Exercises in Probability
Second Edition

Derived from extensive teaching experience in Paris, this second edition now includes 120 exercises in probability. New exercises have been added to reflect important areas of current research in probability theory, including infinite divisibility of stochastic processes and past–future martingales. For each exercise the authors provide detailed solutions as well as references for preliminary and further reading. There are also many insightful notes to motivate the student and set the exercises in context.

Students will find these exercises extremely useful for easing the transition between simple and complex probabilistic frameworks. Indeed, many of the exercises here will lead the student on to frontier research topics in probability. Along the way, attention is drawn to a number of traps into which students of probability often fall. This book is ideal for independent study or as the companion to a course in advanced probability theory.

LOÏC CHAUMONT is Professor in the Laboratoire Angevin de Recherche en Mathématiques (LAREMA) at Université d'Angers.

MARC YOR is Professor in the Laboratoire de Probabilités et Modèles Aléatoires (LPMA) at Université Pierre et Marie Curie (Paris VI). He is a senior member of the Institut Universitaire de France (IUF).

Exercises in Probability

A Guided Tour from Measure Theory to Random Processes,
via Conditioning

Second Edition

Loïc Chaumont
LAREMA, Université d'Angers

Marc Yor
LPMA, Université Pierre et Marie Curie (Paris VI)

CAMBRIDGE
UNIVERSITY PRESS

CAMBRIDGE
UNIVERSITY PRESS

University Printing House, Cambridge CB2 8BS, United Kingdom

Cambridge University Press is part of the University of Cambridge.

It furthers the University's mission by disseminating knowledge in the pursuit of education, learning and research at the highest international levels of excellence.

www.cambridge.org
Information on this title: www.cambridge.org/9781107606555

First edition © Cambridge University Press 2003
Second edition © L. Chaumont and M. Yor 2012

First published 2003
Second edition 2012

A catalogue record for this publication is available from the British Library

Library of Congress Cataloguing in Publication data
Chaumont, L. (Loïc)
Exercises in probability : a guided tour from measure theory to random processes, via conditioning /
Loïc Chaumont, Marc Yor. – 2nd ed.
p.   cm. – (Cambridge series in statistical and probabilistic mathematics)
ISBN 978-1-107-60655-5 (pbk.)
1. Probabilities.   2. Probabilities – Problems, exercises, etc.   I. Yor, Marc.   II. Title.
QA273.25.C492   2012
519.2 – dc23   2012002653

ISBN 978-1-107-60655-5 Paperback

To K. Itô who showed us the big picture.

To P. A. Meyer for his general views and explanations of Itô's work.

Contents

Preface to the second edition

The friendly welcome which the first edition of this book of exercises appears to have received enticed us to give it a second outing, correcting mistakes, adding comments and references, and presenting twenty more exercises, thus bringing the total number of exercises for this second edition to 120. We would also like to point out a few unsolved questions, which are indicated in the text with a circle ∘. (See in particular Exercise **6.29**, which discusses some additive and multiplicative martingale decompositions of Brownian motion, a topic which has fascinated us, but which we have not been able to clear up! So, dear reader, there are still a few challenges in this book, despite the proposed solutions...)

This second edition follows, and whenever possible, reinforces the "guiding principle" of the first edition, that is: to present exercises that are constructed by "stripping to its simplest skeleton" a complex random phenomenon, so that the stripped version may be accessible to a student who has engaged seriously in a first course in probability. To give an example, proving Lévy's second arcsine law for Brownian motion $(B_t, t \leq 1)$, i.e.: $P\left(A \equiv \int_0^1 ds\, \mathbb{1}_{\{B_s > 0\}} \in dx\right) = \frac{dx}{\pi\sqrt{x(1-x)}}$, $(0 < x < 1)$ seems to necessitate, whichever method is employed, some quite sophisticated tools. But, here, we choose to present the first step, namely several representations of A such as:

$$A \stackrel{\text{(law)}}{=} \frac{N^2}{N^2 + N'^2} \stackrel{\text{(law)}}{=} \frac{T}{T + T'} \stackrel{\text{(law)}}{=} \frac{1}{1 + C^2},$$

where (N, N') is a pair of independent reduced Gaussian variables, (T, T') is a pair of independent identically distributed, stable $(1/2)$, positive variables and C is a standard Cauchy variable.

The "opposite" of this stripping exercise is, of course, the "dressing" exercise, meaning that it is also quite natural, and fruitful, in probability theory, to look for an infinite-dimensional probabilistic framework in which a particular finite-dimensional (in fact, often one-dimensional) probabilistic fact may be embedded. Perhaps, a most important example of "dressing" comes with the central limit theorem:

$$\frac{X_1 + \cdots + X_n}{\sqrt{n}} \xrightarrow{\text{(law)}} N,$$

where the $X_i's$ are i.i.d., centered, with variance 1, and N is a reduced Gaussian random variable; this theorem admits the (infinite-dimensional) functional version, known as Donsker's theorem:

$$\left(\frac{X_1 + \cdots + X_{[nt]}}{\sqrt{n}}, \, t \geq 0\right) \xrightarrow{\text{(law)}} (B_t, \, t \geq 0),$$

where $[x]$ is the integer part of x, and $(B_t, \, t \geq 0)$ is one-dimensional Brownian motion.

In the same vein, an infinitely divisible r.v. L may be seen as the value at time 1 of a Lévy process $(L_t, \, t \geq 0)$. The infinite-dimensional "dressing" attitude was advertised in particular by Professor Itô, as we pointed out in the last page of the first (and now second) edition of our book. Further developments of these "stripping–dressing" performances are presented in the paper: *Small and Big Probability Worlds*, [100]. We wish the reader some nice travel through these worlds.

Finally, we would like to advocate the systematic adoption of a probabilistic viewpoint. When reading some maths (e.g. about special functions), and/or witnessing some phenomena, do ask yourself: what does this mean probabilistically? This is often a rewarding attitude...

Angers and Paris, 24th of July 2011

The circle ∘ symbol is appended to a question, or a whole exercise for which we may know only how to start walking in the (right?) direction, but we have not reached the goal. These circled questions are found in Exercises 2.7 and 6.29. We encourage the reader to do better than us!

Preface to the first edition

Originally, the main body of these exercises was developed for, and presented to, the students in the Magistère des Universités Parisiennes between 1984 and 1990; the audience consisted mainly of students from the Écoles Normales, and the spirit of the Magistère was to blend "undergraduate probability" ($\overset{?}{=}$ random variables, their distributions, and so on ...) with a first approach to "graduate probability" ($\overset{?}{=}$ random processes). Later, we also used these exercises, and added some more, either in the Préparation à l'Agrégation de Mathématiques, or in more standard Master courses in probability.

In order to fit the exercises (related to the lectures) in with the two levels alluded to above, we systematically tried to strip a number of results (which had recently been published in research journals) of their random processes apparatus, and to exhibit, in the form of exercises, their random variables skeleton.

Of course, this kind of reduction may be done in almost every branch of mathematics, but it seems to be a quite natural activity in probability theory, where a random phenomenon may be either studied on its own (in a "small" probability world), or as a part of a more complete phenomenon (taking place in a "big" probability world); to give an example, the classical central limit theorem, in which only one Gaussian variable (or distribution) occurs in the limit, appears, in a number of studies, as a one-dimensional "projection" of a central limit theorem involving processes, in which the limits may be several Brownian motions, the former Gaussian variable appearing now as the value at time 1, say, of one of these Brownian motions.

This being said, the aim of these exercises was, and still is, to help a student with a good background in measure theory, say, but starting to learn probability theory, to master the main concepts in basic (?) probability theory, in order that, when reaching the next level in probability, i.e. graduate studies (so called, in France: Diplôme d'Études Approfondies), he/she would be able to recognize, and put aside, difficulties which, in fact, belong to the "undergraduate world", in order to concentrate better on the "graduate world" (of course, this is nonsense, but some analysis of the level of a given difficulty is always helpful...).

Among the main basic concepts alluded to above, we should no doubt list the notions of independence, and conditioning (Chapter 2) and the various modes of convergence of random variables (Chapter 5). It seemed logical to start with a short Chapter 1 where measure theory is deeply mixed with the probabilistic aspects. Chapter 3 is entirely devoted to some exercises on Gaussian variables: of course, no one teaching or studying probability will be astonished, but we have always been struck, over the years, by the number of mistakes which Gaussian type computations seem to lead many students to.

A number of exercises about various distributional computations, with some emphasis on beta and gamma distributions, as well as stable laws, are gathered in Chapter 4, and finally, perhaps as an eye opener, a few exercises involving random processes are found in Chapter 6, where, as an exception, we felt freer to refer to more advanced concepts. However, the different chapters are not autonomous, as it is not so easy – and it would be quite artificial – to separate strictly the different notions, e.g. convergence, particular laws, conditioning, and so on.... Nonetheless, each chapter focusses mainly around the topic indicated in its title.

As often as possible, some comments and references are given after an exercise; both aim at guiding the reader's attention towards the "bigger picture" mentioned above; furthermore, each chapter begins with a "minimal" presentation, which may help the reader to understand the global "philosophy" of this chapter, and/or some of the main tools necessary to solve the exercises there. But, for a more complete collection of important theorems and results, we refer the reader to the list of textbooks in probability – perhaps slightly slanted towards books available in France! – which is found at the end of the volume. Appended to this list, we have indicated on one page some (usually, three) among these references where the notion N is treated; we tried to vary these sources of references.

A good proportion of the exercises may seem, at first reading, "hard", but we hope the solutions – not to be read too quickly before attempting seriously to solve the exercises! – will help; we tried to give almost every ε–δ needed! We have indicated with one star * exercises which are of standard difficulty, and with two stars ** the more challenging ones. We have given references, as much as we could, to related exercises in the literature. Internal references from one exercise to another should be eased by our marking in bold face of the corresponding numbers of these exercises in the *Comments and references, Hint*, and so on....

Our thanks go to Dan Romik, Koichiro Takaoka, and at a later stage, Alexander Cherny, Jan Obloj, Adam Osekowski, for their many comments and suggestions for improvements. We are also grateful to K. Ishiyama who provided us with the picture featured on the cover of our book which represents the graph of densities of the time average of geometric Brownian motion, see Exercise **6.20** for the corresponding discussion.

As a final word, let us stress that we do not view this set of exercises as being "the" good companion to a course in probability theory (the reader may also use the books of exercises referred to in our bibliography), but rather we have tried to present some perhaps not so classical aspects....

Paris and Berkeley, August 2003

Some frequently used notations

a.e.	almost everywhere.
a.s.	almost surely.
r.v.	random variable.
i.i.d.	independent and identically distributed (r.v.s).
Question x of Exercise $a.b$	Our exercises are divided in questions, to which we may refer in different places to compare some results.
*Exercise	Exercise of standard difficulty.
**Exercise	Challenging exercise.
$P_{|\mathcal{A}} \ll Q_{|\mathcal{A}}$	P is absolutely continuous with respect to Q, when both probabilities are considered on the σ-field $\mathcal{A}$. When the choice of $\mathcal{A}$ is obvious, we write only $P \ll Q$.
$\left.\dfrac{dP}{dQ}\right|_{\mathcal{A}}$	denotes the Radon–Nikodym density of P with respect to Q, on the σ-field $\mathcal{A}$, assuming $P_{|\mathcal{A}} \ll Q_{|\mathcal{A}}$, again, $\mathcal{A}$ is suppressed if there is no risk of confusion.
$P \otimes Q$	denotes the tensor product of the two probabilities P and Q.
$X(P)$	denotes the image of the probability P by the r.v. X.
$\nu_n \xrightarrow{w} \nu$	indicates that the sequence of positive measures on $\mathbb{R}$ (or $\mathbb{R}^n$) converges weakly towards ν.
An n-sample $\mathbf{X}_n$ of the r.v. X	denotes an n-dimensional r.v. $(X_1, \ldots, X_n)$, whose components are i.i.d., distributed as X.
ε	Bernoulli (two valued) r.v.
N or G	Standard centered Gaussian variable, with variance 1: $P(N \in dx) = e^{-\frac{x^2}{2}} \dfrac{dx}{\sqrt{2\pi}}, \ (x \in \mathbb{R})$.

T Standard stable(1/2) $\mathbb{R}_+$–valued variable:
$P(T \in dt) = \frac{dt}{\sqrt{2\pi t^3}} \exp\left(-\frac{1}{2t}\right)$, $(t > 0)$.

Z Standard exponential variable: $P(Z \in dt) = e^{-t}\, dt$, $(t > 0)$.

Z_a $(a > 0)$ Standard gamma(a) variable: $P(Z_a \in dt) = t^{a-1} e^{-t} \frac{dt}{\Gamma(a)}$, $(t > 0)$.

$Z_{a,b}$ $(a, b > 0)$ Standard beta(a, b) variable:
$P(Z_{a,b} \in dt) = t^{a-1}(1-t)^{b-1} \frac{dt}{\beta(a,b)}$, $(t \in (0,1))$.

It may happen that, for convenience, we use some different notation for these classical variables.

Chapter 1

Measure theory and probability

Aim and contents

This chapter contains a number of exercises, aimed at familiarizing the reader with some important measure theoretic concepts, such as: Monotone Class Theorem (Williams [98], II.3, II.4, II.13), uniform integrability (which is often needed when working with a family of probabilities, see Dellacherie and Meyer [23]), L^p convergence (Jacod and Protter [40], Chapter 23), conditioning (this will be developed in a more probabilistic manner in the following chapters), absolute continuity (Fristedt and Gray [33], p. 118).

We would like to emphasize the importance for every probabilist to stand on a "reasonably" solid measure theoretic (back)ground for which we recommend, e.g., Revuz [74].

Exercise **1.12** plays a unifying role, and highlights the fact that the operation of taking a conditional expectation is a contraction (in L^2, but also in every L^p) in a strong sense.

---

* 1.1 Some traps concerning the union of σ-fields

1. Show that the union of two σ-fields is never a σ-field unless one is included into the other.

2. Give an example of a filtration $(\mathcal{F}_n)_{n \geq 0}$ which is strictly increasing, i.e. $\mathcal{F}_n \neq \mathcal{F}_{n+1}$, for all n and such that $\cup_{n \geq 0} \mathcal{F}_n$ is not a σ-field.

Comments:

(a) It is clear from the proof that the assertion of question 1 is also valid if we consider only Boole algebras instead of σ-fields.

(b) Question 2 raises the following more general problem: are there examples of strictly increasing filtrations $(\mathcal{F}_n)_{n \geq 0}$, such that the union $\cup_{n \geq 0} \mathcal{F}_n$ is a σ-field? The answer to this question is NO. A proof of this fact can be found in: A. Broughton and B. W. Huff: A comment on unions of sigma-fields. *The American Mathematical Monthly*, **84**, no. 7 (Aug.–Sep., 1977), pp. 553–554. We are very grateful to Rodolphe Garbit and Rongli Liu, who have indicated to us some references about this question.

** 1.2 Sets which do not belong in a strong sense, to a σ-field

Let $(\Omega, \mathcal{F}, P)$ be a complete probability space. We consider two $(\mathcal{F}, P)$ complete sub-σ-fields of $\mathcal{F}$, $\mathcal{A}$ and $\mathcal{B}$, and a set $A \in \mathcal{A}$.

The aim of this exercise is to study the property:

$$0 < P(A|\mathcal{B}) < 1, \quad P \quad \text{a.s.} \tag{1.2.1}$$

1. Show that (1.2.1) holds if and only (iff) there exists a probability Q, which is equivalent to P on $\mathcal{F}$, and such that

     (a) $0 < Q(A) < 1$,   and   (b) $\mathcal{B}$ and A are independent.

Hint. If (1.2.1) holds, we may consider, for $0 < \alpha < 1$, the probability:

$$Q_\alpha = \left\{ \alpha \frac{1_A}{P(A|\mathcal{B})} + (1-\alpha) \frac{1_{A^c}}{P(A^c|\mathcal{B})} \right\} \cdot P .$$

2. Assume that (1.2.1) holds. Define $\mathcal{B}^A = \mathcal{B} \vee \sigma(A)$. Let $0 < \alpha < 1$, and Q be a probability which satisfies (a) and (b) together with:

     (c) $Q(A) = \alpha$, and (d) $\left. \dfrac{dQ}{dP} \right|_{\mathcal{F}}$ is $\mathcal{B}^A$-measurable.

Show then the existence of a $\mathcal{B}$-measurable r.v. Z, which is > 0, P a.s., and such that:

$$E_P(Z) = 1, \quad \text{and} \quad Q = Z \left\{ \alpha \frac{1_A}{P(A|\mathcal{B})} + (1-\alpha) \frac{1_{A^c}}{P(A^c|\mathcal{B})} \right\} \cdot P .$$

Show that there exists a unique probability $\hat{Q}$ which satisfies (a) and (b), together with (c), (d) and (e), where:

     (e) $: \left. \hat{Q} \right|_{\mathcal{B}} = \left. P \right|_{\mathcal{B}} .$

3. We assume, in this and the two next questions, that $\mathcal{A} = \mathcal{B}^A$, but it is not assumed a priori that A satisfies (1.2.1).

    Show then that $A' \in \mathcal{A}$ satisfies (1.2.1) iff the two following conditions are satisfied:

    (f) there exists $B \in \mathcal{B}$ such that: $A' = (B \cap A) \cup (B^c \cap A^c)$, up to a negligible set, and

    (g) A satisfies (1.2.1).

    Consequently, if A does not satisfy (1.2.1), then there exists no set $A' \in \mathcal{A}$ which satisfies (1.2.1).

4. We assume, *in this question and in the next one*, that $\mathcal{A} = \mathcal{B}^A$, and that A satisfies (1.2.1).

    Show that, if $\mathcal{B}$ is not P-trivial, then there exists a σ-field $\mathcal{A}'$ such that $\mathcal{B} \subsetneq \mathcal{A}' \subsetneq \mathcal{A}$, and that no set in $\mathcal{A}'$ satisfies (1.2.1).

5. (i) We further assume that, under P, A is independent of $\mathcal{B}$, and that: $P(A) = \frac{1}{2}$.

    Show that $A' \in \mathcal{A}$ satisfies (1.2.1) iff A' is P-independent of $\mathcal{B}$, and $P(A') = \frac{1}{2}$.

    (ii) We now assume that, under P, A is independent of $\mathcal{B}$, and that: $P(A) = \alpha$, with: $\alpha \neq \left\{ 0, \frac{1}{2}, 1 \right\}$.

    Show that A' (belonging to $\mathcal{A}$, and assumed to be non-trivial) is independent of $\mathcal{B}$ iff $A' = A$ or $A' = A^c$.

    (iii) Finally, we only assume that A satisfies (1.2.1).

    Show that, if $A' (\in \mathcal{A})$ satisfies (1.2.1), then the equality $\mathcal{A} = \mathcal{B}^{A'}$ holds.

Comments and references. The hypothesis (1.2.1) made at the beginning of the exercise means that A does not belong, in a strong sense, to $\mathcal{B}$. Such a property plays an important role in:

J. AZÉMA AND M. YOR: Sur les zéros des martingales continues. *Séminaire de Probabilités XXVI*, 248–306, *Lecture Notes in Mathematics*, **1526**, Springer, 1992.

** 1.3 Some criteria for uniform integrability

Consider, on a probability space $(\Omega, \mathcal{A}, P)$, a set H of r.v.s with values in $\mathbb{R}_+$, which is bounded in L^1, i.e.

$$\sup_{X \in H} E(X) < \infty .$$

Recall that H is said to be uniformly integrable if the following property holds:

$$\sup_{X \in H} \int_{(X > a)} X dP \underset{a \to \infty}{\longrightarrow} 0. \tag{1.3.1}$$

To each variable $X \in H$ associate the positive, bounded measure ν_X defined by:

$$\nu_X(A) = \int_A X dP \qquad (A \in \mathcal{A}).$$

Show that the property (1.3.1) is equivalent to each of the three following properties:

(i) the measures $(\nu_X, X \in H)$ are equi-absolutely continuous with respect to P, i.e. they satisfy the criterion:

$$\forall \varepsilon > 0, \ \exists \delta > 0, \ \forall A \in \mathcal{A}, \ P(A) \leq \delta \Rightarrow \sup_{X \in H} \nu_X(A) \leq \varepsilon, \qquad (1.3.2)$$

(ii) for any sequence (A_n) of sets in $\mathcal{A}$, which decrease to $\emptyset$, then:

$$\lim_{n \to \infty} \left(\sup_{X \in H} \nu_X(A_n) \right) = 0, \qquad (1.3.3)$$

(iii) for any sequence (B_n) of disjoint sets of $\mathcal{A}$,

$$\lim_{n \to \infty} \left(\sup_{X \in H} \nu_X(B_n) \right) = 0. \qquad (1.3.4)$$

Comments and references:

(a) The equivalence between properties (1.3.1) and (1.3.2) is quite classical; their equivalence with (1.3.3) and *a fortiori* (1.3.4) may be less known. These equivalences play an important role in the study of weak compactness in:

C. DELLACHERIE, P.A. MEYER AND M. YOR: Sur certaines propriétés des espaces H^1 et BMO, *Séminaire de Probabilités XII*, 98–113, *Lecture Notes in Mathematics*, **649**, Springer, 1978.

(b) De la Vallée-Poussin's lemma is another very useful criterion for uniform integrability (see Meyer [60]; one may also consult: C. Dellacherie and P.A. Meyer [23]).

   The lemma asserts that $(X_i, \ i \in I)$ is uniformly integrable if and only if there exists a strictly increasing function $\Phi : \mathbb{R}_+ \to \mathbb{R}_+$, such that $\frac{\Phi(x)}{x} \to \infty$, as $x \to \infty$ and $\sup_{i \in I} E[\Phi(X_i)] < \infty$. (Prove that the condition is sufficient!) This lemma is often used (in one direction) with $\Phi(x) = x^2$, i.e. a family $(X_i, \ i \in I)$ which is bounded in L^2 is uniformly integrable. See Exercise **5.9** for an application.

* 1.4 When does weak convergence imply the convergence of expectations?

Consider, on a probability space $(\Omega, \mathcal{A}, P)$, a sequence (X_n) of r.v.s with values in $\mathbb{R}_+$, which are uniformly integrable, and such that:

$$X_n(P) \xrightarrow[n \to \infty]{w} \nu \ .$$

1. Show that ν is carried by $\mathbb{R}_+$, and that $\int \nu(dx)x < \infty$.

2. Show that $E(X_n)$ converges, as $n \to \infty$, towards $\int \nu(dx)x$.

Comments:

(a) Recall that, if $(\nu_n; n \in \mathbb{N})$ is a sequence of probability measures on $\mathbb{R}^d$ (for simplicity), and ν is also a probability on $\mathbb{R}^d$, then:

$$\nu_n \xrightarrow[n\to\infty]{w} \nu \quad \text{if and only if}: \langle \nu_n, f \rangle \xrightarrow[n\to\infty]{} \langle \nu, f \rangle$$

for every bounded, continuous function f.

(b) When $\nu_n \xrightarrow[n\to\infty]{w} \nu$, the question often arises whether $\langle \nu_n, f \rangle \xrightarrow[n\to\infty]{} \langle \nu, f \rangle$ also for some f's which may be either unbounded, or discontinuous. Examples of such situations are dealt with in Exercises **5.4** and **5.9**.

(c) Recall Scheffe's lemma: if (X_n) and X are $\mathbb{R}_+$-valued r.v.s, with $X_n \xrightarrow[n\to\infty]{(P)} X$, and $E[X_n] \xrightarrow[n\to\infty]{} E[X]$, then $X_n \xrightarrow[n\to\infty]{} X$ in $L^1(P)$, hence the $X_n's$ are uniformly integrable, thus providing a partial converse to the statement in this exercise.

* 1.5 Conditional expectation and the Monotone Class Theorem

Consider, on a probability space $(\Omega, \mathcal{F}, P)$, a sub-σ-field $\mathcal{G}$. Assume that there exist two r.v.s, X and Y, with X $\mathcal{F}$-measurable and Y $\mathcal{G}$-measurable such that, for every Borel bounded function $g : \mathbb{R} \to \mathbb{R}_+$, one has:

$$E[g(X) \mid \mathcal{G}] = g(Y) .$$

Prove that: $X = Y$ a.s. *Hint:* Look at the title !

Comments. For a deeper result, see Exercise **1.12**.

** 1.6 L^p-convergence of conditional expectations

Let $(\Omega, \mathcal{F}, P)$ be a probability space and $X \in L^p(\Omega, \mathcal{F}, P)$, $X \geq 0$, for some $p \geq 1$.

1. Let $\mathbb{H}$ be the set of all sub-σ-fields of $\mathcal{F}$. Prove that the family of r.v.s $\{(E[X \mid \mathcal{G}]^p) : \mathcal{G} \in \mathbb{H}\}$ is uniformly integrable. (We refer to Exercise **1.2** for the definition of uniform integrability.)

2. Show that if a sequence of r.v.s (Y_n) , with values in $\mathbb{R}_+$, is such that (Y_n^p) is uniformly integrable and (Y_n) converges in probability to Y, then (Y_n) converges to Y in L^p.

3. Let $(\mathcal{B}_n)$ be a monotone sequence of sub-σ-fields of $\mathcal{F}$. We denote by $\mathcal{B}$ the limit of $(\mathcal{B}_n)$, that is $\mathcal{B} = \vee_n \mathcal{B}_n$ if $(\mathcal{B}_n)$ increases or $\mathcal{B} = \cap_n \mathcal{B}_n$ if $(\mathcal{B}_n)$ decreases. Prove that

$$E(X \,|\, \mathcal{B}_n) \xrightarrow{L^p} E(X \,|\, \mathcal{B}).$$

Hint. First, prove the result in the case $p = 2$.

Comments and references. These three questions are very classical. We present the end result (of question 3.) as an exercise, although it is an important and classical part of the Martingale Convergence Theorem (see the reference hereafter). We wish to emphasize that here, nonetheless, as for many other questions the L^p convergence results are much easier to obtain than the corresponding almost sure one, which is proved in J. Neveu [62] and D. Williams [98].

* 1.7 Measure preserving transformations

Let $(\Omega, \mathcal{F}, P)$ be a probability space, and let $T : (\Omega, \mathcal{F}) \to (\Omega, \mathcal{F})$ be a transformation which preserves P, i.e. $T(P) = P$.

1. Prove that, if $X : (\Omega, \mathcal{F}) \to (\mathbb{R}, \mathcal{B}(\mathbb{R}))$ is almost T-invariant, i.e. $X(\omega) = X(T(\omega))$, P a.s., then, for any bounded function $\Phi : (\Omega \times \mathbb{R}, \mathcal{F} \otimes \mathcal{B}(\mathbb{R})) \to (\mathbb{R}, \mathcal{B}(\mathbb{R}))$, one has:

$$E[\Phi(\omega, X(\omega))] = E[\Phi(T(\omega), X(\omega))]. \qquad (1.7.1)$$

2. Conversely, prove that, if (1.7.1) is satisfied, then, for every bounded function $g : (\mathbb{R}, \mathcal{B}(\mathbb{R})) \to (\mathbb{R}, \mathcal{B}(\mathbb{R}))$, one has:

$$E[g(X) \,|\, T^{-1}(\mathcal{F})] = g(X(T(\omega))), \quad P \text{ a.s.} \qquad (1.7.2)$$

3. Prove that (1.7.1) is satisfied if and only if X is almost T-invariant.

   Hint: Use Exercise **1.5**.

* 1.8 Ergodic transformations

Let $(\Omega, \mathcal{F}, P)$ be a probability space, and let $T : (\Omega, \mathcal{F}) \to (\Omega, \mathcal{F})$ be a transformation which preserves P, i.e. $T(P) = P$.

We denote by $\mathcal{J}$ the invariant σ-field of T, i.e.

$$\mathcal{J} = \{A \in \mathcal{F} : 1_A(T\omega) = 1_A(\omega)\}.$$

T is said to be ergodic if $\mathcal{J}$ is P-trivial.

1. Prove that T is ergodic if the following property holds:

    (a) for every f, g belonging to a vector space $\mathcal{H}$ which is dense in $L^2(\mathcal{F}, P)$,

$$E\left[f(g \circ T^n)\right] \xrightarrow[n \to \infty]{} E(f)E(g),$$

    where T^n is the composition product of T by itself, $(n-1)$ times: $T^n = T \circ T \circ \cdots \circ T$.

2. Prove that, if there exists an increasing sequence $(\mathcal{F}_k)_{k \in \mathbb{N}}$ of sub-σ-fields of $\mathcal{F}$ such that:

    (b) $\vee_k \mathcal{F}_k = \mathcal{F}$,

    (c) for every k, $T^{-1}(\mathcal{F}_k) \subseteq \mathcal{F}_k$,

    (d) for every k, $\bigcap_n (T^n)^{-1}(\mathcal{F}_k)$ is P-trivial,

    then the property (a) is satisfied.

    Consequently, the properties (b)–(c)–(d) imply that T is ergodic.

* 1.9 Invariant σ-fields

Consider, on a probability space $(\Omega, \mathcal{F}, P)$, a measurable transformation T which preserves P, i.e. $T(P) = P$.

Let g be an integrable random variable, i.e. $g \in L^1(\Omega, \mathcal{F}, P)$.

Prove that the two following properties are equivalent:

(i) for every $f \in L^\infty(\Omega, \mathcal{F}, P)$, $E[fg] = E[(f \circ T)g]$,

(ii) g is almost T-invariant, i.e. $g = g \circ T$, P a.s.

Hint: One may use the following form of the ergodic theorem:

$$\frac{1}{n} \sum_{i=1}^n f \circ T^i \xrightarrow[n \to \infty]{L^1} E[f \mid \mathcal{J}],$$

where $\mathcal{J}$ is the invariant σ-field of T.

Comments and references on Exercises **1.7, 1.8, 1.9**:

(a) These are featured at the very beginning of every book on Ergodic Theory. See, for example, K. Petersen [67] and P. Billingsley [8].

(b) Of course, many examples of ergodic transformations are provided in books on Ergodic Theory. Let us simply mention that if (B_t) denotes Brownian motion, then the scaling operation, $B \mapsto \left(\frac{1}{\sqrt{c}} B_c\right)$ is ergodic for $c \neq 1$. Can you prove this result? Actually, the same result holds for the whole class of stable processes, as proved in Exercise **5.17**.

(c) Exercise **1.12** yields a proof of (i) $\Rightarrow$ (ii) which does not use the Ergodic Theorem.

** 1.10 Extremal solutions of (general) moments problems

Consider, on a measurable space $(\Omega, \mathcal{F})$, a family $\Phi = (\varphi_i)_{i \in I}$ of real-valued random variables, and let $c = (c_i)_{i \in I}$ be a family of real numbers.
Define $\mathcal{M}_{\Phi,c}$ to be the family of probabilities P on $(\Omega, \mathcal{F})$ such that:

(a) $\Phi \subset L^1(\Omega, \mathcal{F}, P)$; (b) for every $i \in I$, $E_P(\varphi_i) = c_i$.

A probability measure P in $\mathcal{M}_{\Phi,c}$ is called extremal if whenever $P = \alpha P_1 + (1-\alpha)P_2$, with $0 < \alpha < 1$ and $P_1, P_2 \in \mathcal{M}_{\Phi,c}$, then $P = P_1 = P_2$.

1. Prove that, if $P \in \mathcal{M}_{\Phi,c}$, then P is extremal in $\mathcal{M}_{\Phi,c}$, if and only if the vector space generated by 1 and Φ is dense in $L^1(\Omega, \mathcal{F}, P)$.

2. (i) Prove that, if P is extremal in $\mathcal{M}_{\Phi,c}$, and $Q \in \mathcal{M}_{\Phi,c}$, such that $Q \ll P$, and $\frac{dQ}{dP}$ is bounded, then $Q = P$.

   (ii) Prove that, if P is not extremal in $\mathcal{M}_{\Phi,c}$, and $Q \in \mathcal{M}_{\Phi,c}$, such that $Q \simeq P$, with $0 < \varepsilon \leq \frac{dQ}{dP} \leq C < \infty$, for some ε, $C > 0$, then Q is not extremal in $\mathcal{M}_{\Phi,c}$.

3. Let T be a measurable transformation of $(\Omega, \mathcal{F})$, and define $\mathcal{M}_T$ to be the family of probabilities P on $(\Omega, \mathcal{F})$ which are preserved by T, i.e. $T(P) = P$. Prove that, if $P \in \mathcal{M}_T$, then P is extremal in $\mathcal{M}_T$ if, and only if, T is ergodic under P.

Comments and references:

(a) The result of question 1 appears to have been obtained independently by:

M.A. NAIMARK: Extremal spectral functions of a symmetric operator. *Bull. Acad. Sci. URSS. Sér. Math.*, **11**, 327–344 (1947).

(see e.g. N.I. AKHIEZER: *The Classical Moment Problem and Some Related Questions in Analysis.* Publishing Co., New York, p. 47, 1965), and

R. DOUGLAS: On extremal measures and subspace density II. *Michigan Math. J.*, **11**, 243–246 (1964). Proc. Amer. Math. Soc., **17**, 1363–1365 (1966).

It is often used in the study of indeterminate moment problems, see e.g.

CH. BERG: Recent results about moment problems. *Probability Measures on Groups and Related Structures, XI* (Oberwolfach, 1994), 1–13, World Sci. Publishing, River Edge, NJ, 1995.

CH. BERG: Indeterminate moment problems and the theory of entire functions. *Proceedings of the International Conference on Orthogonality, Moment Problems and Continued Fractions* (Delft, 1994). *J. Comput. Appl. Math.*, **65**, no. 1–3, 27–55 (1995).

(b) Some variants are presented in:

E.B. DYNKIN: Sufficient statistics and extreme points. *Ann. Probab.*, **6**, no. 5, 705–730 (1978).

For some applications to martingale representations as stochastic integrals, see:

M. YOR: Sous-espaces denses dans L^1 et H^1 et représentations des martingales, *Séminaire de Probabilités XII*, 264–309, *Lecture Notes in Mathematics*, **649**, Springer, 1978.

(c) The next exercise gives the most classical example of a non-moments determinate probability on $\mathbb{R}$. It is those particular moments problems which motivated the general statement of Naimark–Douglas.

*1.11 The log normal distribution is moments indeterminate

Let N_{σ^2} be a centered Gaussian variable with variance σ^2. Associate to N_{σ^2} the log normal variable:

$$X_{\sigma^2} = \exp\left(N_{\sigma^2}\right).$$

1. Compute the density of X_{σ^2}; its expression gives an explanation for the term "log normal".

2. Prove that for every $n \in \mathbb{Z}$, and $p \in \mathbb{Z}$,

$$E\left[X_{\sigma^2}^n \sin\left(\frac{p\pi}{\sigma^2} N_{\sigma^2}\right)\right] = 0. \tag{1.11.1}$$

3. Show that there exist infinitely many probability laws μ on $\mathbb{R}_+$ such that:

   (i) for every $n \in \mathbb{Z}$,

   $$\int \mu(dx)\, x^n = \exp\left(\frac{n^2 \sigma^2}{2}\right).$$

   (ii) μ has a bounded density with respect to the law of $\exp\left(N_{\sigma^2}\right)$.

Comments and references:

(a) This exercise and its proof go back to T. Stieltjes' fundamental memoir:

   T.J. STIELTJES: Recherches sur les fractions continues. Reprint of *Ann. Fac. Sci. Toulouse* **9**, (1895), A5–A47. Reprinted in *Ann. Fac. Sci. Toulouse Math.*, **6**, no. 4, A5–A47 (1995).

   There are many other examples of elements of $\mathcal{M}_{\sigma^2}$, including some with countable support; see, e.g., Stoyanov ([86], p. 104).

(b) In his memoir, Stieltjes also provides other similar elementary proofs for different moment problems. For instance, for any $a > 0$, if Z_a denotes a gamma variable, then for $c > 2$, the law of $(Z_a)^c$ is not moments determinate. See, e.g., J.M. Stoyanov [86], § 11.4.

(c) There are some sufficient criteria which bear upon the sequence of moments $m_n = E[X^n]$ of an r.v. X and ensure that the law of X is determined uniquely from the (m_n) sequence. (In particular, the classical *sufficient* Carleman criterion asserts that if $\sum_n (m_{2n})^{-1/2n} = \infty$, then the law of X is moments determinate.) But, these are unsatisfactory in a number of cases, and the search continues. See, for example, the following.

   J. STOYANOV: Krein condition in probabilistic moment problems. *Bernoulli*, **6**, no. 5, 939–949 (2000).

   A. GUT: On the moment problem. *Bernoulli*, **8**, no. 3, 407–421 (2002).

* 1.12 Conditional expectations and equality in law

Let $X \in L^1(\Omega, \mathcal{F}, P)$, and $\mathcal{G}$ be a sub-σ-field of $\mathcal{F}$. The objective of this exercise is to prove that if X and $Y \overset{(\mathrm{def})}{=} E[X \,|\, \mathcal{G}]$ have the same distribution, then X is $\mathcal{G}$ measurable (hence $X = Y$).

1. Prove the result if X belongs to L^2.

2. Prove that for every $a, b \in \mathbb{R}_+$,

$$E[(X \wedge a) \vee (-b) \,|\, \mathcal{G}] = (Y \wedge a) \vee (-b), \tag{1.12.1}$$

   and conclude.

3. Prove the result of Exercise **1.5** using the previous question.

4. In Exercise **1.9**, prove, without using the Ergodic Theorem that (i) implies (ii).

5. Let X and Y belong to L^1, prove that if $E[X\,|\,Y] = Y$ and $E[Y\,|\,X] = X$, then $X = Y$, a.s.

Comments and references:

(a) This exercise, in the generality of question 1, was proposed by A. Cherny. As is clear from questions 2, 3, 4 and 5, it has many potential applications. See also Exercise **2.6** where it is used to prove de Finetti's Representation Theorem of exchangeable sequences of r.v.s.

(b) Question 5 is proposed as Exercise (33.2) on p. 62 in D. Williams' book [99].

* 1.13 Simplifiable random variables

An r.v. Y which takes its values in $\mathbb{R}_+$ is said to be simplifiable if the following property holds:
if $XY \stackrel{\text{(law)}}{=} YZ$, with X and Z taking their values in $\mathbb{R}_+$, and X, resp. Z, are independent of Y, then: $X \stackrel{\text{(law)}}{=} Z$.

1. Prove that, if Y takes its values in $\mathbb{R}_+ \setminus \{0\}$, and if the characteristic function of $(\log Y)$ has only isolated zeros, then Y is simplifiable.

2. Give an example of an r.v. Y which is not simplifiable.

3. Suppose Y is simplifiable, and satisfies: $Y \stackrel{\text{(law)}}{=} AB$, where on the right hand side A and B are independent, and neither of them is a.s. constant.

   Prove that A cannot be factorized as: $A \stackrel{\text{(law)}}{=} YC$, with Y and C independent.

Comments and references:

(a) To prove that two r.v.s are identical in law, it is sometimes very convenient to first multiply both these variables by a third independent r.v. and then to simplify this variable as in the present exercise. Many applications of this idea are shown in Chapter 4, see for instance Exercise **4.18**.

   To simplify an equality in law as above, the positivity of the variables is crucial as the following example shows: let ε_1 and ε_2 be two independent symmetric Bernoulli variables, then $\varepsilon_1\varepsilon_2 \stackrel{\text{(law)}}{=} \varepsilon_1$ does not imply that $\varepsilon_2 = 1$!

   For some variants of these questions, see Durrett [26], p. 107, as well as Feller [28], section XV, 2.a.

(b) An r.v. H is said to be infinitely divisible if for every $n \in \mathbb{N}$, there exist $H_1^{(n)}, \ldots, H_n^{(n)}$ which are i.i.d. and $H \stackrel{\text{(law)}}{=} H_1^{(n)} + \cdots + H_n^{(n)}$. The well-known

Lévy–Khintchin formula asserts that $E[\exp(i\lambda H)] = \exp\psi(\lambda)$, for a function ψ (of a very special form). In particular, the characteristic function of H has no zeros; we may use this result, in the setup of question 1, for Y such that $H = \log Y$ is infinitely divisible.

(c) Exercise 7, p. 295 in A.N. Shiryaev [82] exhibits three independent r.v.s such that $U + V \overset{(\text{law})}{=} W + V$, but U and W do not have the same law. Hence, $\exp(V)$ is not simplifiable.

* 1.14 Mellin transform and simplification

The Mellin transform of (the distribution of) an $\mathbb{R}_+$ valued r.v. Z is the function: $s \mapsto E[Z^s]$, as defined on $\mathbb{R}_+$ (it may take the value $+\infty$).

Let X, Y, Z be three independent r.v.s taking values in $\mathbb{R}_+$, and such that:

(i) $XY \overset{(\text{law})}{=} ZY$,

(ii) for $0 \le s \le \varepsilon$, with some $\varepsilon > 0$, $E[(XY)^s] < \infty$,

(iii) $P(Y > 0) > 0$.

Show that: $X \overset{(\text{law})}{=} Z$.

Comments and references:

(a) The advantage of this last Exercise (and its result) over the previous one is that one needs not worry about the characteristic function of $(\log Y)$. The (small) cost is that we assume X, Y, Z have (small enough) moments. In our applications (e.g. in Chapter 4), we shall be able to use both criteria of Exercises **1.13** and **1.14**.

(b) It may be worth emphasizing here (informally) that the Mellin transform is injective (on the set of probabilities on $\mathbb{R}_+$), whereas its restriction to $\mathbb{N}$ is not. (Please give precise statements and keep them in mind!) See Chapter VI of Widder [96].

* 1.15 There exists no fractional covering of the real line

1. Prove that for any $c \in (0,1)$, there is no Borel set A such that, for every interval I,

$$m(A \cap I) = cm(I), \tag{1.15.1}$$

where m denotes Lebesgue measure on $\mathbb{R}$.

2. Given a Borel set A, describe the pairs (c, d) in $[0, 1]^2$ such that, for every interval I:

$$|m(A \cap I) - cm(I)| \le |m(A \cap I) - dm(I)|.$$

Comments. (a) This (negative) result also holds on any measurable space endowed with any σ-finite measure.

(b) We are grateful to Paul Bourgade who asked us this question in March 2010.

(c) Although we excluded the values 0 and 1 from consideration for c in question 1, it makes sense to consider them as possible values for c and d in question 2.

Solutions for Chapter 1

Solution to Exercise 1.1

1. Let $\mathcal{F}$ and $\mathcal{G}$ be two σ-fields defined on the same space, such that neither of them is included into the other. Then

$$\begin{cases} \text{there exist } A \in \mathcal{F} \text{ and } B \in \mathcal{G}, \\ \text{with } A \notin \mathcal{G} \text{ and } B \notin \mathcal{F}. \end{cases} \tag{1.1.a}$$

Set $A_1 = A \cap B$, $A_2 = A \cap B^c$, $A_3 = A^c \cap B$ and $A_4 = A^c \cap B^c$. If $\mathcal{F} \cup \mathcal{G}$ is a σ-field then each of these four events belongs to either $\mathcal{F}$ or $\mathcal{G}$. If one of these σ-fields, say $\mathcal{F}$, contains three events, then by complementarity, it contains the fourth one thus $B \in \mathcal{F}$ and it contradicts (1.1.a). On the other hand, it is not difficult to see that if both σ-fields contain exactly two events among A_1, A_2, A_3 and A_4, then it is possible to show that either $B \in \mathcal{F}$ or $A \in \mathcal{G}$. Again, this contradicts (1.1.a).

2. Let $(X_n)_{n \geq 0}$ be a sequence of i.i.d, random variables which are uniformly distributed over the interval $[-2, 1]$. Define $S_n = X_0 + \ldots + X_n$ and $\mathcal{F}_n = \sigma\{S_k : k \leq n\}$. We derive from the law of large numbers that $\lim_{n \to \infty} S_n = -\infty$, a.s., so that $0 < \sup_n S_n < \infty$, a.s. Then the set $\{\sup_{n \geq 0} S_n \geq 1/2\}$ is not trivial and it belongs to $\sigma\left(\cup_{n \geq 0} \mathcal{F}_n\right)$. But this set belongs to none of the σ-fields $\mathcal{F}_n$.

Solution to Exercise 1.2

1. Suppose that (1.2.1) holds, then we shall prove that for every $\alpha \in (0, 1)$, the probability Q_α satisfies (a) and (b). First, note that Q_α is equivalent to P. Indeed, let $N \in \mathcal{F}$ be such that $Q_\alpha(N) = 0$, then $E_P\left(\frac{1_{A \cap N}}{P(A|\mathcal{B})}\right) = 0$ and $E_P\left(\frac{1_{A^c \cap N}}{P(A^c|\mathcal{B})}\right) = 0$. Since $0 < P(A|\mathcal{B}) < 1$ and $0 < P(A^c|\mathcal{B}) < 1$, P a.s., the preceding identities imply that $P(A \cap N) = 0$ and $P(A^c \cap N) = 0$. Hence $P(N) = 0$. The converse is obvious. On the other hand, $Q_\alpha(A) = \alpha E_P\left(\frac{1_A}{P(A|\mathcal{B})}\right) = \alpha \in (0, 1)$. To prove the independence,

note that $E_P\left(\frac{1_{A\cap B}}{P(A|\mathcal{B})}\right) = P(B)$, whenever $B \in \mathcal{B}$. Therefore $Q_\alpha(A \cap B) = \alpha P(B)$ and we easily verify that $Q_\alpha(A)Q_\alpha(B) = \alpha P(B)$.

Suppose now that (a) and (b) hold and set $N_0 = \{\omega \in \Omega : P(A|\mathcal{B}) = 0\}$, and $N_1 = \{\omega \in \Omega : P(A|\mathcal{B}) = 1\}$. We have $P(A \cap N_0|\mathcal{B}) = 1_{N_0}P(A|\mathcal{B}) = 0$, thus $P(A \cap N_0) = 0$ and $Q_\alpha(A \cap N_0) = 0$. But $N_0 \in \mathcal{B}$, thus $Q_\alpha(A \cap N_0) = 0 = Q_\alpha(A)Q_\alpha(N_0)$ and $Q_\alpha(N_0) = 0$. This is equivalent to $P(N_0) = 0$. To prove that $P(N_1) = 0$, it suffices to consider $P(A^c \cap N_1|\mathcal{B})$ and to proceed as above.

2. Let Q_α be the probability defined in 1, and set $Z_\alpha = \frac{dQ}{dQ_\alpha}\big|_{\mathcal{F}}$. First, we show that $E_{Q_\alpha}(Z_\alpha|\mathcal{B}) = Z_\alpha$, Q_α a.s. It suffices to prove that for every $\mathcal{F}$-measurable r.v. $X \geq 0$,

$$E_Q(X) = E_{Q_\alpha}(E_{Q_\alpha}(Z_\alpha|\mathcal{B})X).$$

By (d), Z_α is $\mathcal{B}^A$-measurable, thus all we have to prove is:

$$Q(A) = E_{Q_\alpha}(E_{Q_\alpha}(Z_\alpha|\mathcal{B})1_A).$$

It follows from the definition of Q_α that:

$$E_{Q_\alpha}(E_{Q_\alpha}(Z_\alpha|\mathcal{B})1_A) = \alpha E_P\left(E_{Q_\alpha}(Z_\alpha|\mathcal{B})\frac{1_A}{P(A|\mathcal{B})}\right).$$

Furthermore, since Z_α is $\mathcal{B}^A$-measurable,

$$\begin{aligned}
E_{Q_\alpha}(E_{Q_\alpha}(Z_\alpha|\mathcal{B})1_A) &= \alpha E_P(E_{Q_\alpha}(Z_\alpha|\mathcal{B})) \\
&= \alpha E_P(Z_\alpha).
\end{aligned}$$

On the other hand, we have:

$$\begin{aligned}
Q(A) &= E_{Q_\alpha}(Z_\alpha 1_A) \\
&= \alpha E_P\left(Z_\alpha\frac{1_A}{P(A|\mathcal{B})}\right) \\
&= \alpha E_P(Z_\alpha).
\end{aligned}$$

Put $Z = E(Z_\alpha|\mathcal{B})$, then from the above, we verify that $E_P(Z) = 1$. Since Q and Q_α are equivalent, $Z_\alpha > 0$, Q_α a.s. Therefore, $Z > 0$, Q_α a.s., which is equivalent to $Z > 0$, P a.s. We have proven that Z satisfies the required conditions.

Now, let Q and Q_α be two probabilities which satisfy (a) and (b) together with (c), (d) and (e). By (c) and (e), it is obvious that $Q_{|\mathcal{B}^A} = \hat{Q}_{|\mathcal{B}^A}$. This implies that $\frac{dQ}{dP}\big|_{\mathcal{F}} = \frac{d\hat{Q}}{dP}\big|_{\mathcal{F}}$, P a.s. and we conclude that $Q = \hat{Q}$ on the σ-field $\mathcal{F}$.

3. Suppose that (1.2.1) holds. Each element of $\mathcal{B}^A$ is of the form $(B_1 \cap A) \cup (B_2 \cap A^c)$, where B_1 and B_2 belong to $\mathcal{B}$. Indeed, the set $\{(B_1 \cap A) \cup (B_2 \cap A^c) : B_1, B_2 \in \mathcal{B}\}$ is a σ-field which contains $\mathcal{B}^A$ (we leave the proof to the reader). Put $A' = (B_1 \cap A) \cup (B_2 \cap A^c)$, then by (1.2.1), we have

$$0 < 1_{B_1} P(A^c \mid \mathcal{B}) + 1_{B_2} P(A^c \mid \mathcal{B}) < 1, \quad P \text{ a.s.},$$

and thus $B_1 = B_2^c$, up to a negligible set. The converse is obvious.

4. With $\mathcal{A}' = \{A \in \mathcal{A} : (1.2.1) \text{ is not satisfied for } A\}$, it is not difficult to prove that $\mathcal{A}'$ is a σ-field. Moreover, it is clear that $\mathcal{B} \subseteq \mathcal{A}' \subset \mathcal{A}$. Now, let $B \in \mathcal{B}$ be non trivial, then $B \cap A \in \mathcal{A}'$ and $B \cap A \notin \mathcal{B}$, thus $\mathcal{B} \subsetneq \mathcal{A}'$.

5. (i) If A' satisfies (1.2.1) then, by 3, there exists $B \in \mathcal{B}$ such that $A' = (B \cap A) \cup (B^c \cap A^c)$, thus $P(A \mid \mathcal{B}) = 1_B P(A) + 1_{B^c} P(A^c) = 1/2$. The converse is obvious.

(ii) $A' \in \mathcal{A}'$, therefore there exist B_1 and B_2 such that $A' = (B_1 \cap A) \cup (B_2 \cap A^c)$ and A' is independent of $\mathcal{B}$ iff $P(A' \mid \mathcal{B}) = 1_{B_1} P(A) + 1_{B_2} P(A^c)$ is constant. Since A' is non-trivial and $P(A) \notin \{0, 1/2, 1\}$, this holds if and only if $B_1 = \emptyset$ and $B_2 = \Omega$ or $B_1 = \Omega$ and $B_2 = \emptyset$.

(iii) Since $A' \in \mathcal{A}'$, it is clear that $\mathcal{B}^{A'} \subseteq \mathcal{A}$. Moreover, by 3, there exists $B \in \mathcal{B}$ such that $A' = (B \cap A) \cup (B^c \cap A^c)$. Then, we can prove that $A = (B \cap A') \cup (B^c \cap A'^c)$ and $A \in \mathcal{B}^{A'}$, thus $\mathcal{A} \subsetneq \mathcal{B}^{A'}$.

Solution to Exercise 1.3

$(1.3.1) \Longrightarrow (1.3.2)$: Pick $\epsilon > 0$, then on the one hand, according to (1.3.1), there exists $b > 0$ such that:

$$\sup_{X \in H} \nu_X(X > b) \le \epsilon/2.$$

On the other hand, for every $A \in \mathcal{A}$ such that $P(A) \le \epsilon/(2b)$,

$$\sup_{X \in H} \nu_X((X \le b) \cap A) \le b P(A) \le \epsilon/2.$$

Finally, (1.3.2) follows from the inequality:

$$\sup_{X \in H} \nu_X(A) \le \sup_{X \in H} \nu_X((X \le b) \cap A) + \sup_{X \in H} \nu_X(X > b) \le \varepsilon.$$

$(1.3.2) \Longrightarrow (1.3.4)$: If (B_n) is a sequence of disjoint sets in $\mathcal{A}$ then $\sum_{n \ge 0} P(B_n) \le 1$, and thus $\lim_{n \to \infty} P(B_n) = 0$. Therefore, for ε and δ as in (1.3.2), there exists $N \ge 1$ such that $P(B_n) \le \delta$ for every $n \ge N$, hence $\sup_{X \in H} \int_{B_n} X \, dP \le \varepsilon$.

(1.3.4)$\Longrightarrow$(1.3.3): Suppose that (1.3.3) is not verified. Let (A_n) be a sequence of sets in $\mathcal{A}$ which decreases to $\emptyset$ and such that $\lim_{n\to\infty}(\sup_{X\in H}\nu_X(A_n)) = \epsilon > 0$. For every $X \in H$, $\lim_{n\to\infty}\nu_X(A_n) = 0$, thus there exists $X^1 \in H$ and $n_1 > 1$ such that $\nu_{X^1}(A_1\backslash A_{n_1}) \geq \epsilon/2$, put $B_1 = A_1\backslash A_{n_1}$. Furthermore, there exists $n_2 > n_1$ and $X^2 \in H$ such that $\nu_{X^2}(A_{n_1}\backslash A_{n_2}) \geq \epsilon/2$, put $B_2 = A_{n_1}\backslash A_{n_2}$. We can construct, in this manner, a sequence (B_n) of disjoint sets of $\mathcal{A}$ such that, $\sup_{X\in H}\nu_X(B_n) \geq \epsilon/2$ and (1.3.4) is not verified.

(1.3.3)$\Longrightarrow$(1.3.1): Suppose that (1.3.1) does not hold, then there exists $\epsilon > 0$ such that for every $n \in \mathbb{N}$, we can find $X^n \in H$ which verifies $\nu_{X^n}(X^n > 2^n) \geq \epsilon$. Put $A_n = \cup_{p\geq n}(X^p > 2^p)$, then (A_n) is a decreasing sequence of sets of $\mathcal{A}$ such that $\sup_{X\in H}\nu_X(A_n) \geq \epsilon$. Moreover, $\lim_{n\to\infty}P(A_n) = 0$, indeed, for every $n \geq 1$,

$$P(A_n) \leq \sum_{p=n}^{\infty} P(X^p \geq 2^p) \leq C\sum_{p=n}^{\infty}\frac{1}{2^p} \to 0, \quad \text{as } n \to \infty.$$

This proves that (A_n) decreases to a negligible set A. Finally, put $A'_n = A_n\backslash A$, then (A'_n) is a sequence of $\mathcal{A}$ which contradicts (1.3.3).

Solution to Exercise 1.4

1. Since each of the laws $\nu_n \overset{(\text{def})}{=} X_n(P)$ is carried by $\mathbb{R}_+$, then for every bounded, continuous function f which vanishes on $\mathbb{R}_+$, and for every $n \in \mathbb{N}$, we have: $E(f(X_n)) = 0$. By the weak convergence of ν_n to ν, we get $\int f(x)\nu(dx) = 0$ and this proves that ν is carried by $\mathbb{R}_+$.

To prove that $\int x\nu(dx) < \infty$, note that

$$\int x\nu(dx) = \lim_{a\uparrow\infty}\int(x \wedge a)\nu(dx) = \lim_{a\uparrow\infty}\left(\lim_{n\to\infty}\int(x \wedge a)\nu_n(dx)\right) \leq \sup_n E[X_n] < \infty,$$

since the X_ns are uniformly integrable.

2. For any $a \geq 0$, write

$$\left|E[X_n] - \int x\nu(dx)\right| \leq |E[X_n] - E[X_n \wedge a]| + \left|E[X_n \wedge a] - \int(x \wedge a)\nu(dx)\right|$$

$$+ \left|\int(x \wedge a)\nu(dx) - \int x\nu(dx)\right|.$$

Since the r.v.s X_n are uniformly integrable, for any $\varepsilon > 0$, we can find a such that for every n, $|E[X_n] - E[X_n \wedge a]| \leq \varepsilon/3$ and $|\int(x \wedge a)\nu(dx) - \int x\nu(dx)| \leq \varepsilon/3$. Moreover, from the convergence in law, there exists N such that for every $n \geq N$, $|E[X_n \wedge a] - \int(x \wedge a)\nu(dx)| \leq \varepsilon/3$.

Solution to Exercise 1.5

First solution. From the Monotone Class Theorem, the identity:

$$E[g(X) \mid \mathcal{G}] = g(Y)$$

extends to:

$$E[G(X,Y) \mid \mathcal{G}] = G(Y,Y),$$

for every (bounded) Borel function $G : \mathbb{R} \times \mathbb{R} \to \mathbb{R}_+$. Hence taking $G(x,y) = \mathbb{1}_{\{x \neq y\}}$ yields the result.

Second solution. Let $a \geq 0$. From the hypothesis, we deduce:

$$E((X \mathbb{1}_{\{|X| \leq a\}} - Y \mathbb{1}_{\{|Y| \leq a\}})^2 \mid \mathcal{G}) = 0, \quad \text{a.s.}$$

So, $E((X \mathbb{1}_{\{|X| \leq a\}} - Y \mathbb{1}_{\{|Y| \leq a\}})^2) = 0$, hence $X \mathbb{1}_{\{|X| \leq a\}} = Y \mathbb{1}_{\{|Y| \leq a\}}$, a.s, for any $a \geq 0$.

Solution to Exercise 1.6

1. Thanks to the criterion for uniform integrability (1.3.2) in Exercise **1.3**, it suffices to show that the sets: $\{\{E[X \mid \mathcal{G}]^p > a\} : \mathcal{G} \in \mathbb{H}\}$ have small probabilities as $a \to \infty$ uniformly in $\mathbb{H}$. But this follows from

$$P(E[X \mid \mathcal{G}]^p > a) \leq P(E[X^p \mid \mathcal{G}] > a) \leq \frac{1}{a} E[X^p].$$

Note that instead of dealing with only one variable, $X \in L^p$, $X \geq 0$, we might also consider a family $\{X_i, i \in I\}$ of r.v.s such that $\{X_i^p, i \in I\}$ is uniformly integrable. Then, again, the set $\{E[X_i \mid \mathcal{G}]^p, i \in I, \mathcal{G} \in \mathbb{H}\}$ is uniformly integrable.

2. Let $\varepsilon > 0$, then

$$
\begin{aligned}
E[|Y_n - Y|^p] &\leq E[|Y_n - Y|^p \mathbb{1}_{\{|Y_n - Y|^p \leq \varepsilon\}}] + E[|Y_n - Y|^p \mathbb{1}_{\{|Y_n - Y|^p > \varepsilon\}}] \\
&\leq \varepsilon + 2^{p-1}(E[Y_n^p \mathbb{1}_{\{|Y_n - Y|^p > \varepsilon\}}] + E[Y^p \mathbb{1}_{\{|Y_n - Y|^p > \varepsilon\}}]),
\end{aligned}
$$

where the last equality comes from $|x + y|^p \leq 2^{p-1}(|x|^p + |y|^p)$. When n goes to ∞, the terms $E[Y_n^p \mathbb{1}_{\{|Y_n - Y|^p > \varepsilon\}}]$ and $E[Y^p \mathbb{1}_{\{|Y_n - Y|^p > \varepsilon\}}])$ converge to 0, since (Y_n^p) is uniformly integrable and $P(|Y_n - Y|^p > \varepsilon)$ converges to 0 as n goes to ∞.

3. In this question, it suffices to deal with the case $p = 2$. Indeed, suppose the result is true for $p = 2$ and consider the general case where $p \geq 1$.

First suppose that $p \in [2, \infty)$ and let $X \in L^p(\Omega, \mathcal{F}, P)$. This implies that $X \in L^2(\Omega, \mathcal{F}, P)$ and $E(X \mid \mathcal{B}_n) \in L^2(\Omega, \mathcal{F}, P)$ for every $n \in \mathbb{N}$. Since $E(X \mid \mathcal{B}_n) \xrightarrow{L^2}$

$E(X \mid \mathcal{B})$, as $n \to \infty$, the sequence of r.v.s $(E(X \mid \mathcal{B}_n))$ converges in probability to $E(X \mid \mathcal{B})$ and from question 1, the sequence $(E[X \mid \mathcal{B}_n]^p)$ is uniformly integrable. Therefore, from question 2, $(E(X \mid \mathcal{B}_n))$ converges in L^p to $E(X \mid \mathcal{B})$.

If $p \in [1, 2)$ then there exists a sequence $(X_k) \in L^2$ such that $X_k \xrightarrow{L^p} X$, as $k \to \infty$. Assume that for each $k \in \mathbb{N}$, $E(X_k \mid \mathcal{B}_n) \xrightarrow{L^2} E(X_k \mid \mathcal{B})$ as $n \to \infty$; then from Hölder's inequality, $E(X_k \mid \mathcal{B}_n) \xrightarrow{L^p} E(X_k \mid \mathcal{B})$ as $n \to \infty$. Let $\varepsilon > 0$, and k such that $\|X - X_k\|_{L^p} \leq \varepsilon$. There exists n_0 such that for all $n \geq n_0$, $\|E(X_k \mid \mathcal{B}_n) - E(X_k \mid \mathcal{B})\|_{L^p} \leq \varepsilon$, and we have

$$
\begin{aligned}
\|E(X \mid \mathcal{B}_n) - E(X \mid \mathcal{B})\|_{L^p} \leq\ & \|E(X \mid \mathcal{B}_n) - E(X_k \mid \mathcal{B}_n)\|_{L^p} \\
&+\ \|E(X_k \mid \mathcal{B}_n)\ E(X_k \mid \mathcal{B})\|_{L^p} \\
&+\ \|E(X_k \mid \mathcal{B}) - E(X \mid \mathcal{B})\|_{L^p} \leq 3\varepsilon .
\end{aligned}
$$

Now we prove the result in the case $p = 2$. At first, assume that $(\mathcal{B}_n)$ is an increasing sequence of σ-fields. There exist $Y \in L^2(\Omega, \mathcal{B}, P)$ and $Z \in L^2(\Omega, \mathcal{B}, P)^{\perp}$ such that $X = Y + Z$. For every $n \in \mathbb{N}$, $Z \in L^2(\Omega, \mathcal{B}_n, P)^{\perp}$, hence $E(X \mid \mathcal{B}_n) = E(Y \mid \mathcal{B}_n)$. Put $Y_n = E(Y \mid \mathcal{B}_n)$, then for $m \leq n$, $E[(Y_n - Y_m)^2] = E[Y_n^2] - E[Y_m^2]$, so $E[Y_n^2]$ increases and is bounded, so this sequence of reals converges. From Cauchy's criterion, the sequence (Y_n) converge in L^2 towards an r.v. $\tilde{Y}$. Now we show that $\tilde{Y} = Y$: for any $k \leq n$ and $\Gamma_k \in \mathcal{B}_k$, $E[Y_n \mathbb{1}_{\Gamma_k}] = E[Y \mathbb{1}_{\Gamma_k}]$. But the left hand side converges as $n \to \infty$ towards $E[\tilde{Y} \mathbb{1}_{\Gamma_k}] = E[Y \mathbb{1}_{\Gamma_k}]$. Finally, we verify from the Monotone Class Theorem that $\{\Gamma \in \mathcal{B} : E[\tilde{Y} \mathbb{1}_{\Gamma}] = E[Y \mathbb{1}_{\Gamma}]\}$ is equal to the sigma-field $\mathcal{B}$, hence $\tilde{Y} = Y$.

Assume now that $(\mathcal{B}_n)$ decreases. For any integers n and m such that $m > n$, the r.v.s $X - E[X \mid \mathcal{B}_n]$ and $E[X \mid \mathcal{B}_m] - E[X \mid \mathcal{B}_n]$ are orthogonal in $L^2(\Omega, \mathcal{F}, P)$, and from the decomposition: $X - E[X \mid \mathcal{B}_m] = X - E[X \mid \mathcal{B}_n] + E[X \mid \mathcal{B}_n] - E[X \mid \mathcal{B}_m]$, we have

$$
\|X - E[X \mid \mathcal{B}_m]\|_{L^2}^2 = \|X - E[X \mid \mathcal{B}_n]\|_{L^2}^2 + \|E[X \mid \mathcal{B}_n] - E[X \mid \mathcal{B}_m]\|_{L^2}^2 .
$$

This equality implies that the sequence $(\|X - E[X \mid \mathcal{B}_n]\|_{L^2})$ increases with m and is bounded by $2.\|X\|_{L^2}$, hence it converges. Furthermore, the same equality implies that $\|E[X \mid \mathcal{B}_m] - E[X \mid \mathcal{B}_n]\|_{L^2}$ tends to 0 as n and m go to ∞: we proved that $(E[X \mid \mathcal{B}_n])$ is a Cauchy sequence in $L^2(\Omega, \mathcal{F}, P)$, hence it converges. Call Y the limit in L^2 of $(E[X \mid \mathcal{B}_n])$. Note that Y is $\mathcal{B}$-measurable. For any $B \in \mathcal{B}$ and $n \in \mathbb{N}$, we have $E[X \mathbb{1}_B] = E[E[X \mid \mathcal{B}_n] \mathbb{1}_B]$. Letting n go to ∞ on the right hand side, we obtain $E[X \mathbb{1}_B] = E[Y \mathbb{1}_B]$, hence $Y = E[X \mid \mathcal{B}]$.

Solution to Exercise 1.7

1. Since T preserves P, we have:

$$
E[\Phi(\omega, X(\omega))] = E[\Phi(T(\omega), X(T(\omega)))] .
$$

When X is almost T-invariant, the right hand side is $E[\Phi(T(\omega), X(\omega))]$.

2. Let g be such a function. Every $T^{-1}(\mathcal{F})$-measurable function is of the form $\phi \circ T$ where ϕ is an $\mathcal{F}$-measurable function. Therefore, it suffices to prove that for any $\mathbb{R}_+$-valued $\mathcal{F}$-measurable function ϕ,

$$E[\phi(T(\omega))g(X(\omega))] = E[\phi(T(\omega))g(X(T(\omega)))] \,.$$

Since T preserves P, it is equivalent to prove that

$$E[\phi(T(\omega))g(X(\omega))] = E[\phi(\omega)g(X(\omega))] \,.$$

But this identity has already been proved in question 1, with $\Phi(T(\omega), X(\omega)) = \phi(T(\omega))g(X(\omega))$.

3. If (1.7.1) is satisfied, then by question 2, (1.7.2) holds and by Exercise **1.5**, $X = X \circ T$ a.s.

Solution to Exercise 1.8

1. Property (a) implies that for every f and g in L^2,

$$E\left[f(g \circ T^n)\right] \xrightarrow[n\to\infty]{} E(f)E(g) \,. \tag{1.8.a}$$

Indeed, let (f_k) and (g_k) be two sequences of $\mathcal{H}$ which converge respectively towards f and g in L^2 and write

$$|E[f(g \circ T^n)] - E[f]E[g]| \leq |E[f(g \circ T^n)] - E[f_k(g_k \circ T^n)]|$$
$$+ |E[f_k(g_k \circ T^n)] - E[f_k]E[g_k]| + |E[f_k]E[g_k] - E[f]E[g]| \,.$$

Since T preserves P, then for any n,

$$|E[f(g \circ T^n)] - E[f_k(g_k \circ T^n)]| = |E[(f - f_k)(g_k \circ T^n)] + E[f(g \circ T^n - g_k \circ T^n)]|$$
$$\leq E[(f - f_k)^2]^{\frac{1}{2}} E[(g_k \circ T^n)^2]^{\frac{1}{2}} + E[f^2]^{1/2} E[(g \circ T^n - g_k \circ T^n)^2]^{\frac{1}{2}}$$
$$= E[(f - f_k)^2]^{\frac{1}{2}} E[g_k^2]^{\frac{1}{2}} + E[f^2]^{1/2} E[(g - g_k)^2]^{\frac{1}{2}} \,,$$

so that for any ε and n, there exists K, such that for all $k \geq K$, both terms $|E[f(g \circ T^n)] - E[f_k(g_k \circ T^n)]|$ and $|E[f_k]E[g_k] - E[f]E[g]|$ are less than ε. The result is then a consequence of the fact that for any k, the term $|E[f_k(g_k \circ T^n)] - E[f_k]E[g_k]|$ converges towards 0 as $n \to \infty$.

Now let $g \in L^2(\mathcal{I})$, then $g \circ T^n = g$ and (1.8.a) yields $E[fg] = E[f]E[g]$, for any $f \in L^2(\mathcal{F})$. This implies $g = E[g]$, hence $\mathcal{I}$ is trivial.

2. Let $\mathcal{H} = \cup_{k\geq 0}L^2(\Omega, \mathcal{F}_k, P)$, then from Exercise **1.6**, $\mathcal{H}$ is dense in $L^2(\Omega, \mathcal{F}, P)$. Now we prove that $\mathcal{H}$ satisfies property (a).

Let $g \in \mathcal{H}$, then there exists $k \in \mathbb{N}$ such that $g \in L^2(\Omega, \mathcal{F}_k, P)$. Moreover, from (c), $((T^n)^{-1}(\mathcal{F}_k))_{n\geq 0}$ is a decreasing sequence of σ-fields. Let $f \in \mathcal{H}$, then from (d) and Exercise **1.6**, $E[f \,|\, (T^n)^{-1}(\mathcal{F}_k)] \overset{k\to\infty}{\longrightarrow} E[f]$, in $L^2(\Omega, \mathcal{F}, P)$. Put $f_n = E[f \,|\, (T^n)^{-1}(\mathcal{F}_k)]$ and $g_n = g \circ T^n$, then we have,

$$|E[f_n g_n] - E[f]E[g_n]| \leq \|f_n - E[f]\|_{L^2}\|g_n\|_{L^2}$$

and since $E[g_n] = E[g]$ and $\|g \circ T^n\|_{L^2} = \|g\|_{L^2}$, one has the required convergence.

Solution to Exercise 1.9

1. By (ii) and the invariance of P under T, we have:

$$E[fg] = E[f \circ T \cdot g \circ T] = E[f \circ T \cdot g],$$

and thus (ii) implies (i).

Suppose that (i) holds then by applying this property to $f, f \circ T, f \circ T^2, \cdots f \circ T^n$, successively, we get: $E[fg] = E[f \circ T^n \cdot g]$ for every $n \in \mathbb{N}^*$, hence,

$$E[fg] = E\left[\frac{1}{n}\left(\Sigma_{p=1}^n f \circ T^p\right) \cdot g\right].$$

Since $f \in L^\infty$, we can apply Lebesgue's theorem of dominated convergence together with the Ergodic Theorem to get

$$\lim_{n\to\infty} E\left[\frac{1}{n}\left(\Sigma_{p=1}^n f \circ T^p\right) \cdot g\right] = E[E[f \,|\, \mathcal{J}]g].$$

Consequently, one has $E[fg] = E[E[f \,|\, \mathcal{J}]g]$, for every $f \in L^\infty$. This identity is equivalent to

$$\begin{aligned} E[fg] &= E[E[f \,|\, \mathcal{J}]E[g \,|\, \mathcal{J}]] \\ &= E[fE[g \,|\, \mathcal{J}]], \end{aligned}$$

for every $f \in L^\infty$, which implies that $g = E[g \,|\, \mathcal{J}]$, a.s. This last statement is equivalent to $g = g \circ T$, a.s.

Solution to Exercise 1.10

1. Call L the vector space generated by 1 and Φ. First, assume that L is dense in $L^1(\Omega, \mathcal{F}, P)$ and that $P = \alpha P_1 + (1-\alpha)P_2$, for $\alpha \in (0, 1)$ and $P_1, P_2 \in \mathcal{M}_{\Phi, c}$.

We easily derive from the previous relation that L is dense in $L^1(\Omega, \mathcal{F}, P_i)$, $i = 1, 2$. Moreover, it is clear by (b) that P_1 and P_2 agree on L, hence it follows that $P_1 = P_2$.

Conversely, assume that L is not dense in $L^1(\Omega, \mathcal{F}, P)$. Then from the Hahn–Banach theorem, there exists $g \in L^\infty(\Omega, \mathcal{F}, P)$ with $P(g \neq 0) > 0$, such that $\int gf \, dP = 0$, for every $f \in L$. We may assume that $\|g\|_\infty \leq 1/2$, then put $P_1 = (1 - g)P$ and $P_2 = (1 + g)P$. Clearly, P_1 and P_2 belong to $\mathcal{M}_{\Phi,c}$ and we have $P = \frac{1}{2}(P_1 + P_2)$ but these probabilities are not equal to P since $P(g \neq 0) > 0$.

2. (i) From question 1, the vector space $\tilde{\Phi}$ generated by 1 and Φ is dense in $L^1(\Omega, \mathcal{F}, P)$. Since $\frac{dQ}{dP}$ is bounded, $\tilde{\Phi}$ is also dense in $L^1(\Omega, \mathcal{F}, Q)$.
(ii) Under the hypothesis, $L^1(\Omega, \mathcal{F}, Q)$ and $L^1(\Omega, \mathcal{F}, P)$ are identical.

3. This study is a particular moments problem with

$$\Phi \overset{(\text{def})}{=} \{f - f \circ T; f \in b(\Omega, \mathcal{F})\},$$

and the constants $c_f = 0$.

So, $P \in \mathcal{M}_T$ is extremal if and only if $\Phi \cup \{1\}$ spans a dense space in $L^1(P)$, or equivalently the only functions g in $L^\infty(\Omega, \mathcal{F}, P)$ such that:

$$\text{for all } f \in b(\Omega, \mathcal{F}), \quad E[fg] = E[(f \circ T)g] \tag{1.10.a}$$

are the constants. But, in Exercise **1.9**, we proved that (1.10.a) is equivalent to the fact that g is almost T-invariant; thus $P \in \mathcal{M}_T$ is extremal if and only if $\mathcal{I}$ is P-trivial, that is T is ergodic under P.

Solution to Exercise 1.11

1. The density of X_{σ^2} is readily obtained as:

$$\frac{1}{\sqrt{2\pi\sigma^2}} \frac{1}{y} \exp\left(-\frac{(\log y)^2}{2\sigma^2}\right), \quad y > 0.$$

2. We obtain the equality (1.11.1) by applying the formula

$$E[\exp(zN_{\sigma^2})] = \exp\left(\frac{z^2\sigma^2}{2}\right)$$

(which is valid for every $z \in \mathbb{C}$) with $z = n + i\rho$, $\rho = \frac{p\pi}{\sigma^2}$ and $n, p \in \mathbb{Z}$.

3. Let for instance μ be the law of $\exp(N_{\sigma^2})$ under the probability measure

$$\left(1 + \sum_p c_p \sin\left(\frac{p\pi}{\sigma^2} N_{\sigma^2}\right)\right) dP,$$

where (c_p) is any sequence of reals such that $\sum_p |c_p| \leq 1$, then question 1, asserts that (i) is satisfied. Moreover, (ii) is trivially satisfied.

Solution to Exercise 1.12

1. From the hypothesis, we deduce that

$$E[(X - Y)^2] = E[X^2] - E[Y^2] = 0.$$

2. We first note that $E[X \wedge a \,|\, \mathcal{G}] \leq Y \wedge a$, but since $X \wedge a$ and $Y \wedge a$ have the same law, this inequality is in fact an equality. Likewise, we obtain (1.12.1). The same argument as for question 1 now yields:

$$(X \wedge a) \vee (-b) = (Y \wedge a) \vee (-b) \quad \text{a.s.,}$$

and finally, letting a and b tend to $+\infty$, we obtain $X = Y$, a.s.

3. We easily reduce the proof to the case where X and Y are bounded. Then the hypothesis implies simultaneously that X and Y have the same law, and $E[X \,|\, \mathcal{G}] = Y$, so that we can apply the above result.

4. Under the hypothesis (i) of Exercise **1.9**, we deduce

$$E[g \,|\, T^{-1}(\mathcal{F})] = g \circ T.$$

Hence the above result yields $g = g \circ T$, a.s.

5. It is easily deduced from the hypothesis that the identity (1.12.1) is satisfied, hence $X = Y$, a.s.

Solution to Exercise 1.13

1. The main difficulty in this question lies in the fact that we cannot define "$\log X$" and "$\log Z$", since X and Z may actually be zero on non-negligible sets. We rewrite the hypothesis $XY \stackrel{\text{(law)}}{=} ZY$ trivially as

$$\mathbb{1}_{\{X>0\}} XY \stackrel{\text{(law)}}{=} \mathbb{1}_{\{Z>0\}} ZY,$$

and since $P(Y = 0) = 0$, we can write, for any $\lambda \neq 0$:

$$E\left[\mathbb{1}_{\{X>0\}} \exp\left(i\lambda(\log X + \log Y)\right)\right] = E\left[\mathbb{1}_{\{Z>0\}} \exp\left(i\lambda(\log Z + \log Y)\right)\right].$$

From the independence hypothesis, we obtain

$$E\left[\mathbb{1}_{\{X>0\}}\exp\left(i\lambda\log X\right)\right]E\left[\exp\left(i\lambda\log Y\right)\right)]$$
$$= E\left[\mathbb{1}_{\{Z>0\}}\exp\left(i\lambda\log Z\right)\right]E\left[\exp\left(i\lambda\log Y\right)\right)],$$

from which we easily deduce $X \overset{(\text{law})}{=} Z$.

2. Applying the Fourier inverse transform, we can check that the characteristic function of the density $f(x) = \frac{1}{\pi}\frac{1-\cos x}{x^2}$ is given by

$$\varphi(t) = \begin{cases} 1 - |t| & \text{for} \quad |t| \le 1 \\ 0 & \text{for} \quad |t| > 1 \end{cases},$$

and that the characteristic function of the density $g(x) = \frac{1}{\pi}\frac{(1-\cos x)(1+\cos 2x)}{x^2}$ is $\psi(t) = \varphi(t) + \frac{1}{2}(\varphi(t-2) + \varphi(t+2))$, for all $t \in \mathbb{R}$.

Let X and Y be r.v.s with values in $\mathbb{R}_+ \setminus \{0\}$ such that $\log X$ has density g and $\log Y$ has density f. Let Z be an independent copy of Y, then equation $\psi(t)\varphi(t) = \varphi^2(t)$, for all $t \in \mathbb{R}$ ensures that $XY \overset{(\text{law})}{=} ZY$. Nonetheless, X and Z have different laws, so Y is not a simplifiable variable.

3. Assume $A \overset{(\text{law})}{=} YC$, then, we have: $Y \overset{(\text{law})}{=} YCB$, but since Y is simplifiable, we deduce: $1 \overset{(\text{law})}{=} CB$, hence $CB = 1$, a.s. This is impossible since C and B are assumed to be independent, and B is not constant.

Solution to Exercise 1.14

We deduce from the hypothesis that:

$$E[X^s] = E[Z^s],$$

for $0 \le s \le \varepsilon$. Thus, the laws of X and Z have the same Mellin transforms on $[0,\varepsilon]$, hence these laws are equal (see the comments at the end of the statement of this exercise).

———————

A relevant reference: H. Georgii [35] makes use of a number of the arguments employed throughout our exercises, especially in the present Chapter 1. Thus, as further reading, it may be interesting to look at the discussions in [35] related to extremal Gibbs measures.

Solution to Exercise 1.15

1. By monotone class theorem, identity (1.15.1) would imply that, for any $f \in L^1(m)$,

$$\int_A m(dx)f(x) = c \int m(dx)f(x),$$

which implies: $\mathbb{1}_A = c$, $m(dx)$-a.e. This is absurd.

2. Assume that

$$|m(A \cap I) - cm(I)| \leq |m(A \cap I) - dm(I)|.$$

Then it follows that $(m(A \cap I) - cm(I))^2 \leq (m(A \cap I) - dm(I))^2$, and after a few simplifications, we get

$$(c^2 - d^2)m(I) \leq 2m(A \cap I)(c - d). \tag{1.15.a}$$

Assume that $c > d$, then we derive from (1.15.a) that $(c+d)m(I) \leq 2m(A \cap I)$, for every interval I. We deduce from this inequality and the monotone class theorem that for any nonnegative function $f \in L^1(m)$,

$$\int m(dx)f(x)[2\mathbb{1}_A(x) - (c+d)] \geq 0.$$

This implies that $2\mathbb{1}_A(x) \geq c + d$, for m-almost every x.

If $m(A^c) > 0$, then this inequality implies $c + d = 0$ and hence $c = d = 0$.

If $m(A^c) = 0$, then from (1.15.a), one has $c^2 - d^2 \leq 2(c - d)$, hence $c + d \leq 2$, which is true.

Now assume $c \leq d$, then from (1.15.a), we derive that

$$2m(I \cap A) \leq (c+d)m(I),$$

for every interval I. This implies that for m-almost every x, $2\mathbb{1}_A(x) \leq c + d$.

If $m(A) > 0$, then necessarily, we have $c = d = 1$.

If $m(A) = 0$, then c and d can be any constants such that $0 < c \leq d \leq 1$.

We summarize the above discussion in the following table.

	$c > d$	$c \leq d$
$m(A^c) > 0$	$c = d = 0$	
$m(A^c) = 0$	any c, d	
$m(A) > 0$		$c = d = 1$
$m(A) = 0$		any c, d

Chapter 2

Independence and conditioning

> ### "Philosophy" of this chapter

(a) A probabilistic model $\{(\Omega, \mathcal{F}, P); (X_i)_{i \in I}\}$ consists of setting together in a mathematical way different sources of randomness, i.e. the r.v.s $(X_i)_{i \in I}$ usually have some complicated joint distribution

It is always a simplification, and thus a progress, to replace this "linked" family by an "equivalent" family $(Y_j)_{j \in J}$ of independent random variables, where by equivalence we mean the equality of their σ-fields: $\sigma(X_i, i \in I) = \sigma(Y_j, j \in J)$ up to negligible sets.

(b) Assume that the set of indices I splits into $I_1 + I_2$, and that we know the outcomes $\{X_i(\omega); i \in I_1\}$. This modifies deeply our perception of the randomness of the system, which is now reduced to understanding the conditional law of $(X_i)_{i \in I_2}$, given $(X_i)_{i \in I_1}$. This is the main theme of D. Williams' book [64].

(c) Again, it is of great interest, even after this conditioning with respect to $(X_i)_{i \in I_1}$, to be able to replace the family $(X_i)_{i \in I_2}$ by an "equivalent" family $(Y_j)_{j \in J_2}$, which consists of independent variables, conditionally on $(X_i)_{i \in I_1}$.

Note that the terms "independence" and "conditioning" come from our everyday language and are very suitable as translations of the corresponding probabilistic concepts. However, some of our exercises aim at pointing out some traps which may originate from this common language meaning.

(d) The Markov property (in a general framework) asserts the conditional independence of the "past" and "future" σ-fields given the "present" σ-field. It provides an unending source of questions closely related to the topic of this chapter. The elementary articles

F.B. KNIGHT: A remark on Markovian germ fields. *Z. Wahrscheinlichkeitstheorie und Verw. Gebiete*, **15**, 291–296 (1970)

K.L. CHUNG: Some universal field equations. *Séminaire de Probabilités VI*, 90–97, *Lecture Notes in Mathematics*, **258**, Springer, Berlin, 1972.

give the flavor of such studies.

---

* 2.1 Independence does not imply measurability with respect to an independent complement

1. Assume that, on a probability space $(\Omega, \mathcal{F}, P)$, there exist a symmetric Bernoulli variable ε (that is, ε satisfies: $P(\varepsilon = +1) = P(\varepsilon = -1) = \frac{1}{2}$), and an r.v. X, which are independent.

   Show that εX and ε are independent iff X is symmetric (that is: $X \overset{(\text{law})}{=} -X$).

2. Construct, on an adequate probability space $(\Omega, \mathcal{F}, P)$, two independent σ-fields $\mathcal{A}$ and $\mathcal{B}$, and an r.v. Y such that:
   (i) Y is $\mathcal{A} \vee \mathcal{B}$-measurable;   (ii) Y is independent of $\mathcal{B}$;
   (iii) Y is not measurable with respect to $\mathcal{A}$.

3. Construct, on an adequate probability space $(\Omega, \mathcal{F}, P)$, two independent σ-fields $\mathcal{A}$ and $\mathcal{B}$, and a non-constant r.v. Z such that:
   (j) Z is independent of $\mathcal{A}$;   (jj) Z is independent of $\mathcal{B}$;
   (jjj) Z is $\mathcal{A} \vee \mathcal{B}$-measurable.

4. Let $\mathbb{G}$ be a Gaussian subspace of $L^2(\Omega, \mathcal{F}, P)$ which admits the direct sum decomposition: $\mathbb{G} = \mathbb{G}_1 \oplus \mathbb{G}_2$. (Have a brief look at Chapter 3, if necessary....)
   Define $\mathcal{A} = \sigma(\mathbb{G}_1)$, and $\mathcal{B} = \sigma(\mathbb{G}_2)$.

   (a) Show that there is no variable $Y \in \mathbb{G}$, $Y \neq 0$ such that the hypotheses (i)–(ii)–(iii) are satisfied.

   (b) Show that there is no variable $Z \in \mathbb{G}$, $Z \neq 0$, such that the hypotheses (j)–(jj)–(jjj) are satisfied.

Comments and references. This exercise is an invitation to study the notion and properties (starting from the existence) of an independent complement to a given sub-σ-field $\mathcal{G}$ in a probability space $(\Omega, \mathcal{F}, P)$. We refer the reader to the famous article

V.A. ROHLIN: On the fundamental ideas of measure theory. *Mat. Sbornik N.S.*, **25**, no. 67, 107–150 (1949).

(Be aware that this article is written in a language which is closer to Ergodic Theory than to Probability Theory.)

* 2.2 Complement to Exercise 2.1: further statements of independence versus measurability

Consider, on a probability space $(\Omega, \mathcal{F}, P)$, three sub-σ-fields $\mathcal{A}, \mathcal{B}, \mathcal{C}$, which are $(\mathcal{F}, P)$ complete.

Assume that:

$\qquad$ (i) $\;\mathcal{A} \subseteq \mathcal{B} \vee \mathcal{C}$, and (ii) $\;\mathcal{A}$ and $\mathcal{C}$ are independent.

1. Show that if the hypotheses
   (i) $\;\mathcal{A} \subseteq \mathcal{B} \vee \mathcal{C}$, and (ii)$'$ $\mathcal{A} \vee \mathcal{B}$ is independent of $\mathcal{C}$
   are satisfied, then $\mathcal{A}$ is included in $\mathcal{B}$.

2. Show that, if (ii)$'$ is not satisfied, it is not always true that $\mathcal{A}$ is included in $\mathcal{B}$.

3. Show that if, besides (i) and (ii), the property (iii): $\mathcal{B} \subseteq \mathcal{A}$ is satisfied, then: $\mathcal{A} = \mathcal{B}$.

** 2.3 Independence and mutual absolute continuity

Let $(\Omega, \mathcal{F}, P)$ be a probability space and $\mathcal{G}$ be a sub-σ-field of $\mathcal{F}$.

1. Let $\Gamma \in \mathcal{F}$. Prove that the following properties are equivalent:

   (i) Γ is independent of $\mathcal{G}$ under P,

   (ii) for every probability Q on $(\Omega, \mathcal{F})$, equivalent to P, with $\left(\frac{dQ}{dP}\right)$ $\mathcal{G}$ measurable, $Q(\Gamma) = P(\Gamma)$.

2. Let Q be a probability on $(\Omega, \mathcal{F})$ which is equivalent to P and consider the following properties:

   (j) $\frac{dQ}{dP}$ is $\mathcal{G}$ measurable

   (jj) for every set $\Gamma \in \mathcal{F}$ independent of $\mathcal{G}$ under P, $Q(\Gamma) = P(\Gamma)$.

   Prove that (j) implies (jj).

3. Prove that in general, (jj) does not imply (j).

   Hint: Let $\mathcal{G} = \{\emptyset, \Omega, \{a\}, \{a\}^c\}$, assuming that $\{a\} \in \mathcal{F}$, and $P(\{a\}) > 0$. Show that if X, $\mathcal{F}$-measurable, is independent from $\mathcal{G}$, then X is constant. Prove that if $\mathcal{G} \subset \mathcal{F}$, with $\mathcal{G} \neq \mathcal{F}$, then there exists Q which satisfies (jj), but not (j).

Comments and references:

(a) This exercise emphasizes the difficulty of characterizing the set of events which are independent of a given σ-field. Another way to ask question 2, is:
    "Does the set of events which are independent of $\mathcal{G}$ characterize $\mathcal{G}$?"

Corollary 4 of the following paper is closely related to our problem:

M. ÉMERY AND W. SCHACHERMAYER: On Vershik's standardness criterion and Tsirelson's notion of cosiness. *Séminaire de Probabilités XXXV*, 265–305, *Lecture Notes in Mathematics*, **1755**, Springer, Berlin, 2001.

(b) It is tempting to think that if $\mathcal{G}$ has no atoms, then (jj) implies (j). But, this is not true: M. Emery kindly gave us some examples of σ-fields $\mathcal{G}$ without atoms, with $\mathcal{G}$ strictly included in $\mathcal{F}$, and such that any event in $\mathcal{F}$ independent from $\mathcal{G}$ is trivial.

* 2.4 Size-biased sampling and conditional laws

Let $X_1, \ldots, X_n$, be n independent, equidistributed r.v.s which are a.s. strictly positive.

Define $S_n = X_1 + X_2 + \cdots + X_n$, and assume $n \geq 3$.

Let, moreover, J be an r.v. taking values in $\{1, 2, \ldots, n\}$ such that:

$$P(J = j \mid X_1, \ldots, X_n) = X_j/S_n.$$

We define the $(n-1)$-dimensional r.v. $\mathbf{X}^*_{(n-1)} = (X_1^*, \ldots, X_{n-1}^*)$ as follows:

$$X_i^* = \begin{cases} X_i & , \text{ if } i < J \\ X_{i+1} & , \text{ if } i \geq J \end{cases} \qquad (i \leq n-1) \ .$$

Show that, given $S_{n-1}^* \equiv X_1^* + \cdots + X_{n-1}^*$ the r.v.s $\mathbf{X}^*_{(n-1)}$ and X_J are independent, and that, moreover, the conditional law of $\mathbf{X}^*_{(n-1)}$, given $S_{n-1}^* = s$, is identical to the conditional law of $\mathbf{X}_{(n-1)} = (X_1, \ldots, X_{n-1})$, given $S_{n-1} = s$.

Comments and references. This result is the first step in proving the inhomogeneous Markov property for $\left(S_n - \sum_{j=1}^m \tilde{X}_j, m = 1, 2, \ldots, n\right)$, where $(\tilde{X}_1, \ldots, \tilde{X}_n)$ is a size-biased permutation of $(X_1, \ldots, X_n)$, see, e.g., p. 22 in:

M. PERMAN, J. PITMAN AND M. YOR: Size-biased sampling of Poisson point processes and excursions. *Probab. Theory and Related Fields*, **92**, no. 1, 21–39 (1992)

for the definition of such a random permutation. See also:

L. GORDON: Estimation for large successive samples with unknown inclusion probabilities. *Adv. in Appl. Math.*, **14**, no. 1, 89–122 (1993)

and papers cited there for more on size-biased sampling of independent and identically distributed sequences, as well as Lemma 10 and Proposition 11 in:

J. PITMAN: Partition structures derived from Brownian motion and stable subordinators. *Bernoulli*, **3**, no. 1, 79–96 (1997)

for some related results. We thank J. Pitman for his suggestions about this exercise.

** 2.5 Think twice before exchanging the order of taking the supremum and intersection of σ-fields!

Let $(\Omega, \mathcal{F}, P)$ be a probability space, $\mathcal{C}$ a sub-σ-field of $\mathcal{F}$, and $(\mathcal{D}_n)_{n \in \mathbb{N}}$ a decreasing sequence of sub-σ-fields of $\mathcal{F}$.

$\mathcal{C}$ and $\mathcal{D}_n$, for every n, are assumed to be $(\mathcal{F}, P)$ complete.

1. Prove that if, for every n, $\mathcal{C}$ and $\mathcal{D}_1$ are conditionally independent given $\mathcal{D}_n$, then:

$$\bigcap_n (\mathcal{C} \vee \mathcal{D}_n) = \mathcal{C} \vee \left(\bigcap_n \mathcal{D}_n \right) \quad \text{holds.} \tag{2.5.1}$$

2. If there exists a sub-σ-field $\mathcal{E}_n$ of $\mathcal{D}_n$ such that $\mathcal{C}$ and $\mathcal{D}_1$ are conditionally independent given $\mathcal{E}_n$, then $\mathcal{C}$ and $\mathcal{D}_1$ are conditionally independent given $\mathcal{D}_n$.

   Consequently, if $\mathcal{C}$ and $\mathcal{D}_1$ are independent, then (2.5.1) holds.

3. The sequence $(\mathcal{D}_n)_{n \in \mathbb{N}}$ and the σ-field $\mathcal{C}$ are said to be asymptotically independent if, for every bounded $\mathcal{F}$-measurable r.v. X and every bounded $\mathcal{C}$-measurable r.v. C, one has:

$$E[E(X \mid \mathcal{D}_n)C] \xrightarrow[n \to \infty]{} E(X)E(C). \tag{2.5.2}$$

   Prove that the condition (2.5.2) holds iff $\bigcap_n \mathcal{D}_n$ and $\mathcal{C}$ are independent.

4. Let $Y_0, Y_1, \ldots$ be independent symmetric Bernoulli r.v.s For $n \in \mathbb{N}$, define $X_n = Y_0 Y_1 \ldots Y_n$ and set $\mathcal{C} = \sigma(Y_1, Y_2, \ldots)$, $\mathcal{D}_n = \sigma(X_k : k > n)$. Prove that (2.5.1) fails in this particular case.

   Hint: Prove that $\bigcap_n (\mathcal{C} \vee \mathcal{D}_n) = \mathcal{C} \vee \sigma(Y_0)$; but $\mathcal{C} \vee (\bigcap_n \mathcal{D}_n) = \mathcal{C}$, since $\bigcap_n \mathcal{D}_n$ is trivial.

Comments and references:

(a) The need to determine germ σ-fields $\mathcal{G} \overset{\text{def}}{=} \bigcap_n \mathcal{D}_n$ occurs very naturally in many problems in Probability Theory, often leading to 0–1 laws, i.e. $\mathcal{G}$ is trivial.

(b) A number of authors, (including the present authors, separately!), gave wrong proofs of (2.5.1) under various hypotheses. This seems to be one of the worst traps involving σ-fields.

(c) A necessary and sufficient criterion for (2.5.1) to hold is presented in:

H. VON WEIZSÄCKER: Exchanging the order of taking suprema and countable intersection of σ-algebras. *Ann. I.H.P.,* **19**, 91–100 (1983).

However, this criterion is very difficult to apply in any given set-up.

(d) The conditional independence hypothesis made in question 1 above is presented in:

T. LINDVALL AND L.C.G. ROGERS: Coupling of multidimensional diffusions by reflection. *Ann. Prob.,* **14**, 860–872 (1986).

(e) The following papers discuss, in the framework of a stochastic equation, instances where (2.5.1) may hold or fail.

M. YOR: De nouveaux résultats sur l'équation de Tsirelson. *C. R. Acad. Sci. Paris Sér. I Math.,* **309**, no. 7, 511–514 (1989).

M. YOR: Tsirelson's equation in discrete time. *Probab. Theory and Related Fields,* **91**, no. 2, 135–152 (1992).

(f) A simpler question than the one studied in the present exercise is whether the following σ-fields are equal:

$$(\mathcal{A}_1 \vee \mathcal{A}_2) \cap \mathcal{A}_3 \quad \text{and} \quad (\mathcal{A}_1 \cap \mathcal{A}_3) \vee (\mathcal{A}_2 \cap \mathcal{A}_3) \qquad (2.5.3)$$
$$(\mathcal{A}_1 \cap \mathcal{A}_2) \vee \mathcal{A}_3 \quad \text{and} \quad (\mathcal{A}_1 \vee \mathcal{A}_3) \cap (\mathcal{A}_2 \vee \mathcal{A}_3). \qquad (2.5.4)$$

With the help of the Bernoulli variables ε_1, ε_2, $\varepsilon_3 = \varepsilon_1 \varepsilon_2$ and the σ-fields $\mathcal{A}_i = \sigma(\varepsilon_i)$, $i = 1,2,3$, already considered in Exercise **2.2**, one sees that the σ-fields in (2.5.3) and (2.5.4) may not be equal.

* 2.6 Exchangeability and conditional independence: de Finetti's theorem

A sequence of random variables $(X_n)_{n \geq 1}$ is said to be exchangeable if for any permutation σ of the set $\{1, 2, \ldots\}$

$$(X_1, X_2, \ldots) \overset{\text{(law)}}{=} (X_{\sigma(1)}, X_{\sigma(2)}, \ldots).$$

Let $(X_n)_{n\geq 1}$ be such a sequence and $\mathcal{G}$ be its tail σ-field, i.e. $\mathcal{G} = \cap_n \mathcal{G}_n$, with $\mathcal{G}_n = \sigma\{X_n, X_{n+1}, \ldots\}$, for $n \geq 1$.

1. Show that for any bounded Borel function Φ,

$$E[\Phi(X_1)\,|\,\mathcal{G}] \overset{\text{(law)}}{=} E[\Phi(X_1)\,|\,\mathcal{G}_2].$$

2. Show that the above identity actually holds almost surely.

3. Show that the r.v.s $X_1, X_2, \ldots$ are conditionally independent given the tail σ-field $\mathcal{G}$.

Comments and references:

(a) The result proved in this exercise is the famous de Finetti's Theorem, see e.g.

B. DE FINETTI: La prévision: ses lois logiques, ses sources subjectives. *Ann. Inst. H. Poincaré*, **7**, 1–66 (1937).

It essentially says that any sequence of exchangeable random variables is a "mixture" of i.i.d. random variables.

D.J. ALDOUS: Exchangeability and related topics. *École d'été de probabilités de Saint-Flour, XIII–1983*, 1–198, *Lecture Notes in Mathematics*, **1117**, Springer, Berlin, 1985.

O. KALLENBERG: *Foundations of Modern Probability*. Second edition. Springer-Verlag, New York, 2002. (See Theorem 11.10, p. 212).

See also P.A. Meyer's discussion in [40] of the Hewitt–Savage Theorem.

(b) Question 2 is closely related to Exercise **1.5**, and/or to Exercise **1.12**.

(c) De Finetti's Theorem extends to the continuous time setting in the following form: *Any càdlàg exchangeable process is a mixture of Lévy processes.* See Proposition 10.5 of Aldous' course cited above. See also Exercise **6.24** for some applications.

* 2.7 On exchangeable σ-fields

A sequence $(\mathcal{G}_1, \ldots, \mathcal{G}_n)$ of sub-σ-fields of a probability space $(\Omega, \mathcal{F}, P)$ is said to be exchangeable if for all $i, j = 1, 2, \ldots, n$ and all $X \in L^1(\mathcal{P})$,

$$E\left(E(X\,|\,\mathcal{G}_i)\,|\,\mathcal{G}_j\right) = E\left(E(X\,|\,\mathcal{G}_j)\,|\,\mathcal{G}_i\right), \quad \text{a.s.} \tag{2.7.1}$$

1. Find a sequence of sub-σ-fields for which (2.7.1) is not satisfied.
   Hint. You may use the Gaussian vector of Exercise 3.3.

2. (a) Let X be a random variable such that for two σ-fields $\mathcal{G}_1$ and $\mathcal{G}_2$,

$$E(X \mid \mathcal{G}_1) = E(X \mid \mathcal{G}_2).$$

Prove that this common r.v. is also equal to $E(X \mid \mathcal{G}_1 \cap \mathcal{G}_2)$.

(b) Show that the previous assertion admits no converse, i.e. give an example of an r.v. X such that $E(X \mid \mathcal{G}_1)$ is $\mathcal{G}_1 \cap \mathcal{G}_2$-measurable but, nonetheless, $E(X \mid \mathcal{G}_1)$ and $E(X \mid \mathcal{G}_2)$ differ.

3. Show that $(\mathcal{G}_1, \dots, \mathcal{G}_n)$ is exchangeable if and only if $\mathcal{G}_1, \dots, \mathcal{G}_n$ are independent conditionally on $\mathcal{G}_1 \cap \dots \cap \mathcal{G}_n$.

4. Show that if $X_1, \dots, X_n$ are exchangeable random variables in the sense of Exercise **2.6**, then $\sigma(X_1), \dots, \sigma(X_n)$ is exchangeable.

    o Is the converse true?

5. Show that if there exist a filtration $(\mathcal{F}_k)$ and $T_1, \dots, T_n$, n $(\mathcal{F}_k)$-stopping times such that $\mathcal{G}_i = \mathcal{F}_{T_i}$, then $\mathcal{G}_1, \dots, \mathcal{G}_n$ is exchangeable with respect to any probability measure on $(\Omega, \mathcal{F})$.

    o Is the converse true?

Comments and references: It follows from questions 1 and 5 that for any two sub-σ-fields $\mathcal{G}$ and $\mathcal{H}$ of a probability space $(\Omega, \mathcal{F}, P)$, it is not always possible to find a filtration $(\mathcal{F}_t)$ and two stopping times of this filtration S and T such that $\mathcal{F}_S = \mathcal{G}$ and $\mathcal{F}_T = \mathcal{H}$.

M. YOR: Sur certains commutateurs d'une filtration. *Séminaire de Probabilités XV*, 526–528, *Lecture Notes in Mathematics*, **850**, Springer, Berlin, 1981.

Many results on exchangeability are gathered in:

D.J. ALDOUS: Exchangeability and related topics. *École d'été de probabilités de Saint-Flour, XIII—1983*, 1–198, *Lecture Notes in Mathematics*, **1117**, Springer, Berlin, 1985.

* 2.8 Too much independence implies constancy

Let $(\Omega, \mathcal{F}, P)$ be a probability space on which two real valued r.v.s X and Y are defined.

The aim of questions 1, 2, 3 and 4 is to show that the property:

(I) $\begin{cases} (X - Y) \text{ and } X \text{ are independent} \\ (X - Y) \text{ and } Y \text{ are independent} \end{cases}$

can only be satisfied if $X - Y$ is a.s. constant.

1. Prove the result when X and Y have a second moment.

In the following, we make no integrability assumption on either X or Y.

2. Let φ (resp. H), be the characteristic function of X (resp. $X - Y$). Show that, if (I) is satisfied, then the identity:

$$\varphi(x)\left(1 - |H(x)|^2\right) = 0\,, \quad \text{for every } x \in \mathbb{R}\,, \tag{2.8.1}$$

holds.

3. Show that if (2.8.1) is satisfied, then:

$$|H(x)| = 1\,, \quad \text{for } |x| < \varepsilon,\text{ and } \varepsilon > 0,\text{ sufficiently small}\,. \tag{2.8.2}$$

4. Show that if (2.8.2) is satisfied, then $X - Y$ is a.s. constant.

In the same vein, we now discuss how much constraint may be put on the conditional laws of either one of the components of a two dimensional r.v., given the other.

5. Is it possible to construct a pair of r.v.s (X, Y) which satisfy the following property, for all $x, y \in \mathbb{R}$:

(J) $\begin{cases} \text{conditionally on } X = x, & Y \text{ is distributed as } \mathcal{N}(x, 1) \\ \text{conditionally on } Y = y, & X \text{ is distributed as } \mathcal{N}(y, 1) \end{cases}$?

(Here, and in the sequel $\mathcal{N}(a, b)$, denotes a Gaussian variable with mean a, and variance b.)

6. Prove the existence of a pair (X, Y) of r.v.s such that

(K) $\begin{cases} X \text{ is distributed as } \mathcal{N}(a, \sigma^2) \\ \text{conditionally on } X = x, Y \text{ is distributed as } \mathcal{N}(x, 1). \end{cases}$

Compute explicitly the joint law of (X, Y).

Compute the law of X, conditionally on $Y = y$.

Comments and references. See Exercise **3.12** for a unification of questions 5 and 6.

* 2.9 A double paradoxical inequality

Give an example of a pair of random variables X, Y taking values in $(0, \infty)$ such that:

$\begin{cases} \text{(i)} & E[X\,|\,Y] < \infty, \quad E[Y\,|\,X] < \infty, \quad \text{a.s.,} \\ \text{(ii)} & E[X\,|\,Y] > Y, \quad E[Y\,|\,X] > X, \quad \text{a.s..} \end{cases}$

Hint. Assume that $P\left(\frac{Y}{X} \in \{\frac{1}{2}, 2\}\right) = 1$, and $P(X \in \{1, 2, 4, \ldots\}) = 1$. More precisely, denote $p_n = P(X = 2^n, Y = 2^{n-1})$, $q_n = P(X = 2^n, Y = 2^{n+1})$ and find necessary and sufficient conditions on (p_n), (q_n) for (i) and (ii) to be satisfied. Finally, find for which values of a, the preceding discussion applies when: $p_n = q_n =$ (const.) a^n.

Comments and references. That (i) and (ii) may be realized for a pair of non-integrable variables is hinted at in D. Williams ([97], p.401), but this exercise has been suggested to us by B. Tsirel'son.

For another variant, see Exercise (33.2) on p. 62 in D. Williams [99]. See also question 5 of Exercise **1.12** for a related result.

* 2.10 Euler's formula for primes and probability

Let $\mathcal{N}$ denote the set of positive integers n, $n \neq 0$, and $\mathcal{P}$ the set of prime numbers (1 does not belong to $\mathcal{P}$). We write: $a|b$ if a divides b.

To a real number $s > 1$, we associate $\zeta(s) = \sum\limits_{k=1}^{\infty} \frac{1}{k^s}$ and we define the probability P_s on $\mathcal{N}$ by the formula:

$$P_s(\{n\}) = \frac{1}{\zeta(s)n^s} \ .$$

1. Define, for any $p \in \mathcal{P}$, the random variables ρ_p by the formula:

$$\rho_p(n) = 1_{\{p|n\}} \ .$$

Show that the r.v.s $(\rho_p, p \in \mathcal{P})$ are independent under P_s, and prove Euler's identity:

$$\frac{1}{\zeta(s)} = \prod_{p \in \mathcal{P}} \left(1 - \frac{1}{p^s}\right).$$

Hint. $\frac{1}{\zeta(s)} = P_s(\{1\})$.

2. We write the decomposition of $n \in \mathcal{N}$ as a product of powers of prime numbers, as follows:

$$n = \prod_{p \in \mathcal{P}} p^{\alpha_p(n)},$$

thereby defining the r.v.s $(\alpha_p, p \in \mathcal{P})$.

Prove that, for any $p \in \mathcal{P}$, the variable α_p is geometrically distributed, with parameter q, that is: $P_s(\alpha_p = k) = q^k(1 - q)(k \in \mathbb{N})$.

Compute q. Prove that the variables $(\alpha_p, p \in \mathcal{P})$ are independent.

Comments and references. This is a very well-known exercise involving probabilities on the integers; for a number of variations on this theme, see for example, the following.

P. DIACONIS AND L. SMITH: Honest Bernoulli excursions. *J. App. Prob.*, **25**, 464–477 (1988).

S.W. GOLOMB: A class of probability distributions on the integers. *J. Number Theory*, **2**, 189–192 (1970).

M. KAC: *Statistical Independence in Probability, Analysis and Number Theory. The Carus Mathematical Monographs*, **12**. Published by the Mathematical Association of America. Distributed by John Wiley and Sons, Inc., New York (1959).

PH. NANOPOULOS: Loi de Dirichlet sur $\mathbb{N}^*$ et pseudo-probabilités. *C. R. Acad. Sci. Paris Sér. A-B*, **280**, no. 22, Aiii, A1543–A1546 (1975).

M. SCHROEDER: *Number Theory in Science and Communications.* Springer Series in Information Sciences, Springer, 1986.

G.D. LIN AND C.-Y. HU: The Riemann zeta distribution. *Bernoulli*, **7**, no. 5, 817–828 (2001).

This exercise is also discussed in D. Williams [97].

*2.11 The probability, for integers, of being relatively prime

Let $(A_j)_{j \leq m}$ be a finite sequence of measurable events of a probability space $(\Omega, \mathcal{F}, P)$ and define $\rho_k = \sum_{1 \leq i_1 < \cdots < i_k \leq m} P(A_{i_1} \cap \ldots \cap A_{i_k})$.

1. Show that

$$P(\cup_{k=1}^m A_k) = \sum_{k=1}^m (-1)^{k-1} \rho_k \,. \tag{2.11.1}$$

2. We denote by $k \wedge l$ the g.c.d. of two integers k and l. Let $n \geq 1$ and N_1, N_2 be two independent r.v.s which are uniformly distributed on $\{1, \ldots, n\}$. We denote by Q_n the law of (N_1, N_2) and define the events $A = \{N_1 \wedge N_2 = 1\}$, $A_p = \{p|N_1 \text{ and } p|N_2\}$, $p \geq 1$. We denote by $p_1, \ldots, p_l$ the prime numbers less than or equal to n.

   (a) Show that for any positive integer m,

$$Q_n(\cup_{k=1}^m A_{p_k}) = \sum_{k=1}^m (-1)^{k-1} \mu_k \,, \tag{2.11.2}$$

   where $\mu_k = n^{-2} \sum_{1 \leq i_1 < \ldots < i_k \leq n} [n/(p_{i_1} \ldots p_{i_k})]^2$.

(b) Deduce that for any fixed m,

$$\lim_n Q_n(\cup_{k=1}^m A_{p_k}) = 1 - \prod_{k=1}^m \left(1 - \frac{1}{p_k^2}\right).$$

(c) Show that

$$\lim_n Q_n(A) = \frac{1}{\zeta(2)} = \frac{6}{\pi^2},$$

where ζ were introduced in the previous exercise.

3. Generalize the above results to the case where $N_1, \ldots, N_k$ are independent and uniformly distributed on $\{1, \ldots, n\}$.

Comments and references. The reader will find deep links between probability theory and number theory developed in:

G.H. HARDY AND E.M. WRIGHT: *An Introduction to the Theory of Numbers.* Fifth edition. The Clarendon Press, Oxford University Press, New York, 1979.

G. TENENBAUM: *Introduction to Analytic and Probabilistic Number Theory. Cambridge Studies in Advanced Mathematics,* **46**, Cambridge University Press, Cambridge, 1995.

G. TENENBAUM: *Exercices Corrigés de Théorie Analytique et Probabiliste des Nombres.* With the collaboration of Jie Wu. Cours Spécialisés, 2. Société Mathématique de France, Paris, 1996.

J. KUBILIUS: *Probabilistic Methods in the Theory of Numbers. Translations of Mathematical Monographs,* **11**, American Mathematical Society, Providence, RI, 1964.

* 2.12 Completely independent multiplicative sequences of $\mathbb{U}$-valued random variables

1. Show that there exists a sequence $(X_n)_{n \geq 2}$ taking values on the unit circle, $\mathbb{U}$ (torus), and such that:

   (a) for every m, n, $X_m X_n = X_{mn}$, a.s.,

   (b) for every m, X_m is uniform on the torus,

   (c) if n_1, $n_2, \ldots n_k$ are coprime numbers, then X_{n_1}, $X_{n_2}, \ldots, X_{n_k}$ are independent.

Then prove that this sequence is unique in law.

From now on, we consider only such a sequence.

2. Give a necessary and sufficient condition on two integers m and n for X_m and X_n to be independent.

   o More generally, solve the same question with a finite number of integers.

Comments and references. We are grateful to Eric Saias for telling us about the notion of a completely multiplicative sequence of $\mathbb{U}$-valued random variables. This notion is found on p. 64, Chapter 4, of:

J. STEUDING: *Value-Distribution of L-Functions. Lecture Notes in Mathematics,* **1877**, Springer, Berlin, 2007.

* 2.13 Bernoulli random walks considered at some stopping time

Let $(X_n, n \geq 1)$ and $(Y_n, n \geq 1)$ be two sequences of r.v.s which take values in $\{0, 1\}$.

Assume that all the variables $(X_n, Y_m; n, m \geq 1)$ are independent, and that, for any $n \geq 1$, one has:
$$P(X_n = 1) = p \quad \text{and} \quad P(Y_n = 1) = q ,$$
where p and q are fixed, and $0 < p, q < 1$.

1. Show that the r.v.s $Z_n = X_n Y_n$, $n \geq 1$, are independent, and identically distributed. Compute their common distribution.

2. Define $S_n = \sum_{m=1}^{n} X_m$, and $T_n = \sum_{m=1}^{n} Z_m$.

   What are the laws of S_n and T_n?

3. Define $\tau(\omega) = \inf \{n \geq 1 : T_n(\omega) = 1\}$, with the convention: $\inf(\emptyset) = \infty$. We write simply S_τ for $S_{\tau(\omega)}(\omega)$.

   Show that τ and S_τ are random variables. What is the law of τ?

4. Show that, for $n \geq 2$ and $1 \leq k \leq n$:

$$P\left(X_k = 1 \mid \tau = n\right) = P\left(X_k = 1 \mid Z_k = 0\right) = \frac{p(1 - q)}{1 - pq} .$$

5. Show that:

$$P\left(\bigcap_{i=1}^{n}(X_i = x_i) \mid \tau = n\right) = \prod_{i=1}^{n} P\left(X_i = x_i \mid \tau = n\right) ,$$

and deduce therefrom the value of $P(S_\tau = k \mid \tau = n)$.

6. Compute $E(S_\tau \mid \tau = n)$, and then $E(S_\tau)$. Check that:

$$E[S_\tau] = E[\tau]E[X_1], \quad \text{and} \qquad (2.13.1)$$
$$E[T_\tau] = E[\tau]E[Z_1].$$

Comments. This is a simple exercise whose main aim is to help the reader check his/her familiarity when working with a randomly indexed sequence of random variables.

The r.v. τ is a stopping time with respect to the increasing family of σ-fields $\mathcal{F}_n = \sigma\{X_k, Y_k; k \leq n\}$, and the final equalities (2.13.1) may be obtained as consequences of the optional stopping theorem applied to the $(\mathcal{F}_n)$-martingales: $(S_n - nE[X_1])$ and $(T_n - nE[Z_1])$.

** 2.14 cosh, sinh, the Fourier transform and conditional independence

Let W_-, W_+, L be three r.v.s such that:

(i) L takes its values in $\mathbb{R}_+$;

(ii) for all $a \geq 0$, b, $c \in \mathbb{R}$, $E\left[\exp(-aL + ibW_- + icW_+)\right] = f(2a + |b|, c)$, where:
$f(u,v) = \left(\cosh v + \dfrac{u}{v}\sinh v\right)^{-1}$, if $v \neq 0$, and $f(u,0) = (1+u)^{-1}$.

1. What is the law of L? What is the law of $W \equiv W_+ + W_-$?

2. Show that W_- and W_+ are independent, conditionally on L. What is the conditional law of W_-, given that $W_+ = x$ and $L = \ell$?

   Compute explicitly $E\left[\exp(icW_+) \mid L = \ell\right]$.

Comments and references. The triplet (L, W_-, W_+) may be realized as follows: consider two independent one-dimensional Brownian motions $(\beta_t, t \geq 0))$ and $(\gamma_t, t \geq 0)$, starting from 0, and define:

$$L = \ell_\sigma, \ W_- = \int_0^\sigma d\gamma_s \, \mathbb{1}_{\{\beta_s < 0\}}, \ W_+ = \int_0^\sigma d\gamma_s \, \mathbb{1}_{\{\beta_s > 0\}},$$

where $\sigma = \inf\{t : \beta_t = 1\}$ and $(\ell_t, t \geq 0)$ is the local time of β at 0. This triplet appears naturally in the limit laws for the (small and large) winding numbers of planar Brownian around a given point; see for example:

J.W. PITMAN AND M.YOR: Asymptotic laws of planar Brownian motion. *Ann. Prob.*, **14**, no. 3, 733–779 (1986)

or Chapter 5 in M. Yor [101], or Chapter 10 in Revuz-Yor [75].

** 2.15 cosh, sinh, and the Laplace transform

Let A_-, A_+ and L be three $\mathbb{R}_+$-valued r.v.s whose joint distribution is characterized by:

$$E\left[\exp-\left(aL + \frac{1}{2}\lambda^2 A_- + \frac{1}{2}\mu^2 A_+\right)\right] = \left(\cosh(\mu) + \frac{2a+\lambda}{\mu}\sinh(\mu)\right)^{-1}, \quad \lambda, \mu \geq 0.$$

(2.15.1)

1. Prove that A_+ and A_- are conditionally independent given L. Prove that

$$A_- \stackrel{\text{(law)}}{=} \frac{Z^2}{N^2},$$

where Z and N are independent; Z has an exponential law with parameter 1 and N is a centered Gaussian variable with variance 1.

Hint. Use the result of question 7 of Exercise **4.2**.

2. Prove that the triplet (L, W_+, W_-) studied in Exercise **2.14** satisfies:

$$(L, W_+, W_-) \stackrel{\text{(law)}}{=} (L, \sqrt{A_+}N_+, \sqrt{A_-}N_-),$$

(2.15.2)

where on the right hand side, the triplet (L, A_+, A_-), and N_+, and N_- are independent and N_+ and N_- are centered Gaussian variables, with variance 1.

3. Deduce the conditional independence of W_+ and W_- given L (as shown in Exercise **2.14**) from that of A_+ and A_- given L.

4. Give an explicit formula for the joint density of $(H = \frac{L}{2}, A_-)$. Deduce the density of A_- in an integral form. Give the conditional density of H given $A_- = u$. Give the conditional Laplace transform of A_+ given A_-.

5. Fix λ and μ. Give a series development of the density of $(\lambda^2 A_- + \mu^2 A_+)$.

Comments:

(a) Using the same notation as in the comments for Exercise **2.14**, we may realize the triplet (L, A_-, A_+) as: $\left(l_\sigma, \int_0^\sigma ds\,\mathbb{1}_{\{B_s<0\}}, \int_0^\sigma ds\,\mathbb{1}_{\{B_s>0\}}\right)$. See also Exercise **6.18** for another occurrence in terms of Brownian motion of these two or three-dimensional distributions.

(b) This exercise points out at an apparently paradoxical situation: question 3 allows us to describe "explicitly" the distribution of any linear combination of A_- and A_+ (with positive coefficients); hence, we "know" the law of (A_-, A_+) (already through the Laplace transform formula (**2.15.1**)). But it is notorious that the conditional law of A_+ given $(A_- = a)$ is almost "impossible" to invert explicitly! (See Exercise **6.18** for some further discussion.)

** 2.16 Conditioning and changes of probabilities

Let $(\Omega, \mathcal{F})$ be a measurable space and $\mathcal{G}$ be a sub-σ-field of $\mathcal{F}$. Let P and Q be two probabilities which are mutually absolutely continuous on $\mathcal{F}$. We denote by X_0 the Radon–Nikodym density of Q with respect to P and $\mathcal{F}$.

1. Show that the two following properties are satisfied:

   (a) $0 < E_P(X_0 \mid \mathcal{G}) < \infty$ P a.s.,

   and

   (b) for every $\mathcal{F}$-measurable r.v. f, with values in $\mathbb{R}_+$,

   $$E_P(fX_0 \mid \mathcal{G}) = E_Q(f \mid \mathcal{G})E_P(X_0 \mid \mathcal{G}).$$

2. The aim of the rest of the exercise is to describe all $\mathcal{F}$-measurable r.v.s X', which take values in $\mathbb{R}_+$, and which satisfy:

   (a') $E_P(X' \mid \mathcal{G}) < \infty$ P a.s.,

   and

   (b') for every $\mathcal{F}$-measurable r.v. f, with values in $\mathbb{R}_+$,

   $$E_P(fX' \mid \mathcal{G}) = E_Q(f \mid \mathcal{G})E_P(X' \mid \mathcal{G}).$$

   Prove that, if $X' = \gamma X_0$, where γ is a $\mathcal{G}$-measurable r.v., with values in $\mathbb{R}_+$, then X' satisfies (a') and (b').

3. Assume that $X' = \gamma X_0$, with γ a $\mathcal{G}$-measurable r.v. with values in $\mathbb{R}_+$. Compute γ in terms of $E_P(X' \mid \mathcal{G})$ and of $E_P(X_0 \mid \mathcal{G})$.

4. Prove that, if X' satisfies (a') and (b'), then, there exists γ, a $\mathcal{G}$-measurable r.v. with values in $\mathbb{R}_+$ such that: $X' = \gamma X_0$.

5. Find a necessary and sufficient condition, in terms of X_0, such that, for any $\mathcal{F}$-measurable r.v. f, with values in $\mathbb{R}_+$,

   $$E_P(f \mid \mathcal{G}) = E_Q(f \mid \mathcal{G}) \ .$$

* 2.17 Radon–Nikodym density and the Acceptance–Rejection Method of von Neumann

Consider, on a measurable space $(\Omega, \mathcal{F})$, two probabilities P and Q, with $Q << P$, and $\frac{dQ}{dP} = f$.

Enlarge the probability space $(\Omega, \mathcal{F}, P)$ in $(\Omega' = \Omega \times [0, 1], \mathcal{F} \otimes \mathcal{B}_{[0,1]}, P = P \otimes \lambda)$, where λ denotes the Lebesgue measure on $[0, 1]$. Denote by $\omega' = (\omega, u)$ the generic point of Ω', and $U(\omega') = u$.

1. Assume that $P(f \leq M) = 1$, for some $M > 0$. Then prove that, for any $\Phi : (\Omega, \mathcal{F}) \to (\mathbb{R}_+, \mathcal{B}_{\mathbb{R}_+})$, one has:

$$E[\Phi \mid f \geq MU] = E_Q[\Phi].$$

2. We no longer assume that f is essentially bounded; nonetheless, prove that, with the same notations as above:

$$E_Q[\Phi] = \lim_{M \to \infty} E[\Phi \mid f \geq MU].$$

Comments and references. This exercise originates from Dan Romik's Ph.D. thesis. A relevant reference is:

R.Y. RUBINSTEIN: *Simulation and the Monte Carlo Method.* Wiley Series in Probability and Mathematical Statistics. John Wiley & Sons, Inc., New York, 1981.

** 2.18 Negligible sets and conditioning

Let (X, Y) be a pair of $\mathbb{R}_+$-valued random variables, and assume that

(i) (X, Y) has a jointly continuous density;

(ii) if $g(y)$ denotes the density of Y, then: $g(y) \sim c y^\alpha$, as $y \to 0$, for some $c > 0$, and some $\alpha \geq 0$.

Furthermore, consider an $\mathbb{R}_+$-valued variable S which is independent of the pair (X, Y), and satisfies: $E\left[S^{-(\alpha+1)}\right] < \infty$.

1. Prove that for any bounded Borel function $f : \mathbb{R}_+ \to \mathbb{R}_+$, one has

$$E[f(SX) \mid SY = 0] = \frac{1}{E\left[S^{-(\alpha+1)}\right]} E\left[S^{-(\alpha+1)} f(SX) \mid Y = 0\right], \qquad (2.18.1)$$

where both conditional expectations may be defined as:

$$E[F \mid Z = 0] \stackrel{\text{def}}{=} \lim_{\varepsilon \to 0} \frac{E[F \mathbb{1}_{\{Z \leq \varepsilon\}}]}{P(Z \leq \varepsilon)},$$

with obvious notations for F and Z.

2. Prove that, if Y satisfies (ii) with $\alpha = 0$, and $S \stackrel{(\text{law})}{=} \sqrt{T}$, where T is an exponential variable with parameter $1/2$, then

$$E[f(SX) \mid SY = 0] = E[f(|N|X) \mid Y = 0], \qquad (2.18.2)$$

where N denotes a Gaussian, centered, variable, which is independent of the pair (X, Y), and has variance 1.

3. Prove that, if Y satisfies (ii) with $\alpha \geq 0$, and if $a - (\alpha + 1) > 0$ then, if $S \stackrel{(\text{law})}{=} Z_a$, one has

$$E[f(SX) \mid SY = 0] = E[f(Z_{a-(\alpha+1)}X) \mid Y = 0], \qquad (2.18.3)$$

where $Z_{a-(\alpha+1)}$ denotes a gamma variable with parameter $a - (\alpha + 1)$, which is independent of the pair (X, Y).

Comments and references. Either of the formulas (2.18.1), (2.18.2), (2.18.3) may look strange (unintuitive?) at first sight; for instance, one is tempted *a priori* to think that the left hand side of (2.18.2) is equal to

$$E[f(SX) \mid Y = 0],$$

but this differs from the correct right hand side of (2.18.2)! For a closely related discussion, see D. Williams' warning ([97], pp. 260–261) about ill-defined quantities $P(X \in dx \mid X = Y)$, which could mean either $P\left(X \in dx \mid \frac{X}{Y} = 1\right)$, or $P(X \in dx \mid X - Y = 0)$, and so on... .

Here, one is clearly confronted with the difficulty of conditioning with respect to negligible sets; about this subject, P. Lévy wrote an article:

P. LÉVY: Une hiérarchie des probabilités plus ou moins nulles, application à certains nuages de points. *L'Enseignement Math.*, (2) **15**, 217–225 (1969).

This is also found in:

P. LÉVY: *Œuvres de Paul Lévy.* Published under the direction of Daniel Dugué with the collaboration of Paul Deheuvels and Michel Ibéro. Gauthier-Villars, Paris, 1980.

* 2.19 Gamma laws and conditioning

Let X and L be two r.v.s which take their values in $\mathbb{R}_+$. Assume furthermore that:

$$E[X] < \infty \quad \text{and} \quad P(L \in dl) = e^{-l}\, dl\,. \qquad (2.19.1)$$

1. Show that the three following properties are equivalent.

(a) There exists $a > 0$ such that for any $\lambda \geq 0$:

$$E\left[X \exp(-\lambda L)\right] = \frac{a}{1 + \lambda a}.$$

(b) There exists $\alpha \geq 0$ such that $E\left[X \mid L = \ell\right] = \exp(-\ell\alpha)$, a.e. *(dl)*.

(c) For every $k \in \mathbb{N}$,

$$E\left[X\frac{L^k}{k!}\right] = (E[X])^{k+1} .$$

Show that, moreover, if those properties are satisfied, then the following relationship between a and α holds:

$$a = \frac{1}{1+\alpha}.$$

2. Prove that the variables $X = (1 + \varphi(L)\varepsilon)\exp(-\alpha L)$, with a Borel function φ such that $-\frac{1}{2} \leq \varphi \leq \frac{1}{2}$, and ε is a symmetric Bernoulli variable independent of L, satisfy the equivalent properties (a), (b) and (c).

3. Find the appropriate variant of question 1 when the property (2.19.1) is replaced by:

$$P(L \in d\ell) = e^{-\ell}\frac{\ell^{\gamma-1}d\ell}{\Gamma(\gamma)} , \qquad (2.19.2)$$

for some $\gamma > 0$.

4. Assume that the equivalent properties stated in question 1 hold. If $\nu(dt)$ is a positive σ-finite measure on $\mathbb{R}_+$, we define:

$$\mathcal{L}_\nu(\ell) = \int_{\mathbb{R}_+} \nu(dt)e^{-t\ell} \quad \text{and} \quad \mathcal{S}_\nu(m) = \int_{\mathbb{R}_+} \nu(dt)\frac{m}{1+tm} .$$

Then, prove that:

$$E[X\mathcal{L}_\nu(L)] = \mathcal{S}_\nu(E[X]) .$$

*2.20 Random variables with independent fractional and integer parts

Let Z be a standard exponential variable, i.e. $P(Z \in dt) = e^{-t}\,dt$. Define $\{Z\}$ and $[Z]$ to be respectively the fractional part, and the integer part of Z.

1. Prove that $\{Z\}$ and $[Z]$ are independent, and compute their distributions explicitly.

2. Consider X a positive random variable whose law is absolutely continuous. Let $P(X \in dt) = \varphi(t)\,dt$. Find a density φ such that:

   (i) $\{X\}$ and $[X]$ are independent;

   (ii) $\{X\}$ is uniformly distributed on $[0, 1]$.

Hint. Use the previous question and make an adequate change of probability from $e^{-t}\,dt$ to $\varphi(t)\,dt$.

3. We make the same hypothesis and use the same notations as in question 2. Characterize the densities φ such that $\{X\}$ and $[X]$ are independent.

* 2.21 Two characterizations of the simple random walk

We call *skip-free process* any sequence $S = (S_n,\, n \geq 0)$ of random variables such that $S_0 = 0$ and for all $n \geq 1$, $S_n - S_{n-1} \in \{-1, +1\}$. All skip-free processes considered in this exercise will be assumed to satisfy $T_k \stackrel{\text{(def)}}{=} \inf\{i : S_i = k\} < \infty$, a.s., for all integer k. Then define the path transformation:

$$\Theta^k(S) \stackrel{\text{(def)}}{=} \left(S_n \mathbb{1}_{\{n \leq T_k\}} + (2k - S_n)\mathbb{1}_{\{n > T_k\}},\, n \geq 0\right),$$

which amounts to a reflection of S around the axis $y = S_k$ by time T_k. Recall that a skip-free process is also a Bernoulli random walk if in addition the r.v.s $(S_n - S_{n-1},\, n \geq 1)$ are independent and symmetric.

1. Show that any skip-free process that is a martingale is a simple random walk.

2. Let M be a martingale such that $M_0 = 0$ and $M_n - M_{n-1} \in \{-1, 0, +1\}$, a.s. and define its quadratic variation by $[M]_n = \sum_{k=0}^{n-1}(M_{k+1} - M_k)^2$, $n \geq 1$, $[M]_0 = 0$. Assume that $\lim_{n \to +\infty}[M]_n = +\infty$, a.s. and define the inverse of $[M]$ by $\tau_n = \inf\{k : [M]_k = n\}$, $n \geq 0$. Show that the process $S^M = (M_{\tau_n},\, n \geq 0)$ is a Bernoulli random walk.

3. Prove that, for any two elements s and s' of the set

$$\Lambda^n = \{s = (s_0, \ldots, s_n) : s_0 = 0, s_k - s_{k-1} \in \{-1, +1\}\}$$

with $s \neq s'$, there exist integers $a_1, a_2, \ldots a_k$ such that

$$s' = \Theta^{a_k}\Theta^{a_{k-1}}\ldots\Theta^{a_1}(s).$$

4. We say that a skip-free process S satisfies the reflection principle at level k if the following identity in law is satisfied:

$$S \stackrel{\text{(law)}}{=} \Theta^k(S).$$

Show that any skip-free process which satisfies the reflection principle at any level $k \geq 0$ is a simple random walk.

Comments and references. The first question states a discrete time equivalent of P. Lévy's characterization of Brownian motion: *any continuous martingale M whose quadratic variation is given by* $\langle M \rangle_t = t$ *for all* $t \geq 0$ *is a Brownian motion.*

Question 2 brings out the discrete equivalent of Dambis–Dubins–Schwarz Theorem : *If M is a continuous martingale such that* $\langle M \rangle_\infty = \infty$ *a.s., then with* $\tau_t = \inf\{s : \langle M \rangle_s = t\}$*, the process* $(M(\tau_t), t \geq 0)$ *is a Brownian motion.*

In question 4, it is actually enough for a skip-free process to satisfy the reflection principle at levels $k = 0, 1$ and 2 to be a simple random walk. However, the validity of the reflection principle at levels 0 and 1 is not enough. These results are proved in:

L. CHAUMONT AND L. VOSTRIKOVA: Reflection principle and Ocone martingales. *Stochastic Process. Appl.*, **119**, no. 10, 3816–3833 (2009),

where an equivalent characterization is obtained in continuous time, for Brownian motion. We also refer to the following recent work:

J. BROSSARD AND C. LEURIDAN: Characterising Ocone local martingales with reflections. Preprint, 2011.

Solutions for Chapter 2

Solution to Exercise 2.1

1. Let f and g be two bounded measurable functions on $(\Omega, \mathcal{F}, P)$. Then,

$$E(f(\varepsilon X)g(\varepsilon)) = \frac{1}{2}[E(f(X)g(1)) + E(f(-X)g(-1))]$$

and the variables εX and ε are independent if and only if

$$E(f(X))g(1) + E(f(-X))g(-1) = E(f(X))(g(1) + g(-1))$$

for every f and g as above. If X is symmetric then this identity holds. Conversely, let g such that $g(1) = 0$ and $g(-1) \neq 0$, then the above identity shows that X is symmetric.

2. Let $(\Omega, \mathcal{F}, P)$ be a probability space on which there exist $A, B \in \mathcal{F}$ such that A, B are independent and $P(A) = P(B) = 1/2$. Defining $X = \mathbb{1}_A - \mathbb{1}_{A^c}$, $\varepsilon = \mathbb{1}_B - \mathbb{1}_{B^c}$, $\mathcal{A} = \sigma(X) = \{\emptyset, \Omega, A, A^c\}$ and $\mathcal{B} = \sigma(\varepsilon) = \{\emptyset, \Omega, B, B^c\}$, then $\mathcal{A}$ and $\mathcal{B}$ are independent sub-σ-fields of $\mathcal{F}$. Moreover, the r.v.s X and ε are independent symmetric Bernoulli variables, so from question 1, $Y = \varepsilon X$ and ε are independent. Hence Y is $\mathcal{A} \vee \mathcal{B}$-measurable and independent of $\mathcal{B}$, but Y is not measurable with respect to $\mathcal{A}$ since $\sigma(Y) = \{\emptyset, \Omega, (A \cap B) \cup (A^c \cap B^c), (A \cap B^c) \cup (A^c \cap B)\}$ and neither of the sets $(A \cap B) \cup (A^c \cap B^c)$, and $(A \cap B^c) \cup (A^c \cap B)$ can be equal to A.

3. Take $Z = Y = \varepsilon X$, $\mathcal{A}$, and $\mathcal{B}$ defined in the previous question. By applying question 1 to the pair of variables $(\varepsilon X, \varepsilon)$ and then to the pair of variables $(\varepsilon X, X)$, we obtain that Z (which is obviously $\mathcal{A} \vee \mathcal{B}$-measurable), is both independent of $\mathcal{A}$ and independent of $\mathcal{B}$.

4. (a) Let $Y_1 \in \mathbb{G}_1$ and $Y_2 \in \mathbb{G}_2$ be such that $Y = Y_1 + Y_2$ and assume that Y is independent of Y_2. Since Y and Y_2 are centered, the variables $Y_1 + Y_2$ and Y_2 are orthogonal. Therefore, $Y_2 = 0$ a.s. and (iii) does not hold (we have supposed that $Y \neq 0$).

(b) We answer the last point in the same manner: let $Z_1 \in \mathbb{G}_1$ and $Z_2 \in \mathbb{G}_2$ such that $Z = Z_1 + Z_2$. If Z is independent of both Z_1 and Z_2 then $Z_1 = 0$ and $Z_2 = 0$ a.s.

Comments on the solution. Since any solution to question 3 also provides a solution to question 2, it is of some interest to modify question 2 as follows:

(i) and (ii) are unchanged; (iii) is changed in (iii)': Y is neither measurable with respect to $\mathcal{A}$, nor independent from $\mathcal{A}$.

(Solution: take $\mathcal{A} = \sigma(G)$, $\mathcal{B} = \sigma(\varepsilon)$, where G is Gaussian, centered, ε is Bernoulli, independent from G, then $Y = G\varepsilon$ solves the modified question 2.)

Solution to Exercise 2.2

1. All variables considered below are assumed to be bounded (so that no integrability problem arises). Since $\mathcal{A}$, $\mathcal{B}$ and $\mathcal{C}$ are complete, it suffices to prove that for every $\mathcal{A}$-measurable r.v., X, one has $E(X \mid \mathcal{B}) = X$, a.s. Let Y and Z be two r.v.s respectively $\mathcal{B}$-measurable and $\mathcal{C}$-measurable then

$$
\begin{aligned}
E(E(X \mid \mathcal{B} \vee \mathcal{C})YZ) &= E(XYZ) \\
&= E(XY)E(Z), \quad \text{by (ii)}', \\
&= E(E(X \mid \mathcal{B})Y)E(Z) \\
&= E(E(X \mid \mathcal{B})YZ), \quad \text{since } \mathcal{B} \text{ and } \mathcal{C} \text{ are independent.}
\end{aligned}
$$

Since the equality between the extreme terms holds for every pair of r.v.s Y and Z as above, the Monotone Class Theorem implies that $E(X \mid \mathcal{B}) = E(X \mid \mathcal{B} \vee \mathcal{C})$ a.s. But X is $\mathcal{B} \vee \mathcal{C}$-measurable, by (i), thus: $E(X \mid \mathcal{B}) = X$, a.s..

2. See question 2 of Exercise **2.1**.

3. This follows from question 1 of the present exercise.

Solution to Exercise 2.3

1. (i) $\Rightarrow$ (ii): let Q be a probability measure which is equivalent to P. Since Γ is independent of $\mathcal{G}$, and $\frac{dQ}{dP}$ is $\mathcal{G}$ measurable then

$$
Q(\Gamma) = E_P\left[\frac{dQ}{dP}\mathbb{1}_\Gamma\right] = E_P\left[\frac{dQ}{dP}\right]P(\Gamma) = P(\Gamma).
$$

(ii) $\Rightarrow$ (i): the property (ii) is clearly equivalent to:

for every bounded $\mathcal{G}$ measurable function ϕ, $E_P[\phi\mathbb{1}_\Gamma] = E_P[\phi]P(\Gamma)$, which amounts to Γ being independent of $\mathcal{G}$.

2. $(j) \Rightarrow (jj)$: this follows from $(i) \Rightarrow (ii)$.

3. Suppose that X is independent of the event $\{a\}$, then for any bounded measurable function f:

$$E[f(X)\mathbb{1}_{\{a\}}] = E[f(X)]P(\{a\})$$

but, almost surely $f(X)\mathbb{1}_{\{a\}} = f(X(a))\mathbb{1}_{\{a\}}$, so that

$$E[f(X)\mathbb{1}_{\{a\}}] = f(X(a))P(\{a\}),$$

hence $E[f(X)] = f(X(a))$ for any f, bounded and measurable. This proves that $X = X(a)$, a.s.

Solution to Exercise 2.4

Let f be a bounded Borel function defined on $\mathbb{R}^{n-1}$ and g, h be two bounded real valued Borel functions, then:

$$E(f(X^*_{(n-1)})g(X_J)h(S^*_{n-1})) = E(E(f(X^*_{(n-1)}))g(X_J)\,|\,S^*_{n-1})h(S^*_{n-1})). \qquad (2.4.a)$$

On the other hand, put

$$X^{(j)}_{(n-1)} = (X_1, X_2, \ldots, X_{j-1}, X_{j+1}, \ldots, X_n)$$
$$S^{(j)}_{n-1} = \sum_{1 \le i \le n,\, i \ne j} X_i,$$

then from the hypotheses:

$$E(f(X^*_{(n-1)})g(X_J)h(S^*_{n-1})) = \sum_{j=1}^{n} E(f(X^{(j)}_{(n-1)})g(X_j)h(S^{(j)}_{n-1})1_{\{J=j\}})$$

$$= \sum_{j=1}^{n} E\left(f(X^{(j)}_{(n-1)})g(X_j)h(S^{(j)}_{n-1})\frac{X_j}{S^{(j)}_{n-1} + X_j} \right).$$

Since $X^{(j)}_{(n-1)}$ and X_j are independent, then:

$$E(f(X^*_{(n-1)})g(X_J)h(S^*_{n-1})) = \sum_{j=1}^{n} E\left(E(f(X^{(j)}_{(n-1)})\,|\,S^{(j)}_{n-1})g(X_j)h(S^{(j)}_{n-1})\frac{X_j}{S^{(j)}_{n-1} + X_j} \right).$$

Put $E(f(X^{(j)}_{(n-1)}) \,|\, S^{(j)}_{n-1}) = k(S^{(j)}_{n-1})$, then from above,

$$E(f(X^*_{(n-1)})g(X_J)h(S^*_{n-1})) = E(k(S^*_{(n-1)})g(X_J)h(S^*_{n-1})).$$

This implies:

$$\begin{aligned} E(f(X^*_{(n-1)})g(X_J)h(S^*_{n-1})) &= E(k(S^*_{(n-1)})E(g(X_J)\,|\,S^*_{n-1})h(S^*_{n-1})) \\ &= E(E(f(X^*_{(n-1)}\,|\,S^*_{(n-1)})E(g(X_J)\,|\,S^*_{n-1})h(S^*_{n-1})). \end{aligned}$$

(2.4.b)

Comparing (2.4.a) and (2.4.b), we have:

$$E(f(X^*_{(n-1)})g(X_J)\,|\,S^*_{n-1}) = E(f(X^*_{(n-1)})\,|\,S^*_{n-1})E(g(X_J)\,|\,S^*_{n-1}),$$

thus, given S^*_{n-1}, $X^*_{(n-1)}$ and X_J are independent. Now, putting $E(f(X^*_{(n-1)})\,|\,S^*_{n-1}) = k_1(S^*_{n-1})$ and $E(f(X_{(n-1)})\,|\,S_{n-1}) = k_2(S_{n-1})$ then we shall show that $k_1(S^*_{n-1}) = k_2(S_{n-1})$ a.s. By the same arguments as above, we have:

$$E(f(X^*_{(n-1)})g(S^*_{n-1})) = \sum_{j=1}^{n} E\left(E(f(X^{(j)}_{(n-1)}\,|\,S^{(j)}_{n-1})g(S^{(j)}_{n-1})\frac{X_j}{S_n}\right).$$

Since the law of $X^{(j)}_{(n-1)}$ is the same as the law of $X_{(n-1)}$,

$$\begin{aligned} E(f(X^*_{(n-1)})g(S^*_{n-1})) &= \sum_{j=1}^{n} E\left(k_2(S^{(j)}_{n-1})g(S^{(j)}_{n-1})\frac{X_j}{S_n}\right) \\ &= E(k_2(S^*_{n-1})g(S^*_{n-1})). \end{aligned}$$

But also, by definition, $E(f(X^*_{(n-1)})g(S^*_{n-1})) = E(k_1(S^*_{n-1})g(S^*_{n-1}))$, which proves the result.

Solution to Exercise 2.5

1. In this solution, we set: $\mathcal{D} = \cap_n \mathcal{D}_n$. It suffices to show that for every bounded, $\mathcal{C} \vee \mathcal{D}_1$-measurable r.v. X:

$$E(X \,|\, \cap_n (\mathcal{C} \vee \mathcal{D}_n)) = E(X \,|\, \mathcal{C} \vee \mathcal{D}), \quad \text{a.s.} \tag{2.5.a}$$

We may consider variables X of the form: $X = fg$, f and g being bounded and respectively $\mathcal{C}$-measurable and $\mathcal{D}_1$-measurable. Let X be such a variable, then on the one hand, since $\mathcal{C} \vee \mathcal{D}_n$ decreases, Exercise **1.5** implies:

$$E(X \,|\, \mathcal{C} \vee \mathcal{D}_n) \xrightarrow{L^1} E(X \,|\, \cap_n (\mathcal{C} \vee \mathcal{D}_n)), \quad (n \to \infty). \tag{2.5.b}$$

On the other hand, since $\mathcal{C}$ and $\mathcal{D}_1$ are independent conditionally on $\mathcal{D}_n$, then

$$E(X \mid \mathcal{C} \vee \mathcal{D}_n) = f E(g \mid \mathcal{C} \vee \mathcal{D}_n) = f E(g \mid \mathcal{D}_n) .$$

But again, from Exercise **1.6**, $E(g \mid \mathcal{D}_n) \xrightarrow{L^1} E(g \mid \mathcal{D})$ as n goes to ∞. Moreover, $\mathcal{C}$ and $\mathcal{D}_1$ being independent conditionally on $\mathcal{D}_n$ for each n, these σ-fields are independent conditionally on $\mathcal{D}$, hence

$$E(X \mid \mathcal{C} \vee \mathcal{D}_n) \xrightarrow{L^1} f E(g \mid \mathcal{D}) = E(X \mid \mathcal{C} \vee \mathcal{D}) , \qquad (2.5.c)$$

as n goes to ∞. We deduce (2.5.a) from (2.5.b) and (2.5.c).

2. With f, g, h such that f is $\mathcal{C}$-measurable, g is $\mathcal{D}_1$-measurable and h is $\mathcal{D}_n$-measurable, we want to show that:

$$E(E(f \mid \mathcal{D}_n) E(g \mid \mathcal{D}_n) h) = E(fgh) . \qquad (2.5.d)$$

We shall show the following stronger equality:

$$E(E(f \mid \mathcal{D}_n) E(g \mid \mathcal{D}_n) h \mid \mathcal{E}_n) = E(fgh \mid \mathcal{E}_n) . \qquad (2.5.e)$$

Let k be $\mathcal{E}_n$-measurable; then, from the hypothesis and since gh is $\mathcal{D}_1$-measurable,

$$\begin{aligned}
E(fghk) &= E(E(f \mid \mathcal{E}_n) E(gh \mid \mathcal{E}_n) k) \\
&= E(E(f \mid \mathcal{E}_n) E(E(g \mid \mathcal{D}_n) h \mid \mathcal{E}_n) k) .
\end{aligned}$$

By applying the hypothesis again:

$$E(fghk) = E(E(f E(g \mid \mathcal{D}_n) h \mid \mathcal{E}_n) k) .$$

Finally, by conditioning on $\mathcal{D}_n$:

$$E(fghk) = E(E(E(f \mid \mathcal{D}_n) E(g \mid \mathcal{D}_n) h \mid \mathcal{E}_n) k) .$$

Since, on the other hand, for every k, $\mathcal{E}_n$-measurable,

$$E(fghk) = E(E(fgh \mid \mathcal{E}_n) k) ,$$

one deduces that (2.5.e) holds. (This result holds in that particular case because $\mathcal{D}_n \subset \mathcal{D}_1$ but it is not true in general.)

3. If (2.5.2) holds then it is obvious that $\mathcal{D}$ and $\mathcal{C}$ are independent. Suppose now that $\mathcal{D}$ and $\mathcal{C}$ are independent and let X be bounded and $\mathcal{F}$-measurable and C be bounded and $\mathcal{C}$-measurable, then from Exercise **1.6**,

$$E(E(X \mid \mathcal{D}_n) C) \longrightarrow E(E(X \mid \mathcal{D}) C) = E(X) E(C) , \quad (n \to \infty) .$$

4. First, we have $\cap_n(\mathcal{C} \vee \mathcal{D}_n) \subset \mathcal{C} \vee \sigma(Y_0)$ since for each n, $\mathcal{C} \vee \mathcal{D}_n \subset \mathcal{C} \vee \sigma(Y_0)$. Moreover, it is obvious that Y_0 is $\cap_n(\mathcal{C} \vee \mathcal{D}_n)$-measurable, so $\mathcal{C} \vee \sigma(Y_0) \subset \cap_n(\mathcal{C} \vee \mathcal{D}_n)$.

On the other hand, we easily check that $X_1, X_2, \ldots$ are independent, hence from Kolmogorov's 0–1 law, the σ-field $\cap_n \mathcal{D}_n$ is trivial, and $\mathcal{C} \vee (\cap_n \mathcal{D}_n) = \mathcal{C}$.

Comments on the solution. The counterexample given in question 4 is presented in Exercise 4.12 on p. 48 of D. Williams [98].

Solution to Exercise 2.6

1. It follows from the exchangeability property that for any $n \geq 2$,

$$(X_1, X_2, X_3, \ldots) \stackrel{(\text{law})}{=} (X_1, X_n, X_{n+1}, \ldots),$$

hence for any bounded Borel function Φ,

$$E[\Phi(X_1) \,|\, \mathcal{G}_n] \stackrel{(\text{law})}{=} E[\Phi(X_1) \,|\, \mathcal{G}_2].$$

We conclude by applying the result of Exercise **1.6** which asserts that $E[\Phi(X_1) \,|\, \mathcal{G}_n]$ converges in law towards $E[\Phi(X_1) \,|\, \mathcal{G}]$, as n goes to ∞.

2. This is a direct consequence of Exercise **1.12**.

3. Question 2 shows that X_1 and $\mathcal{G}_2$ are conditionally independent given $\mathcal{G}$. Similarly, for any n, X_n and $\mathcal{G}_{n+1}$ are conditionally independent given $\mathcal{G}$, so, by iteration, all r.v.s of the sequence $(X_1, X_2, \ldots)$ are conditionally independent given $\mathcal{G}$.

Solution to Exercise 2.7

Without loss of generality, we may restrict ourselves to the case $n = 2$.

1. Let (X_1, X_2, X_3) be a centered Gaussian vector such that $E(X_1^2) = E(X_2^2) = E(X_3^2) = 1$ and $E(X_1 X_2) = a \neq 0$, $E(X_1 X_3) = b \neq 0$, $E(X_2 X_3) = 0$. Define $\mathcal{G}_1 = \sigma(X_1)$ and $\mathcal{G}_2 = \sigma(X_2)$. Then $E(X_3 \,|\, \mathcal{G}_1) = bX_1$, $E(X_1 \,|\, \mathcal{G}_2) = aX_2$ and $E(X_3 \,|\, \mathcal{G}_2) = 0$, so that $E(E(X_3 \,|\, \mathcal{G}_1) \,|\, \mathcal{G}_2) = abX_2$ and $E(E(X_3 \,|\, \mathcal{G}_2) \,|\, \mathcal{G}_1) = 0$. Hence the σ-fields $\mathcal{G}_1$ and $\mathcal{G}_2$ are not exchangeable.

2. (a) It suffices to note that the random variable $E(X \,|\, \mathcal{G}_1) = E(X \,|\, \mathcal{G}_2)$ is $\mathcal{G}_1 \cap \mathcal{G}_2$-measurable.

(b) Let X be $\mathcal{G}_2$ measurable, centered and independent from $\mathcal{G}_1$. Then $E(X \,|\, \mathcal{G}_1) = 0$, but $E(X \,|\, \mathcal{G}_2) = X$.

3. Suppose first that $\mathcal{G}_1$ and $\mathcal{G}_2$ are exchangeable. Let X_1 and X_2 be two random variables which are respectively $\mathcal{G}_1$-measurable and $\mathcal{G}_2$-measurable. Then

$$
\begin{aligned}
E(X_1 X_2 \mid \mathcal{G}_1 \cap \mathcal{G}_2) &= E(E(X_1 \mid \mathcal{G}_2) X_2 \mid \mathcal{G}_1 \cap \mathcal{G}_2) \\
&= E\left(E\left(E(X_1 \mid \mathcal{G}_1) \mid \mathcal{G}_2\right) X_2 \mid \mathcal{G}_1 \cap \mathcal{G}_2\right) \\
&= E\left(E\left(X_1 \mid \mathcal{G}_1 \cap \mathcal{G}_2\right) X_2 \mid \mathcal{G}_1 \cap \mathcal{G}_2\right) \\
&= E\left(X_1 \mid \mathcal{G}_1 \cap \mathcal{G}_2\right) E(X_2 \mid \mathcal{G}_1 \cap \mathcal{G}_2),
\end{aligned}
$$

where the third equality follows from question 2(a) applied to

$$
X = E(E(X_1 \mid \mathcal{G}_1) \mid \mathcal{G}_2) = E(E(X_1 \mid \mathcal{G}_2) \mid \mathcal{G}_1).
$$

Conversely, suppose that $\mathcal{G}_1$ and $\mathcal{G}_2$ are independent conditionally on $\mathcal{G}_1 \cap \mathcal{G}_2$. Let X be any square integrable random variable. Set $X_{i,j} \overset{(\mathrm{def})}{=} E\left(E(X \mid \mathcal{G}_i) \mid \mathcal{G}_j\right)$, for $i = 1, 2$ and $j = 1, 2$, then note that

$$
\begin{aligned}
E[X_{i,j}^2] &= E[E\left(E(X \mid \mathcal{G}_i) \mid \mathcal{G}_j\right) E\left(E(X \mid \mathcal{G}_i) \mid \mathcal{G}_j\right)] \\
&= E[E(X \mid \mathcal{G}_i) E\left(E(X \mid \mathcal{G}_i) \mid \mathcal{G}_j\right)] \\
&= E\left[E[E(X \mid \mathcal{G}_i) E\left(E(X \mid \mathcal{G}_i) \mid \mathcal{G}_j\right) \mid \mathcal{G}_i \cap \mathcal{G}_j]\right] \\
&= E\left[E(X \mid \mathcal{G}_i \cap \mathcal{G}_j)^2\right],
\end{aligned}
$$

where the last equality follows from the conditional independence between $\mathcal{G}_i$ and $\mathcal{G}_j$ given $\mathcal{G}_i \cap \mathcal{G}_j$. Then applying this conditional independence again together with the above identity, we may write

$$
\begin{aligned}
E[(X_{i,j} - X_{j,i})^2] &= E[X_{i,j}^2] + E[X_{j,i}^2] - 2E[X_{i,j} X_{j,i}] \\
&= E\left[E(X \mid \mathcal{G}_i \cap \mathcal{G}_j)^2\right] + E\left[E(X \mid \mathcal{G}_i \cap \mathcal{G}_j)^2\right] - 2E\left[E(X \mid \mathcal{G}_i \cap \mathcal{G}_j)^2\right] \\
&= 0,
\end{aligned}
$$

and identity (2.7.1) follows for any square integrable random variable. It can be extended to any integrable random variable by classical arguments.

4. Suppose that X_1 and X_2 are two random variables such that $(X_1, X_2) \overset{(\mathrm{law})}{=} (X_2, X_1)$. From question 3 it suffices to show that X_1 and X_2 are independent conditionally on $\sigma(X_1) \cap \sigma(X_2)$. Let f be a bounded Borel function, then from the exchangeablity property,

$$
E(f(X_1) \mid \sigma(X_1) \cap \sigma(X_2)) \overset{(\mathrm{law})}{=} E(E(f(X_2) \mid \sigma(X_1) \cap \sigma(X_2)).
$$

But the right hand side of this equality equals $E(E(f(X_2) \mid \sigma(X_1)))$ and, from the exchangeability, this term is equal in law to $E(E(f(X_1) \mid \sigma(X_2)))$, so that we have the identity in law:

$$
E(f(X_1) \mid \sigma(X_1) \cap \sigma(X_2)) \overset{(\mathrm{law})}{=} E(E(f(X_1) \mid \sigma(X_2))).
$$

From Exercise **1.12** we deduce that, actually,

$$E(f(X_1) \mid \sigma(X_1) \cap \sigma(X_2)) = E(E(f(X_1) \mid \sigma(X_2))), \quad \text{a.s.}$$

Let g be another bounded Borel function, then from this identity, we have

$$
\begin{aligned}
E[f(X_1)g(X_2) \mid \sigma(X_1) \cap \sigma(X_2)] &= E[E[f(X_1) \mid \sigma(X_2)]g(X_2) \mid \sigma(X_1) \cap \sigma(X_2)] \\
&= E[E[f(X_1) \mid \sigma(X_1) \cap \sigma(X_2)]g(X_2) \mid \sigma(X_1) \cap \sigma(X_2)] \\
&= E[f(X_1) \mid \sigma(X_1) \cap \sigma(X_2)]E[g(X_2) \mid \sigma(X_1) \cap \sigma(X_2)],
\end{aligned}
$$

which proves the result.

5. It follows from the well-known fact that, for any bounded random variable X and any probability measure P on $(\Omega, \mathcal{F})$, one has $E(E(X \mid \mathcal{F}_{T_1}) \mid \mathcal{F}_{T_2})) = E(X \mid \mathcal{F}_{T_1 \wedge T_2})$.

Solution to Exercise 2.8

1. When X and Y are square integrable r.v.s, assumption (I) implies $\text{Var}(X - Y) = E[(X - Y)^2] - E[X - Y]^2 = 0$, hence $X - Y$ is a.s. constant.

2. Let $x \in \mathbb{R}$, then

$$
\begin{aligned}
\varphi(x)|H(x)|^2 &= E[e^{ixX}]E[e^{-ix(X-Y)}]E[e^{ix(X-Y)}] \\
&= E[e^{ixY}]E[e^{ix(X-Y)}] = E[e^{ixX}] = \varphi(x).
\end{aligned}
$$

The second equality follows from the independence between X and $X - Y$, while the third one, follows from the independence between Y and $X - Y$.

3. This follows immediately from the continuity of φ, and the fact that $\varphi(0) = 1$. Indeed, since $\varphi(x) \neq 0$, for $|x| < \varepsilon$, ε sufficiently small, from (2.8.1) we have $|H(x)| = 1$, for $|x| < \varepsilon$.

4. We shall show that if for an r.v. Z, its characteristic function $\psi(x) = E[e^{ixZ}]$ satisfies $|\psi(x)| = 1$ for $|x| \leq \varepsilon$, for some $\varepsilon > 0$, then Z is a.s. constant.
   Indeed, consider an independent copy Z' of Z. We have: $E[\exp ix(Z - Z')] = 1$, for $|x| < \varepsilon$, so that: $1 - \cos(x(Z - Z')) = 0$, or equivalently: $x(Z - Z') \in 2\pi\mathbb{Z}$, a.s. This implies $Z - Z' = 0$, a.s, since if $|Z(\omega) - Z'(\omega)| > 0$, then $|Z(\omega) - Z'(\omega)| \geq \frac{2\pi}{|x|} \to \infty$, as $x \to 0$. Now, trivially, $Z = Z'$ a.s. is equivalent to Z being a.s. constant.

5. *First solution.*

Under the assumption (J), the law of $Y - X$ given $(X = x)$ would be $\mathcal{N}(0, 1)$, hence $Y - X$ would be independent of X. Likewise, under the assumption (J), $Y - X$ would be independent of Y. But, this could only happen if $Y - X$ is constant, which is in contradiction with the fact (also implied by (J)) that $Y - X$ is $\mathcal{N}(0, 1)$. Thus (J) admits no solution.

Second solution.

Under the assumption (J), we have:

$$P[X \in dx]\, e^{-\frac{(x-y)^2}{2}}\, dy = P[Y \in dy]\, e^{-\frac{(x-y)^2}{2}}\, dx\,, \quad x, y \in \mathbb{R}\,.$$

This implies that

$$P[X \in dx]\, dy = P[Y \in dy]\, dx\,, \quad x, y \in \mathbb{R}\,.$$

But this identity cannot hold because $P[X \in dx]$ and $P[Y \in dy]$ are finite measures whereas the Lebesgue measures dx and dy are not.

6. The bivariate r.v. (X, Y) admits a density which is given by :

$$P[X \in dx, Y \in dy] = \frac{e^{-\frac{(x-a)^2}{2\sigma^2}}}{\sqrt{2\pi\sigma^2}} \frac{e^{-\frac{(y-x)^2}{2}}}{\sqrt{2\pi}}\, dx\, dy\,, \quad x, y \in \mathbb{R}\,.$$

So, the distribution of Y is given by:

$$\int_{\{x \in \mathbb{R}\}} P[X \in dx, Y \in dy] = \frac{dy}{2\pi\sigma} \int_{\{x \in \mathbb{R}\}} e^{-\frac{(x-a)^2}{2\sigma^2}} e^{-\frac{(y-x)^2}{2}}\, dx$$

$$= \frac{dy}{2\pi} \int_{\{x \in \mathbb{R}\}} e^{-\frac{1+\sigma^2}{2\sigma^2}\left[x - \frac{a+\sigma^2 y}{1+\sigma^2}\right]^2}\, dx\, e^{-\frac{(y-a)^2}{2(1+\sigma^2)}}$$

$$= \frac{dy}{\sqrt{2\pi(1+\sigma^2)}} e^{-\frac{(y-a)^2}{2(1+\sigma^2)}}\,, \quad y \in \mathbb{R}\,.$$

We conclude that Y is distributed as $\mathcal{N}(a, 1 + \sigma^2)$ and the conditional density of X given $Y = y$ is:

$$f_{X/Y=y}(x) = \sqrt{\frac{1+\sigma^2}{2\pi\sigma^2}} e^{-\frac{1+\sigma^2}{2\sigma^2}\left[x - \frac{a+\sigma^2 y}{1+\sigma^2}\right]^2}\,, \quad x \in \mathbb{R}\,,$$

which is the density of the law $\mathcal{N}(\frac{a+\sigma^2 y}{1+\sigma^2}, \frac{\sigma^2}{1+\sigma^2})$.

Solution to Exercise 2.9

We use the notations defined in the hint of the statement. We are looking for X and Y such that

$$E\left[\frac{Y}{X} \mid X\right] > 1\,, \quad E\left[\frac{X}{Y} \mid Y\right] > 1\,.$$

First, for any $n \geq 0$,

$$
E\left[\frac{Y}{X} \mid X = 2^n\right] = \frac{1}{2^n}\left(2^{n-1}\frac{P(Y = 2^{n-1}, X = 2^n)}{P(X = 2^n)} + 2^{n+1}\frac{P(Y = 2^{n+1}, X = 2^n)}{P(X = 2^n)}\right)
$$

$$
= \frac{1}{2}\frac{p_n}{p_n + q_n} + 2\frac{q_n}{p_n + q_n}.
$$

The latter expression is greater than 1 whenever

$$
q_n > \frac{1}{2}p_n \tag{2.8.a}
$$

for any n. Now, the other conditional expectation, is equal to:

$$
E\left[\frac{X}{Y} \mid Y = 2^n\right] = \frac{1}{2^n}\left(2^{n-1}\frac{P(X = 2^{n-1}, Y = 2^n)}{P(Y = 2^n)} + 2^{n+1}\frac{P(X = 2^{n+1}, Y = 2^n)}{P(Y = 2^n)}\right)
$$

$$
= \frac{1}{2}\frac{q_{n-1}}{q_{n-1} + p_{n+1}} + 2\frac{p_{n+1}}{q_{n-1} + p_{n+1}}.
$$

From above the necessary and sufficient condition for $E\left[\frac{X}{Y} \mid Y = 2^n\right]$ to be greater than 1 is

$$
p_{n+1} > \frac{1}{2}q_{n-1}. \tag{2.8.b}
$$

When $p_n = q_n$, conditions (2.8.a) and (2.8.b) reduce to: $p_n > \frac{1}{2}p_{n-2}$. This is satisfied by any sequence (p_n) of the form $p_n(= q_n) = Ca^n$, for $0 < C < 1$ and $\frac{1}{\sqrt{2}} < a < 1$.

Note that conditions (2.8.a) and (2.8.b) imply $E[X] = \sum_{n=0}^{\infty} 2^n(p_n + q_n) = \infty$. Actually, it is easily shown that condition (ii) in the statement never occurs when both X and Y are integrable.

Solution to Exercise 2.10

1. It suffices to check that

$$
P_s[\rho_{p_1} = 1, \ldots \rho_{p_k} = 1] = \Pi_{j=1}^k P_s[\rho_{p_j} = 1], \tag{2.9.a}
$$

for any finite sub-sequence $(\rho_{p_1}, \ldots, \rho_{p_k})$, $k \geq 1$ of $(\rho_p)_{p \in \mathcal{P}}$. Indeed, for such a sub-sequence, we have:

$$
P_s[\rho_{p_1} = 1, \ldots \rho_{p_k} = 1] = P_s[\cap_{j=1}^k \{lp_j : l \in \mathcal{N}\}]
$$

$$
= P_s[\{lp_1 \ldots p_k : l \in \mathcal{N}\}]
$$

$$
= \sum_{l \in \mathcal{N}} \frac{1}{\zeta(s)(lp_1 \ldots p_k)^s}
$$

$$
= \frac{1}{(p_1 \ldots p_k)^s},
$$

since $\sum_{l \in \mathcal{N}} \frac{1}{\zeta(s)l^s} = P_s(\mathcal{N}) = 1$. Moreover, this identity implies $P_s(\rho_{p_j} = 1) = 1/p_j^s$, for each $j = 1, \ldots, k$, and (2.9.a) is proven.

To prove Euler's identity, first note that $\cap_{p \in \mathcal{N}}\{\rho_p = 0\} = \{1\}$. Then, we have $P_s(\{1\}) = P_s[\cap_{p \in \mathcal{P}}\{\rho_p = 0\}]$, that is:

$$\frac{1}{\zeta(s)} = \Pi_{p \in \mathcal{P}} P_s[\rho_p = 0]$$

$$= \Pi_{p \in \mathcal{P}} \left(1 - \frac{1}{p^s}\right).$$

2. It is not difficult to see that

$$\{\alpha_p = k\} = p^k\{\rho_p = 0\},$$

where we use the notation: $p^k\{\rho_p = 0\} = \{p^k j : j \in \mathcal{N}, \rho_p(j) = 0\}$. Hence

$$P_s[\alpha_p = k] = P_s[p^k\{\rho_p = 0\}] = \frac{1}{p^{ks}} P_s[\rho_p = 0] = \frac{1}{p^{ks}}\left(1 - \frac{1}{p^s}\right).$$

The latter identity shows that α_p is geometrically distributed with parameter $q = 1/p^s$. Let $k_1, \ldots, k_n$ be any sequence of integers and $p_1, \ldots, p_n$ be any sequence of prime numbers, then

$$\{\alpha_{p_1} = k_1, \ldots, \alpha_{p_n} = k_n\} = p_1^{k_1} \ldots p_n^{k_n}\{\rho_{p_1} = 0, \ldots, \rho_{p_n} = 0\}.$$

And

$$P_s[p_1^{k_1} \ldots p_n^{k_n}\{\rho_{p_1} = 0, \ldots, \rho_{p_n} = 0\}] = \frac{1}{p_1^{k_1 s} \ldots p_n^{k_n s}} P[\rho_{p_1} = 0, \ldots, \rho_{p_n} = 0].$$

Hence, the independence between $\alpha_{p_1}, \ldots, \alpha_{p_n}$ follows from the independence between $\rho_{p_1}, \ldots, \rho_{p_n}$.

Solution to Exercise 2.11

1. First note that

$$\mathbb{1}_{\cup_{k=1}^m A_k} = \mathbb{1}_{(\cap_{k=1}^m A_k^c)^c} = 1 - \mathbb{1}_{\cap_{k=1}^m A_k^c} = 1 - \prod_{k=1}^m (1 - \mathbb{1}_{A_k}).$$

Developing the last term, we have:

$$1 - \prod_{k=1}^m (1 - \mathbb{1}_{A_k}) = \sum_{k=1}^m (-1)^{k-1} \sum_{1 \le i_1 < \cdots < i_k \le m} \mathbb{1}_{A_{i_1} \cap \cdots \cap A_{i_k}}.$$

Formula (2.11.1) is obtained by identifying the expectations of the above equalities.

2. (a) Comparing (2.11.1) with (2.11.2), it suffices to show: $Q_n(A_{p_{i_1}} \cap \cdots \cap A_{p_{i_k}}) = n^{-2}[n/(p_{i_1} \ldots p_{i_k})]^2$. Moreover, for any k: $A_{p_{i_1}} \cap \cdots \cap A_{p_{i_k}} = A_{p_{i_1} p_{i_2} \cdots p_{i_k}}$. Since N_1 and N_2 are independent, $Q_n(A_p) = Q_n(p|N_1)^2$ for any p. Therefore, it remains to check that for any integer p, $Q_n(p|N_1) = \frac{1}{n}[\frac{n}{p}]$, which is easy to do.

(b) For any $k \geq 1$, $n^{-2}[n/(p_{i_1} \ldots p_{i_k})]^2 \to \frac{1}{(p_{i_1} \cdots p_{i_k})^2}$, as $n \to \infty$, and since m is fixed, for $k \leq m$, $\mu_k \to \sum_{1 \leq i_1 < \ldots < i_k \leq i_m} \frac{1}{(p_{i_1} \cdots p_{i_k})^2}$, as $n \to \infty$. Hence

$$\lim_n Q_n(\cup_{k=1}^m A_{p_k}) = \sum_{k=1}^m (-1)^{k-1} \sum_{1 \leq i_1 < \ldots < i_k \leq i_m} \frac{1}{(p_{i_1} \cdots p_{i_k})^2}$$
$$= 1 - \prod_{k=1}^m \left(1 - \frac{1}{p_k^2}\right).$$

(c) We first observe that $Q_n(A) = 1 - Q_n(\cup_{k=1}^n A_{p_k})$. Now, to deal with $Q_n(\cup_{k=1}^n A_{p_k})$ as $n \to \infty$, fix an integer m, let $n \geq m$ and write:

$$\cup_{k=1}^m A_{p_k} \subset \cup_{k=1}^n A_{p_k} \subset (\cup_{k=1}^m A_{p_k}) \cup (\cup_{k>m} A_{p_k}).$$

This leads to the inequality

$$Q_n(\cup_{k=1}^m A_{p_k}) \leq Q_n(\cup_{k=1}^n A_{p_k}) \leq Q_n(\cup_{k=1}^m A_{p_k}) + \frac{1}{n^2} \sum_{k>m} \left[\frac{n}{p_k}\right]^2,$$

and since $\frac{1}{n^2} \sum_{k>m} \left[\frac{n}{p_k}\right]^2 \leq \sum_{k>m} \frac{1}{p_k^2}$, we have, for any m

$$1 - \prod_{k=1}^m \left(1 - \frac{1}{p_k^2}\right) \leq \lim_n Q_n(\cup_{k=1}^n A_{p_k}) \leq 1 - \prod_{k=1}^m \left(1 - \frac{1}{p_k^2}\right) + \sum_{k>m} \frac{1}{p_k^2},$$

as $n \to \infty$. We obtain the required equality by letting m go to ∞.

3. It suffices to set $A = \{N_1 \wedge \cdots \wedge N_k = 1\}$, $A_p = \{p|N_1, \ldots, p|N_k\}$ and to note that from the independence of $N_1, \ldots, N_k$, we have $Q_n(A_p) = \frac{1}{n^k}[p/n]^k$. The rest of the proof is identical to the case $k = 2$ and we obtain

$$\lim_n Q_n(A) = \frac{1}{\zeta(k)}.$$

Solution to Exercise 2.12

1. Let $\mathcal{P}$ be the set of prime numbers and let $(X_p)_{p\in\mathcal{P}}$ be a sequence of independent and uniformly distributed random variables on the torus. Then for any integer n, with factorization $n = \Pi_i p_i^{\alpha_i}$, ($\alpha_i > 0$, for all i), we set $X_n = \Pi_i X_{p_i}^{\alpha_i}$. Then condition (a) is clearly satisfied. Moreover, for any $k \in \mathbb{Z}$, if $k \neq 0$, we have $E[X_n^k] = \Pi_i E[X_{p_i}^{\alpha_i k}] = 0$, hence X_n follows the uniform law on the torus and (b) is satisfied. Finally, in order to show condition (c), it suffices to check that, for any k, X_{n_k} is independent of the vector $(X_{n_1}, X_{n_2}, \ldots, X_{n_{k-1}})$. Indeed, if the factorization of n_k is $n_k = \Pi_i p_i^{\alpha_i}$, then none of the p_is appears in the factorizations of $n_1, \ldots n_{k-1}$, therefore X_{n_k} is independent of $(X_{n_1}, X_{n_2}, \ldots, X_{n_{k-1}})$. Then by induction, we have proved that $X_{n_1}, X_{n_2}, \ldots, X_{n_k}$ are independent.

2. First note that two random variables X_n and X_m are independent if and only if:

$$E[X_n^k X_m^{k'}] = 0, \quad \text{for all } k \text{ and } k' \text{ in } \mathbb{Z}\setminus\{0\}. \tag{2.12.a}$$

Suppose that p is a prime factor of n with degree α, but p is not a prime factor of m. Let k and k' in $\mathbb{Z}\setminus\{0\}$. Then we have $X_n^k X_m^{k'} = X_p^{k\alpha} Y$ where Y is a random variable independent from X_p. Hence $E[X_n^k X_m^{k'}] = 0$. So a sufficient condition for X_n and X_m to be independent is that there exists a prime factor of n (at least one) which is not a prime factor of m.

Now, suppose that n and m have the same prime factors: $n = \Pi_i p_i^{\alpha_i}$ and $m = \Pi_i p_i^{\beta_i}$. Then $E[X_n^k X_m^{k'}] = \Pi_i E[X_{p_i}^{k\alpha_i + k'\beta_i}]$ and this expression is different from 0 if and only if $k\alpha_i + k'\beta_i = 0$, for all i. Hence X_n and X_m are independent whenever α_i/β_i is not constant.

In conclusion, two random variables X_n and X_m are independent if either there is a prime factor of n which is not a prime factor of m (or vice versa) or n and m have decompositions $n = \Pi_i p_i^{\alpha_i}$ and $m = \Pi_i p_i^{\beta_i}$ and α_i/β_i is not constant (in i). Conversely, we easily check that, if X_n and X_m are independent, then one of the above conditions is satisfied.

Solution to Exercise 2.13

1. Set $K_n = (X_n, Y_n)$, $n \geq 1$. The independence between the r.v.s $(X_n, Y_m : n, m \geq 1)$ implies in particular that the r.v.s (K_n) are independent. Since for each $n \geq 1$, Z_n is a Borel function of K_n, the r.v.s (Z_n) are independent. It is easy to check that for each $n \geq 1$, Z_n is Bernoulli distributed with parameter pq.

2. For every $n \geq 1$, let G_{X_n}, G_{Z_n}, G_{S_n} and G_{T_n}, be respectively the generating functions of X_n, Z_n, S_n and T_n. It follows from the independence hypothesis

that
$$G_{X_n}(s) = 1 - p + ps, \quad \text{and} \quad G_{Z_n}(s) = 1 - pq + pqs,$$
$$\text{hence:} \quad G_{S_n}(s) = (1 - p + ps)^n \quad \text{and} \quad G_{T_n}(s) = (1 - pq + pqs)^n.$$

Therefore S_n and T_n have binomial laws with respective parameters (n, p) and (n, pq).

3. Let $n \geq 1$, and note that $\{\tau \geq n\} = \cap_{p=1}^{n-1}\{T_p = 0\}$. Since the T_ps are r.v.s, the event $\cap_{p=1}^{n-1}\{T_p = 0\}$ belongs to $\mathcal{F}$. This shows that τ is an r.v. Now, we deal with S_τ. We have: $\{S_\tau = n\} = \cup_{p=1}^{\infty}\{S_p = n, \tau = p\} \in \mathcal{F}$, since S_p, $p \geq 1$ and τ are r.v.s. Hence, S_τ is an r.v. For every $n \geq 1$, the equalities

$$P[\tau \geq n] = P[\cap_{p=1}^{n-1}\{Z_p = 0\}] = (1 - pq)^{n-1},$$

show that τ is geometrically distributed with parameter pq.

4. From the independence between the r.v.s Z_n, we obtain for $n \geq 2$ and $1 \leq k \leq n$:

$$P[X_k = 1, \tau = n]$$
$$= P[Z_1 = 0, \ldots, Z_{k-1} = 0, X_k = 1, Y_k = 0, Z_{k+1} = 0, \ldots, Z_{n-1} = 0, Z_n = 1]$$
$$= p(1 - q)(1 - pq)^{n-2}pq.$$

Then it follows that,

$$P[X_k = 1 \mid \tau = n] = \frac{p(1 - q)(1 - pq)^{n-2}pq}{(1 - pq)^{n-1} - (1 - pq)^n}$$
$$= \frac{(1 - q)p}{1 - pq}.$$

On the other hand, the independence between X_k and Y_k yields:

$$P[X_k = 1 \mid Z_k = 0] = \frac{P[X_k = 1, Y_k = 0]}{P[Z_k = 0]}$$
$$= \frac{(1 - q)p}{1 - pq}.$$

5. First note that if $x_n = 0$, then both sides are equal to 0. Note also that the identity is trivial for $n = 1$, so, suppose that $n \geq 2$. Developing the right hand side gives:

$$P[\cap_{i=1}^{n}\{X_i = x_i\} \mid \tau = n]$$
$$= \frac{P[X_1 = x_1, \ldots, X_n = x_n, Z_1 = 0, \ldots, Z_{n-1} = 0, Z_n = 1]}{(1 - pq)^{n-1}pq}$$
$$= \frac{P[X_1 = x_1, \ldots, X_n = x_n, x_1Y_1 = 0, \ldots, x_{n-1}Y_{n-1} = 0, X_n = 1, Y_n = 1]}{(1 - pq)^{n-1}pq}$$
$$= \frac{P[X_1 = x_1] \ldots P[X_n = x_n]P[x_1Y_1 = 0] \ldots P[x_{n-1}Y_{n-1} = 0]P[X_n = 1]P[Y_n = 1]}{(1 - pq)^{n-1}pq}.$$

Now, we compute the term $P[X_i = x_i \mid \tau = n]$, for $1 \le i \le n$:

$$P[X_i = x_i \mid \tau = n] = [(1 - pq)^{n-1}pq]^{-1}P[Z_1 = 0, \ldots, Z_{i-1} = 0, X_i = x_i, x_i Y_i = 0,$$
$$Z_{i+1} = 0, \ldots, Z_{n-1} = 0, X_n = 1, Y_n = 1]$$
$$= \frac{P[X_i = x_i]P[x_i Y_i = 0]}{1 - pq}.$$

The result follows from the above identity:

- If $k > n$, then $P[S_\tau = k \mid \tau = n] = 0$.

- If $k \le n$, then

$$P[S_\tau = k \mid \tau = n] = P[S_n = k \mid \tau = n]$$
$$= \sum_{\{x_i : \sum_{i=1}^n x_i = k\}} P[\cap_{i=1}^n \{X_i = x_i\} \mid \tau = n]$$
$$= \binom{n}{k}\left(\frac{p(1-q)}{1-pq}\right)^k\left(\frac{1-p}{1-pq}\right)^{n-k}.$$

6. From above, we have

$$E[S_\tau \mid \tau = n] = \sum_{k=1}^n kP[S_\tau = k \mid \tau = n] = \frac{np(1-q)}{1-pq},$$

then it follows that

$$E[S_\tau] = \sum_{n=1}^\infty P[\tau = n]E[S_\tau \mid \tau = n] = \sum_{n=1}^\infty \frac{(1-pq)^n nqp^2(1-q)}{1-pq}$$
$$= \sum_{n=0}^\infty n(1-pq)^{n-1}p^2q(1-q) = \frac{1-q}{q}.$$

By definition, $Z_\tau = X_\tau Y_\tau = 1$, which implies $X_\tau = Y_\tau = 1$. Thus we have: $E[T_\tau] = 1$. On the other hand,

$$E[\tau]E[X_1] = E[\tau]p = \left(\frac{1}{pq}\right)p = \frac{1}{q},$$

and $E[\tau]E[Z_1] = (1/pq)\,pq = 1$.

Solution to Exercise 2.14

1. Taking $b = c = 0$ in (ii) gives $E[e^{-aL}] = \frac{1/2}{1/2+a}$, for every $a \ge 0$, hence, L is exponentially distributed with parameter $1/2$. Now set $a = 0$ and $b = c$ in (ii), then

we obtain the characteristic function of W:

$$E[e^{ibW}] = \left(\cosh b + \frac{|b|}{b}\sinh b\right)^{-1}, \quad \text{if } b \neq 0,$$
$$= 1, \quad \text{if } b = 0,$$

that is

$$E[e^{ibW}] = e^{-|b|}.$$

Hence, W is Cauchy distributed with parameter 1.

2. Set $\varphi(L) = E[e^{ibW_- + icW_+} \mid L]$. The r.v. $\varphi(L)$ satisfies:

$$E[e^{-aL}\varphi(L)] = E[e^{-aL + ibW_- + icW_+}],$$

for every $a \geq 0$. But from the expression of the law of L obtained in question 1. and (ii), we have:

$$E[e^{-aL + ibW_- + icW_+}] = E\left[e^{-aL}\frac{c}{\sinh c}e^{-\left(\frac{c\coth c}{2} - \frac{1}{2} + \frac{|b|}{2}\right)L}\right],$$

thus $\varphi(L) = \frac{c}{\sinh c}e^{-\left(\frac{c\coth c}{2} - \frac{1}{2} + \frac{|b|}{2}\right)L}$ and it is clear, from this expression for $\varphi(L)$, that it can be written as:

$$\varphi(L) = E[e^{ibW_-} \mid L]E[e^{icW_+} \mid L],$$

for all $b, c \in \mathbb{R}$, where:

$$E[e^{ibW_-} \mid L] = \exp\left(-\frac{|b|}{2}L\right) \tag{2.14.a}$$

$$E[e^{icW_+} \mid L] = \frac{c}{\sinh c}\exp -\frac{1}{2}(c\coth c - 1)L. \tag{2.14.b}$$

Thus, W_+ and W_- are conditionally independent given L.

From the conditional independence proved above, we have for all $l \in \mathbb{R}_+$ and $x \in \mathbb{R}$:

$$E[e^{ibW_-} \mid L = l, W_+ = x] = E[e^{ibW_-} \mid L = l],$$

moreover, from (2.14.a), we obtain:

$$E[e^{ibW_-} \mid L = l] = e^{-\frac{l}{2}|b|}, \quad b \in \mathbb{R}.$$

Hence, conditionally on $W_+ = x$ and $L = l$, W_- is Cauchy distributed with parameter $l/2$.

From (2.14.b), we have

$$E[e^{icW_+} \mid L = l] = \frac{c}{\sinh c}e^{-\left(\frac{c\coth c}{2} - \frac{1}{2}\right)l}, \quad c \neq 0.$$

Comments on the solution. For $l = 0$, the latter Fourier transform can be inverted whereas for $l \neq 0$, there is no known explicit expression for the corresponding density. (However, computations in A.N. Borodin and P. Salminen [10] show that an extremely complicated "closed form" formula might be obtained.) This difficulty has been observed in:

J. BASS AND P. LÉVY: Propriétés des lois dont les fonctions caractéristiques sont $1/\mathrm{ch}\,z$, $z/\mathrm{sh}\,z$, $1/\mathrm{ch}^2 z$. *C. R. Acad. Sci. Paris*, **230**, 815–817 (1950).

One can find series developments for the corresponding densities in A.N. Borodin and P. Salminen [10]. See also:

P. BIANE, J. PITMAN AND M. YOR: Probability laws related to the Jacobi theta and Riemann zeta functions, and Brownian excursions. *Bull. (N.S.) Amer. Math. Soc.*, **38**, no. 4, 435–465 (2001).

Solution to Exercise 2.15

1. This is quite similar to the computations made in question 2 of Exercise **2.14**, and one finds:

$$
E\left[\exp - \left(\frac{\lambda^2}{2}A_- + \frac{\mu^2}{2}A_+\right) \mid L\right] = E\left[\exp - \frac{\lambda^2}{2}A_- \mid L\right] E\left[\exp - \frac{\mu^2}{2}A_+ \mid L\right]
$$

$$
= \exp\left(-\frac{\lambda}{2}L\right) \frac{\mu}{\sinh \mu} \exp\left(-\frac{1}{2}(\mu \coth \mu - 1)L\right).
$$

From question 7 of Exercise **4.2**, if T is (s)-distributed (that is $P(T \in dt) = \frac{dt}{\sqrt{2\pi t^3}} \exp\left(-\frac{1}{2t}\right)$, $t > 0$), then its Laplace transform is:

$$
E\left[\exp\left(-\frac{\lambda^2}{2}T\right)\right] = \exp(-\lambda), \quad \lambda \geq 0,
$$

and the law of T is characterized by the identity

$$
T \stackrel{(\text{law})}{=} \frac{1}{N^2},
$$

where N is a centered Gaussian variable with variance 1. Hence, the Laplace transform of A_- may be expressed as:

$$
E\left[\exp\left(-\frac{\lambda^2}{2}A_-\right)\right] = \frac{1}{1 + \lambda} = \int_0^\infty dv \, e^{-v} e^{-v\lambda}
$$

$$
= \int_0^\infty dv \, e^{-v} E\left(e^{-\frac{\lambda^2 v^2}{2}T}\right)
$$

$$
= E\left[\exp\left(-\frac{\lambda^2 Z^2}{2}T\right)\right] = E\left[\exp\left(-\frac{\lambda^2}{2}\frac{Z^2}{N^2}\right)\right]
$$

and question 1 follows.

2. The identity in law (2.15.2) follows from the double equality:

$$E[\exp(-aL + i\lambda W_+ + i\mu W_-)] = E\left[\exp\left(-aL - \frac{\lambda^2}{2}A_+ - \frac{\mu^2}{2}A_-\right)\right]$$

$$= E\left[\exp\left(-aL + i\lambda N_+\sqrt{A_+} + i\mu N_-\sqrt{A_-}\right)\right].$$

3. This follows from:

$$E[\exp i(\lambda W_+ + \mu W_-) \mid L = l] = E\left[\exp\left(-\frac{\lambda^2}{2}A_+ - \frac{\mu^2}{2}A_-\right) \mid L = l\right].$$

4. Write the Laplace transform of (A_-, A_+) as

$$\frac{1}{\cosh\mu + \frac{\lambda}{\mu}\sinh\mu} = \int_0^\infty dt \exp -t\left(\cosh\mu + \frac{\lambda}{\mu}\sinh\mu\right)$$

$$= \frac{\mu}{\sinh\mu}\int_0^\infty dt\, e^{-t\frac{\mu\cosh\mu}{\sinh\mu}}\int_0^\infty du\, e^{-\frac{\lambda^2 u}{2}}\frac{t}{\sqrt{2\pi u^3}}e^{-\frac{t^2}{2u}}.$$

By letting μ converge towards 0 in the above expression, we obtain the following different expressions for the density g of A_-:

$$g(u) = \int_0^\infty dt\, e^{-t}\frac{t}{\sqrt{2\pi u^3}}\exp\left(-\frac{t^2}{2u}\right) = \int_0^\infty dt\, e^{-t}\left(\frac{1 - e^{-t^2/2u}}{\sqrt{2\pi u}}\right)$$

$$= \frac{e^{u/2}}{\sqrt{u}}E[(N - \sqrt{u})^+] = \frac{1}{\sqrt{2\pi u}} - e^{u/2}P(N \geq \sqrt{u}),$$

where N denotes a standard normal variable. These expressions are closely related to the Hermite functions h_{-1} and h_{-2}; see e.g. Lebedev [51].

Note also that from (2.15.1) the variable L is exponentially distributed with parameter $\frac{1}{2}$ and from question 1, the law of A_- given $L = l$ is the same as the law of $\frac{l^2}{4}T$. The law of L given A_- is then obtained from Bayes' formula, since we know the law of A_- given L, the law of L and the law of A_-. So, with $H \stackrel{\text{(def)}}{=} \frac{L}{2}$ we find:

$$P(H \in dh \mid A_- = u) = \frac{1}{g(u)\sqrt{2\pi u^3}}dh\, he^{-\left(\frac{h^2}{2u}+h\right)}.$$

We then use the conditional independence of A_+ and A_- given L, to write:

$$E\left[\exp\left(-\frac{\mu^2}{2}A_+\right)\mid A_-=u\right]=E\left[E\left[\exp-\frac{\mu^2}{2}A_+\mid H\right]\mid A_-=u\right]$$

$$=\int P(H\in dh\mid A_-=u)E\left[\exp-\frac{\mu^2}{2}A_+\mid H=h\right]$$

$$=\frac{1}{g(u)\sqrt{2\pi u^3}}\int_0^\infty dh\,h\exp-\left(\frac{h^2}{2u}+h\right)\frac{\mu}{\sinh\mu}\exp-h(\mu\coth\mu-1)$$

$$=\left(\frac{\mu}{\sinh\mu}\right)\frac{1}{g(u)\sqrt{2\pi u^3}}\int_0^\infty dt\,t\,e^{-(\frac{t^2}{2}+t\sqrt{u}\coth\mu)}$$

$$=\left(\frac{\mu}{\sinh\mu}\right)\frac{g(u(\mu\coth\mu)^2)}{g(u)}.$$

5. This can be done by first establishing a series development of the Laplace transform of (A_-,A_+) in terms of e^{-s}:

$$E\left[\exp\left(-\frac{s^2}{2}(\lambda^2 A_-+\mu^2 A_+)\right)\right]=\frac{1}{\cosh(s\mu)+\frac{\lambda}{\mu}\sinh(s\mu)}$$

$$=\frac{2}{e^{s\mu}\left(1+\frac{\lambda}{\mu}\right)+e^{-s\mu}\left(1-\frac{\lambda}{\mu}\right)}$$

$$=\frac{2e^{-s\mu}}{\left(1+\frac{\lambda}{\mu}\right)\left(1+\rho e^{-2s\mu}\right)},$$

where $\rho=\frac{1-\frac{\lambda}{\mu}}{1+\frac{\lambda}{\mu}}$. Note that since $|\rho|<1$, we get

$$E\left[\exp\left(-\frac{s^2}{2}(\lambda^2 A_-+\mu^2 A_+)\right)\right]=\frac{2}{1+\frac{\lambda}{\mu}}e^{-s\mu}\sum_{n=0}^\infty(-\rho)^n e^{-2s\mu n}$$

$$=\frac{2}{1+\frac{\lambda}{\mu}}\sum_{n=0}^\infty(-\rho)^n e^{-s\mu(2n+1)}$$

$$=\frac{2}{1+\frac{\lambda}{\mu}}\sum_{n=0}^\infty(-\rho)^n\int_0^\infty\frac{dt}{\sqrt{2\pi t^3}}a_n e^{-\frac{a_n^2}{2t}}e^{-\frac{s^2 t}{2}},$$

where $a_n=(2n+1)\mu$. Hence, we obtain:

$$P(\lambda^2 A_-+\mu^2 A_+\in dt)=\frac{2}{1+\frac{\lambda}{\mu}}\sum_{n=0}^\infty(-\rho)^n\frac{dt}{\sqrt{2\pi t^3}}a_n e^{-\frac{a_n^2}{2t}}.$$

Solution to Exercise 2.16

1. We first recall that for Q absolutely continuous with respect to P, and X_0 the Radon–Nikodym density of Q with respect to P, then $E_P(X_0)=E_Q(1)=1$, and

$X_0 \geq 0\,P$a.s., since $Q(X_0 < 0) = E_P[X_0 1_{\{X_0 < 0\}}] \geq 0$. Moreover, $X_0 \in L_+^1(P)$ and $E_P[X_0] = 1$. Indeed, $Q(X_0 < 0) = E_P[X_0 \mathbb{I}_{\{X_0 < 0\}}] \geq 0$ implies $X_0 \geq 0$, P-a.s. The equality $E_P[X_0] = 1$ follows from the definition of X_0.

(a) From above, for any sub-σ-field $\mathcal{G}$ of $\mathcal{F}$, we have $0 \leq E_P[X_0 \mid \mathcal{G}] < +\infty$, P-a.s. Now, let $Y_0 \,(= 1/X_0)$ be the Radon–Nikodym density of P with respect to Q, and put $C = \{E_P[X_0 \mid \mathcal{G}] = 0\}$. We have

$$P(C) = E_Q[Y_0 \mathbb{I}_C] = E_Q[E_Q[Y_0 \mid \mathcal{G}]\mathbb{I}_C]$$
$$= E_P[E_P[X_0 \mid \mathcal{G}]E_Q[Y_0 \mid \mathcal{G}]\mathbb{I}_C] = 0\,,$$

which implies that $E_P[X_0 \mid \mathcal{G}] > 0$, P-a.s.

(b) Let $A \in \mathcal{G}$, then

$$E_Q[f\mathbb{I}_A] = E_Q[E_Q[f \mid \mathcal{G}]\mathbb{I}_A] = E_P[E_Q[f \mid \mathcal{G}]X_0 \mathbb{I}_A]$$
$$= E_P[E_Q[f \mid \mathcal{G}]E_P[X_0 \mid \mathcal{G}]\mathbb{I}_A]\,.$$

But one also has,

$$E_Q[f\mathbb{I}_A] = E_P[fX_0\mathbb{I}_A] = E_P[E_P[fX_0 \mid \mathcal{G}]\mathbb{I}_A]\,.$$

Identifying the right hand sides of both equations above, we obtain

$$E_P[E_P[fX_0 \mid \mathcal{G}]\mathbb{I}_A] = E_P[E_Q[f \mid \mathcal{G}]E_P[X_0 \mid \mathcal{G}]\mathbb{I}_A]\,,$$

for any $A \in \mathcal{G}$, which establishes identity (b).

2. *First solution*:

We have : $E_P[fX' \mid \mathcal{G}] = \gamma E_P[fX_0 \mid \mathcal{G}]$ and $E_P[X' \mid \mathcal{G}] = \gamma E_P[X_0 \mid \mathcal{G}]$, hence (a') and (b') follow respectively from (a) and (b).

Second solution:

For any $A \in \mathcal{G}$, we have

$$E_Q[f\gamma\mathbb{I}_A] = E_P[E_Q[f \mid \mathcal{G}]\gamma X_0 \mathbb{I}_A]$$
$$= E_P[E_Q[f \mid \mathcal{G}]E_P[\gamma X_0 \mid \mathcal{G}]\mathbb{I}_A]\,.$$

Writing the left hand side as

$$E_Q[f\gamma\mathbb{I}_A] = E_P[E_P[f\gamma X_0 \mid \mathcal{G}]\mathbb{I}_A]$$

gives (b').

3. Since γ is $\mathcal{G}$-measurable, then $E_P[X' \mid \mathcal{G}] = \gamma E_P[X_0 \mid \mathcal{G}]$. So, from (a), we may write $\gamma = E_P[X_0 \mid \mathcal{G}]^{-1}E_P[X' \mid \mathcal{G}]$.

4. From 1(b) we have $E_Q[f \,|\, \mathcal{G}] = E_P[X_0 \,|\, \mathcal{G}]^{-1} E_P[fX_0 \,|\, \mathcal{G}]$. Set $\gamma = E_P[X_0 \,|\, \mathcal{G}]^{-1}$ $E_P[X' \,|\, \mathcal{G}]$, then 2(b') gives

$$E_P[fX' \,|\, \mathcal{G}] = E_P[fX_0 \,|\, \mathcal{G}]\gamma = E_P[f\gamma X_0 \,|\, \mathcal{G}].$$

Identifying the expectations of each member of the above equality, it follows that $E_P[fX'] = E_P[f\gamma X_0]$, for any $\mathcal{F}$-measurable function f, so that $X' = \gamma X_0$, P-a.s.

5. From 1(b), a sufficient condition is that X_0 is $\mathcal{G}$-measurable. Conversely, suppose that for any $\mathcal{F}$-measurable function f, $E_P[f \,|\, \mathcal{G}] = E_Q[f \,|\, \mathcal{G}]$, then from 1(b), we have

$$E_P[fX_0] = E_P[E_P[f \,|\, \mathcal{G}]E_P[X_0 \,|\, \mathcal{G}]] = E_P[fE_P[X_0 \,|\, \mathcal{G}]].$$

It follows that $X_0 = E_P[X_0 \,|\, \mathcal{G}]$, P-a.s.

Solution to Exercise 2.17

We may directly answer question 2. Indeed, for any $M > 0$,

$$E[\Phi \,|\, f \geq MU] = \frac{E_P\left[\Phi\left(\frac{f}{M} \wedge 1\right)\right]}{E_P\left[\frac{f}{M} \wedge 1\right]} = \frac{E_P\left[\Phi(f \wedge M)\right]}{E_P\left[f \wedge M\right]},$$

which converges, as $M \to \infty$, towards $E_P[\Phi f]$, by monotone convergence.

Solution to Exercise 2.18

1. First note that under the conditions of the statement, the functions $y \mapsto E[f(SX) \,|\, Y = y]$ and $z \mapsto E[f(SX) \,|\, SY = z]$ are continuous on $\mathbb{R}_+$ and the conditional expectation $E[f(SX) \,|\, SY = 0]$ may be defined as

$$E[f(SX) \,|\, SY = 0] = \lim_{\varepsilon \to 0} \frac{E[f(SX)\mathbb{1}_{\{0 \leq SY \leq \varepsilon\}}]}{P(0 \leq SY \leq \varepsilon)}.$$

Let $\lambda > 0$ and write

$$E[f(SX)\mathbb{1}_{\{0 \leq SY \leq \varepsilon\}}] = E[f(SX)\mathbb{1}_{\{0 \leq SY \leq \varepsilon\}}\mathbb{1}_{\{S < \lambda\varepsilon\}}] + E[f(SX)\mathbb{1}_{\{0 \leq SY \leq \varepsilon\}}\mathbb{1}_{\{S \geq \lambda\varepsilon\}}].$$
$$(2.18.\mathrm{a})$$

Since $E[S^{-(\alpha+1)}] < +\infty$, for the first term of the right hand side, we have

$$E[f(SX)\mathbb{1}_{\{0 \leq SY \leq \varepsilon\}}\mathbb{1}_{\{S < \lambda\varepsilon\}}] \leq AP(S < \lambda\varepsilon) \leq A(\lambda\varepsilon)^{\alpha+1} \int_{[0,\lambda\varepsilon)} \frac{P(S \in ds)}{s^{\alpha+1}} = \mathrm{o}(\varepsilon^{\alpha+1}),$$
$$(2.18.\mathrm{b})$$

where A is such that $f \leq A$ on $\mathbb{R}_+$. Now, to deal with the second term, define $k(s, y) = E[f(sX) | Y = y]$, for any $s \geq 0$ and write

$$E[f(SX)\mathbb{1}_{\{0 \leq SY \leq \varepsilon\}}\mathbb{1}_{\{S \geq \lambda\varepsilon\}}] = \int_{[\lambda\varepsilon,\infty)} P(S \in ds) \int_0^{\varepsilon/s} dy\, k(s,y)g(y).$$

From (ii) in the statement of the exercise and by continuity of the function $k(s, y)$, we have

$$\int_0^{\varepsilon/s} dy\, k(s,y)g(y) = \left(c + \varphi\left(\frac{\varepsilon}{s}\right)\right) k(s,0)(\varepsilon/s)^{\alpha+1}, \qquad (2.18.c)$$

where $\varphi(\varepsilon) \to 0$ as $\varepsilon \to 0$. The above expression becomes

$$E[f(SX)\mathbb{1}_{\{0 \leq SY \leq \varepsilon\}}\mathbb{1}_{\{S \geq \lambda\varepsilon\}}] = \int_{[\lambda\varepsilon,\infty)} P(S \in ds)c(\varepsilon/s)^{\alpha+1}k(s,0)$$
$$+ \varepsilon^{\alpha+1}\int_{[\lambda\varepsilon,\infty)} P(S \in ds)\varphi\left(\frac{\varepsilon}{s}\right)\frac{1}{s^{\alpha+1}}k(s,0).$$

Since $\varphi\left(\frac{\varepsilon}{s}\right)$ is bounded in ε/s on the interval $[\lambda\varepsilon, \infty)$, by dominated convergence the expression $\int_{[\lambda\varepsilon,\infty)} P(S \in ds)\varphi\left(\frac{\varepsilon}{s}\right)\frac{1}{s^{\alpha+1}}k(s,0)$ tends to 0 as ε goes to 0, so that we can write

$$E[f(SX)\mathbb{1}_{\{0 \leq SY \leq \varepsilon\}}\mathbb{1}_{\{S \geq \lambda\varepsilon\}}] = \int_{[\lambda\varepsilon,\infty)} P(S \in ds)c(\varepsilon/s)^{\alpha+1}k(s,0) + o(\varepsilon^{\alpha+1}). \quad (2.18.d)$$

Finally from (ii), $P(0 \leq Y \leq \varepsilon) \sim c\varepsilon^{\alpha+1}$, as ε goes to 0, so from (2.18.a), (2.18.b) and (2.18.d), we have

$$\lim_{\varepsilon \to 0} \frac{E[f(SX)\mathbb{1}_{\{0 \leq SY \leq \varepsilon\}}]}{P(0 \leq Y \leq \varepsilon)} = \int_{\mathbb{R}_+} P(S \in ds)\frac{k(s,0)}{s^{\alpha+1}} = E\left[\frac{1}{S^{\alpha+1}}f(SX) \mid Y = 0\right],$$

(Note that in the above equalities, the integral $\int_{\mathbb{R}_+} P(S \in ds)\frac{k(s,0)}{s^{\alpha+1}}$ should be $\int_{(0,\infty)} P(S \in ds)\frac{k(s,0)}{s^{\alpha+1}}$ but this makes no difference, since the assumption, $E[S^{-(\alpha+1)}] < +\infty$ implies: $P(S = 0) = 0$.) We conclude by noticing that from the same arguments as above,

$$\lim_{\varepsilon \to 0} \frac{P(0 \leq SY \leq \varepsilon)}{P(0 \leq Y \leq \varepsilon)} = E\left[\frac{1}{S^{\alpha+1}}\right].$$

2. First, applying the result of question 1, we obtain

$$E[f(SX) | SY = 0] = \frac{1}{E[T^{-1/2}]}E[T^{-1/2}f(\sqrt{T}X) | Y = 0].$$

We get (2.18.2) by computing the right hand side of the above equality:

$$\frac{1}{E[T^{-1/2}]}E[T^{-1/2}f(\sqrt{T}X) | Y = 0]$$
$$= \sqrt{\frac{2}{\pi}}\int_0^\infty dt\, \frac{1}{2}e^{-t/2}E[t^{-1/2}f(\sqrt{t}X) | Y = 0]$$
$$= \sqrt{\frac{2}{\pi}}\int_0^\infty dy\, e^{-y^2/2}E[f(yX) | Y = 0] = E[f(|N|X) | Y = 0].$$

3. Again, from question 1, we have

$$E[f(SX) \mid SY = 0] = \frac{1}{E[Z_a^{-(\alpha+1)}]} E[Z_a^{-(\alpha+1)} f(Z_a X) \mid Y = 0].$$

We refer to Exercise **4.3** for the expression of the density of Z_a. This gives:

$$\frac{1}{E[Z_a^{-(\alpha+1)}]} E[Z_a^{-(\alpha+1)} f(Z_a X) \mid Y = 0]$$

$$= \frac{\Gamma(a)}{\Gamma(a - (\alpha+1))} \int_0^\infty dz \frac{1}{\Gamma(a)} z^{a-1} e^{-z} E[z^{-(\alpha+1)} f(zX) \mid Y = 0]$$

$$= \int_0^\infty dz \frac{1}{\Gamma(a - (\alpha+1))} z^{a-1-(\alpha+1)} e^{-z} E[f(zX) \mid Y = 0] = E[f(Z_{a-(\alpha+1)}X) \mid Y = 0].$$

Solution to Exercise 2.19

1. First recall that the Laplace transform of L is given by

$$E[\exp -\lambda L] = \frac{1}{\lambda + 1}, \quad \lambda > 0. \tag{2.19.a}$$

Now suppose that (a) holds, then from (2.19.a), we have

$$E[E[X \mid L]e^{-\lambda L}] = aE[\exp -\lambda a L],$$

for any $\lambda > 0$. This can also be written as

$$\int_0^\infty \left(E[X \mid L = l]e^{-l} \right) e^{-\lambda l} \, dl = \int_0^\infty \left(e^{-a^{-1}l} \right) e^{-\lambda l} \, dl.$$

The functions between parentheses in the above integrals have the same Laplace transform, so they are a.e. equal, hence, we have:

$$E[X \mid L = l] = \exp -(a^{-1} - 1)l, \quad \text{a.e. } (dl).$$

and thus: $a = (1 + \alpha)^{-1}$.
   If (b) holds, then we may write:

$$E[X \exp -\lambda L] = E[E[X \mid L] \exp -\lambda L] = E[\exp -(\alpha + \lambda)L],$$

and from (2.19.a), this last expression equals $\frac{a}{1+\lambda a}$. So we obtained (a).
   Suppose that (a) holds. By analytical continuation, we can show that $E[X \exp (\lambda L)] = \frac{a}{1-\lambda a}$ holds for any $\lambda < \frac{1}{a}$. Hence, for any $\lambda < \frac{1}{a}$, from Fubini's Theorem, and the series development of the exponential function, we have

$$E[X \exp (\lambda L)] = E\left[X \sum_{k \geq 0} \frac{(\lambda L)^k}{k!} \right] = \sum_{k \geq 0} \lambda^k E\left[X \frac{L^k}{k!} \right].$$

Then we obtain (c) by writing : $\sum_{k \geq 0} \lambda^k a^{k+1} = \frac{a}{1-\lambda a}$ and identifying this expression with the previous one.

Conversely, if (c) holds, then, for any $0 \leq \mu < \frac{1}{E[X]}$,

$$E[X \exp(\mu L)] = \sum_{k \geq 0} \mu^k E\left[X \frac{L^k}{k!}\right] = \frac{E[X]}{1 - \mu E[X]},$$

from Fubini's Theorem. By analytical continuation, this expression also holds for $\lambda = -\mu \geq 0$, so (a) holds.

2. For variables of the type of X, it is easy to show that $E[X \mid L] = \exp(-\alpha L)$, i.e. (b) is satisfied.

3. Let us show that under condition (2.19.2), the following properties are equivalent.

(a) There exists $a > 0$ such that for any $\lambda \geq 0$:

$$E[X \exp(-\lambda L)] = \left(\frac{a}{1 + \lambda a}\right)^{\gamma}.$$

(b) There exists $\alpha \geq 0$ such that $E[X \mid L = \ell] = \exp(-\ell \alpha)$.

(c) For every $k \geq 1$,

$$E\left[X L^k\right] = \gamma(\gamma + 1) \dots (\gamma + k - 1) E[X]^{\frac{k+\gamma}{\gamma}}.$$

Moreover, if those properties are satisfied, then the following relationship between a and α holds:

$$a = \frac{1}{1 + \alpha}.$$

Under condition (2.19.2), we say that L is gamma distributed with parameter γ (see Exercise **4.3**). It is not difficult to check that its Laplace transform is given by

$$E[\exp{-\lambda L}] = \left(\frac{1}{\lambda + 1}\right)^{\gamma}, \quad \lambda > 0. \tag{2.19.b}$$

We verify (a)$\Longleftrightarrow$(b) in the same manner as in question 1. To check (a)$\Longleftrightarrow$(c), we use the same arguments together with the series development: $(1 + x)^{-\gamma} = 1 - \gamma x + \gamma(\gamma + 1)\frac{x^2}{2} + \dots + \gamma(\gamma + 1) \dots (\gamma + k - 1)\frac{x^k}{k!} + \dots$, which holds for any $x \geq 0$.

4. We start from the equality

$$E\left[X \frac{(tL)^k}{k!}\right] = t^k E[X]^{k+1}, \quad t \in \mathbb{R}_+, \ k \geq 0.$$

Integrating with respect to $\nu(dt)$, we have:

$$\int_{\mathbb{R}_+} \nu(dt) E\left[X\frac{(tL)^k}{k!}\right] = \int_{\mathbb{R}_+} \nu(dt) E[X](tE[X])^k.$$

Finally we obtain the result by performing the sum over $k \geq 0$ and using Fubini's Theorem.

Solution to Exercise 2.20

1. For any pair of bounded Borel functions f and g, we write:

$$E[f(\{Z\})g([Z])] = \sum_{n=0}^{\infty} \int_n^{n+1} du\, e^{-u} f(u-n)g(n)$$

$$= \sum_{n=0}^{\infty} e^{-n} \int_0^1 dv\, f(v)e^{-v}g(n).$$

Hence $P(\{Z\} \in dv) = \frac{e^{-v}}{1-e^{-1}}\mathbb{I}_{\{v\in[0,1]\}}\,dv$, $P([Z] = n) = e^{-n}(1 - e^{-1})$, $n \in \mathbb{N}$ and $\{Z\}$ and $[Z]$ are independent.

2. We take $\varphi(t) \stackrel{(\text{def})}{=} e^{-[t]}(1-e^{-1}) = (1-e^{-1})e^{\{t\}}e^{-t}$. Then with $Q = (1-e^{-1})e^{\{Z\}}P$, we obtain:

$$E_Q[f(\{Z\})g([Z])] = (1 - e^{-1}) \sum_{n=0}^{\infty} e^{-n}g(n) \int_0^1 dv\, f(v).$$

3. Let X be an r.v. with density φ then

$$E[f(\{X\})g([X])] = \sum_{n=0}^{\infty} \int_n^{n+1} dt\, f(t-n)g(n)\varphi(t)$$

$$= \sum_{n=0}^{\infty} \left(\int_0^1 du\, f(u)\varphi(u+n)\right) g(n).$$

From above, the independence between $\{X\}$ and $[X]$ is equivalent to

$$\sum_{n=0}^{\infty} \left(\int_0^1 du\, f(u)\varphi(u+n)\right) g(n)$$

$$= \left(\sum_{n=0}^{\infty} \int_0^1 du\, f(u)\varphi(u+n)\right) \left(\sum_{n=0}^{\infty} \int_0^1 du\, g(n)\varphi(u+n)\right),$$

for any pair of bounded Borel functions f and g. This is also equivalent to

$$\int_0^1 du\, f(u)\varphi(u+n) = \left(\sum_{k=0}^{\infty} \int_0^1 du\, f(u)\varphi(k+u)\right) \int_0^1 dv\, \varphi(v+n),$$

for any bounded Borel function f that is

$$\varphi(u+n) = \left(\sum_{k=0}^{\infty} \varphi(k+u) \right) \int_0^1 dv\, \varphi(v+n), \quad \text{a.e. } (du).$$

Then, φ must satisfy: for every $n \in \mathbb{N}$,

$$\varphi(u+n) = \psi(u)c_n, \quad u \in (0,1), \text{ a.e.,} \tag{2.20.a}$$

for a positive sequence (c_n) in l^1, and $\psi \geq 0$, $\int_0^1 du\, \psi(u) < \infty$. Conversely, it follows from the above arguments that if φ satisfies (2.20.a), then the independence of $\{X\}$ and $[X]$ holds.

Comments on the solution.

(a) More generally, B. Tsirel'son told us about the following result. Given every pair of probabilities on $\mathbb{R}_+$, μ and ν, which admit densities with respect to Lebesgue measure, there exists a joint law on $\mathbb{R}_+^2$, that of (A, B) such that:

   (i) A is distributed as μ,

   (ii) B is distributed as ν,

   (iii) $B - A$ belongs to $\mathbb{Q}$, a.s.

   In our example: B is exponentially distributed and $A = \{B\}$.

(b) Note how the property (2.20.a) is weaker than the "loss of memory property" for the exponential variable, which is equivalent to:

$$\varphi(u+s) = C\varphi(u)\varphi(s), \quad u, s \geq 0.$$

Solution to Exercise 2.21

1. It is easy to check that $\{-1, +1\}$-valued r.v.s $\varepsilon_1, \ldots, \varepsilon_n$ are i.i.d. and symmetric if and only if, for any subsequence $1 \leq n_1 \leq \ldots \leq n_k \leq n$,

$$E(\varepsilon_{n_1} \varepsilon_{n_2} \ldots \varepsilon_{n_k}) = 0.$$

But the identity

$$E[(S_{n_1} - S_{n_1-1})(S_{n_2} - S_{n_2-1}) \ldots (S_{n_k} - S_{n_k-1})] = 0$$

is readily verified from the martingale property of S.

2. Call $\mathcal{F} = (\mathcal{F}_n)_{n \geq 0}$ the natural filtration generated by M, that is $\mathcal{F}_n = \sigma(M_k, k \leq n)$. Since $[M]_n$ is an $\mathcal{F}$-adapted process, from the optional stopping theorem, S^M is

a martingale with respect to the filtration $(\mathcal{F}_{\tau(n)})_{n \geq 0}$ and since its increments belong to $\{-1, +1\}$, we reach our conclusion from question 1.

3. In order to prove this fact, let $\bar{s} \overset{(\mathrm{def})}{=} (0, 1, \ldots, n)$ and observe that, for any sequence s of Λ^n with $s \neq \bar{s}$, there are integers $0 \leq b_1 \leq b_2 \leq \ldots \leq b_p \leq n - 1$ such that

$$\bar{s} = \Theta^{b_p} \Theta^{b_p - 1} \ldots \Theta^{b_1}(s).$$

Indeed, let $b_1 = \sup\{i \geq 0 : s_i = i\}$ and for $j \geq 2$, if $\Theta^{b_j - 1} \Theta^{b_j - 2} \ldots \Theta^{b_1}(s) \neq \bar{s}$, then let $b_j = \sup\{i \geq 0 : \Theta^{b_j - 1} \Theta^{b_i - 2} \ldots \Theta^{b_1}(s)_i = i\}$. This procedure stops at the index $p = 1, \ldots, n$ such that $\Theta^{b_p} \Theta^{b_p - 1} \ldots \Theta^{b_1}(s) = \bar{s}$.

Now let $s' \in \Lambda^n$. If $s' = \bar{s}$, then the result is proved from above. If $s' \neq \bar{s}$, then let $c_1, \ldots, c_m$ be integers such that

$$\bar{s} = \Theta^{c_l} \Theta^{c_l - 1} \ldots \Theta^{c_1}(s'),$$

and note that the transformations Θ^a verify $\Theta^a \Theta^a(x) = x$, for all $x \in \Lambda^n$. Then we have $\Theta^{c_1} \Theta^{c_2} \ldots \Theta^{c_l}(\bar{s}) = s'$, so that

$$s' = \Theta^{c_1} \Theta^{c_2} \ldots \Theta^{c_l} \Theta^{b_p} \Theta^{b_p - 1} \ldots \Theta^{b_1}(s),$$

and the result is proved.

4. Let $s, s' \in \Lambda^n$ with $s \neq s'$, and consider the sequence of integers $a_1, a_2, \ldots, a_k$ given in question 3. such that

$$s' = \Theta^{a_k} \Theta^{a_k - 1} \ldots \Theta^{a_1}(s). \tag{2.21.a}$$

Denote by S^n the restricted path $(S_0, S_1, \ldots, S_n)$. Since $\Theta^{a_k} \Theta^{a_k - 1} \ldots \Theta^{a_1}(S^n) \overset{(\mathrm{law})}{=} S^n$, we have

$$P(S^n = s) = P(\Theta^{a_k} \Theta^{a_k - 1} \ldots \Theta^{a_1}(S^n) = s).$$

Applying (2.21.a), we see that this term is equal to

$$P(\Theta^{a_k} \Theta^{a_k - 1} \ldots \Theta^{a_1}(S^n) = \Theta^{a_k} \Theta^{a_k - 1} \ldots \Theta^{a_1}(s')).$$

Since $\Theta^{a_k} \Theta^{a_k - 1} \ldots \Theta^{a_1}$ is a one to one transformation of Λ^n, we have

$$P(S^n = s) = P(S^n = s').$$

Therefore, S^n is uniformly distributed over Λ^n, hence it is a Bernoulli random walk.

Chapter 3

Gaussian variables

Basic Gaussian facts

(a) A Gaussian space $\mathbb{G}$ is a sub-Hilbert space of $L^2(\Omega, \mathcal{F}, P)$ which consists of centered Gaussian variables (including 0).

A most useful property is that, if $X_1, \ldots, X_k \in \mathbb{G}$, they are independent if and only if they are orthogonal, i.e.

$$E[X_i X_j] = 0, \quad \text{for all } i, j \text{ such that } i \neq j.$$

(b) It follows that if $\mathbb{H}$ is a subspace of $\mathbb{G}$, and if $X \in \mathbb{G}$, then:

$$E[X \mid \sigma(\mathbb{H})] = \text{proj}_{\mathbb{H}}(X).$$

For these classical facts, see e.g. Neveu ([64], Chapter 2), and/or Hida–Hitsuda ([39], chapter 2).

(c) The above facts explain why linear algebraic computations play such an important role in the study of Gaussian vectors, and/or processes, e.g. the conditional law of a Gaussian vector with respect to another is Gaussian and can be obtained without manipulating densities (see, e.g., solution to Exercise **3.12**).

(d) However, dealing with nonlinear functionals of Gaussian vectors may necessitate other (nonlinear!) techniques; see e.g. Exercise **3.11**.

(e) The following non-decomposability property of the Gaussian distribution, due to H. Cramer, plays an important role in our solution of Exercise 6.29. *If X and $Y* *are two independent centered variables such that their sum is Gaussian, then each of them is Gaussian.* See, e.g., Lukacs [55], p. 243, for this result as well as its counterpart, due to Raikov, for the Poisson law.

(f) Further fine results on Gaussian random functions are found in Fernique [29], Lifschits [54] and Janson [41].

---

* 3.1 Constructing Gaussian variables from, but not belonging to, a Gaussian space

Let X and Y be two centered, independent, Gaussian variables.

1. Show that, if ε is an r.v. which takes only the values $+1$ and -1 and which is measurable with respect to Y, then εX is a Gaussian variable which is independent of Y. In particular, εX and ε are independent.

2. Show that the sub-σ-field Σ of $\sigma(X, Y)$ which is generated by the r.v.s Z which satisfy:

   (i) *Z is Gaussian* and   (ii) *Z is independent* of Y

   is the σ-field $\sigma(X, Y)$ itself.

3. Prove an analogous result to that of question 2, when Y is replaced by a Gaussian space $\mathbb{H}$ which is assumed to be independent of X.

Comments. The results of this exercise – question 2, say – highlight the difference between the Gaussian space generated by X and Y, i.e. the two-dimensional space $\Sigma = \{\lambda X + \mu Y; \lambda, \mu \in \mathbb{R}\}$, and the space $L^2(\sigma(X, Y))$, which contains many other Gaussian variables than the elements of Σ.

* 3.2 A complement to Exercise 3.1

Let $(\Omega, \mathcal{F}, P)$ be a probability space, and let $\mathbb{G}$ $(\subseteq L^2(\Omega, \mathcal{F}, P))$ be a Gaussian space.

Let Z be a $\sigma(\mathbb{G})$-measurable r.v., which is, moreover, assumed to be Gaussian, and centered.

1. Show that, if the condition

   (γ) *the closed vector space (in $L^2(\Omega, \mathcal{F}, P)$) which is generated by Z and $\mathbb{G}$ is a Gaussian space*

   holds, then Z belongs to $\mathbb{G}$.

2. In the case $\dim(\mathbb{G}) = 1$, construct an r.v. Z, which is $\sigma(\mathbb{G})$-measurable, Gaussian, centered, and such that (γ) is not satisfied.

Comments. This exercise complements Exercise **3.1**, in that it shows that many variables Z constructed in question 2 of Exercise **3.1** did not satisfy (γ) with $\mathbb{G} = \{\lambda X + \mu Y : \lambda, \mu \in \mathbb{R}\}$.

* 3.3 Gaussian vectors and orthogonal projections

Consider (X_1, X_2, X_3) a Gaussian vector which consists of three reduced Gaussian variables such that: $E(X_1 X_2) = a$, $E(X_1 X_3) = b$, with a and b satisfying

$$|a| < 1, \quad |b| < 1, \quad a \neq 0, \quad b \neq 0. \tag{3.3.1}$$

We also denote: $c = E[X_2 X_3]$.

1. (a) Prove that there exists a two-dimensional Gaussian vector (X_1', X_1'') such that
   (i) X_1' and X_1'' are reduced Gaussian variables,
   (ii) the vector (X_1', X_1'') is independent of X_1,
   (iii) $X_2 = aX_1 + \sqrt{1 - a^2} X_1'$ and $X_3 = bX_1 + \sqrt{1 - b^2} X_1''$.

   (b) Express the correlation $\rho = E(X_1' X_1'')$ in terms of a, b and c.

2. Prove that, for any $a \neq 0$, $|a| < 1$, there exists a unique $b \neq 0$, $|b| < 1$ such that the two Gaussian variables X_2 and X_3 are independent.

** 3.4 On the negative moments of norms of Gaussian vectors

We consider, on a probability space, two $\mathbb{C}$-valued r.v.s $Z = X + iY$ and $Z' = X' + iY'$ such that:

(a) the vector (X, Y, X', Y') is Gaussian, and centered;

(b) the variables X and Y have variance 1, and are independent;

(c) the variables X' and Y' have variance 1, and are independent.

In the sequel, $\mathbb{R}^2$ and $\mathbb{C}$ are often identified, and $\langle z, z' \rangle$ denotes the Euclidean scalar product of z and z', elements of $\mathbb{C} \simeq \mathbb{R}^2$, and $|z| = \langle z, z \rangle^{1/2}$.

1. Show that $E\left(\dfrac{1}{|Z|^p}\right) < \infty$ if and only if $p < 2$.

2. Let A be the covariance matrix of Z and Z', that is

$$\forall \theta_1, \theta_2 \in \mathbb{R}^2, \quad E\left[\langle \theta_1, Z \rangle \langle \theta_2, Z' \rangle\right] = \langle \theta_1, A\theta_2 \rangle.$$

Show that there exists a Gaussian vector ξ, with values in $\mathbb{R}^2$, independent of Z, such that: $Z' = A^*Z + \xi$, where A^* denotes the transpose of A.

  Show that $\xi = 0$ a.s. if and only if A is an orthogonal matrix, that is: $A^*A = \mathrm{Id}$.

3. Show that if $(I_2 - A^*A)$ is invertible, and if $p < 2$, then:

  (i) $E\left(\dfrac{1}{|Z'|^p}\Big| Z\right) \leq C$, for a certain constant C;

  (ii) $E\left(\dfrac{1}{|Z|^p}\dfrac{1}{|Z'|^p}\right) < \infty$.

Comments and references. It is interesting to compare the fact that the last expression is finite for $p < 2$, under the hypothesis of question 3, with what happens for $Z = Z'$, that is:

$$E\left(\frac{1}{|Z|^{2p}}\right) < \infty \quad \text{if, and only if}: p < 1.$$

The finiteness result (ii) is a key point in the proof of the asymptotic independence of the winding numbers, as time goes to infinity, for certain linearly correlated planar Brownian motions, as shown in:

M. YOR: Etude asymptotique des nombres de tours de plusieurs mouvements browniens complexes corrélés. In: Festschrift volume in honor of F. Spitzer: *Random Walks, Brownian Motion and Interacting Particle Systems*, eds: R. Durrett, H. Kesten, 441–455, Birkhäuser, 1991.

** 3.5 Quadratic functionals of Gaussian vectors and continued fractions

Let $p \in \mathbb{N}$, and $\beta_0, \beta_1, \ldots, \beta_{p+1}$ be a sequence of $(p+2)$ independent, Gaussian r.v.s with values in $\mathbb{R}^d$. For each i, the d components of $\beta_i \equiv (\beta_{i,j}; j = 1, 2, \ldots, d)$ are themselves independent and centered, and there exists $c_i > 0$ such that:

$$E\left[\beta_{i+1,j}^2\right] = \frac{1}{c_i} \qquad (i = -1, 0, \ldots, p, \quad j = 1, 2, \ldots, d).$$

We write $\langle x, y \rangle$ for the scalar product of two vectors in $\mathbb{R}^d$.

1. Let U be an orthogonal $d \times d$ matrix, with real coefficients.

   Show that the r.v.s $\sum_{i=0}^{p} \langle \beta_i, \beta_{i+1} \rangle$ and $\sum_{i=0}^{p} \langle \beta_i, U\beta_{i+1} \rangle$ have the same law.

   Hint. For every $k \in \mathbb{N}$, U^k is an orthogonal matrix.

2. In this question, we assume $d = 2$.

   Define $\mathcal{S}_k = \sum_{i=k}^{p} \langle \beta_i, \beta_{i+1} \rangle$.

   (In the sequel, it may be helpful to use the convention: $\beta_{-1} = \beta_{p+2} = 0$, and $\mathcal{S}_{-1} = \mathcal{S}_0$, $\mathcal{S}_{p+1} = 0$.)

   Prove, using descending iteration starting from $n = p$, that for every n, there exist two functions h_n and k_n such that:

   $$E\left[\exp(ix\,\mathcal{S}_n) \mid \beta_n = m\right] = h_n(x) \exp\left(-\frac{|m|^2}{2} k_n(x)\right). \qquad (3.5.1)$$

   Prove a recurrence relation between, on one hand, h_n, h_{n+1} and k_n, and on the other hand, k_n and k_{n+1}.

   Deduce therefrom the formulae

   $$k_n(x) = \frac{x^2}{c_n+} \frac{x^2}{c_{n+1}+} \frac{x^2}{c_{n+2}+} \cdots \frac{x^2}{c_p} \equiv \cfrac{x^2}{c_n + \cfrac{x^2}{c_{n+1} + \cfrac{x^2}{\ddots}}} \qquad (3.5.2)$$

   and

   $$E\left[\exp(ix\mathcal{S}_0)\right] = \prod_{n=-1}^{p} \left(\frac{c_n k_n(x)}{x^2}\right). \qquad (3.5.3)$$

3. In this question, we assume $d = 1$. Define $a_i = \frac{1}{2\sqrt{c_{i-1}c_i}}(i = -1, 0, \ldots, p)$.

Show that $\mathcal{S}_0$ has the same distribution as: $\langle g, Ag \rangle$, where $g = (g_0, \ldots, g_{p+1})$ is a random vector which consists of centered, independent Gaussian variables,

with variance 1, and $A = \begin{pmatrix} 0 & a_0 & 0 & \ldots & 0 & 0 \\ a_0 & 0 & a_1 & \ldots & 0 & 0 \\ \cdot & \cdot & \cdot & \cdot & \cdot & \cdot \\ \cdot & \cdot & \cdot & \cdot & \cdot & \cdot \\ \cdot & \cdot & \cdot & \cdot & \cdot & \cdot \\ 0 & 0 & \ldots & a_{p-1} & 0 & a_p \\ 0 & 0 & \ldots & 0 & a_p & 0 \end{pmatrix}$.

Deduce therefrom that $\mathcal{S}_0$ is distributed as $\sum_{n=0}^{p+1} d_n g_n^2$, where (d_n) is the sequence of eigenvalues of A.

4. In this question, we assume $d = 2$. Prove the formula:
$$E\left[\exp(ix\,\mathcal{S}_0)\right] = \left(\frac{i}{2x}\right)^{p+2} \frac{1}{\det\left(A + \frac{i}{2x}I_2\right)} ,$$

5. Write $D_{a_0,\ldots,a_p}(\lambda) = \det(A + \lambda I_d)$.

Prove a recurrence relation between $D_{a_0,\ldots,a_p}$, $D_{a_1,\ldots,a_p}$, and $D_{a_2,\ldots,a_p}$.

How can one understand, in terms of linear algebra, the identity of formulae (b) and (c)?

6. Give a necessary and sufficient condition in terms of the sequence of real numbers $(\alpha_n, n \geq 0)$ such that

$$\sum_{i=0}^{n} \alpha_i g_i g_{i+1} \qquad \text{converges in } L^2,$$

where (g_i) is a sequence of independent, centered, reduced Gaussian variables.

Comments and references:

(a) For more details, we refer to the article from which the above exercise is drawn:

PH. BIANE AND M. YOR: Variations sur une formule de Paul Lévy. *Ann. Inst. H. Poincaré Probab. Statist.*, **23**, no. 2, suppl., 359–377 (1987)

and its companion:

C. DONATI-MARTIN AND M. YOR: Extension d'une formule de Paul Lévy pour la variation quadratique du mouvement Brownien plan. *Bull. Sci. Math.*, **116**, no. 3, 353–382 (1992).

(b) Both articles and, to a lesser extent, the present exercise, show how some classical continued fractions (some originating with Gauss) are related to series expansions of quadratic functionals of, say, Brownian motion.

(c) In question 6, using the Martingale Convergence Theorem (see e.g Exercise **1.6**), it is possible to show that, $\sum_{i=0}^{n} \alpha_i g_i g_{i+1}$ converges almost surely as n goes to $+\infty$.

** 3.6 Orthogonal but non-independent Gaussian variables

1. We denote by $z = x + iy$ the generic point of $\mathbb{C} = \mathbb{R} + i\mathbb{R}$. Prove that, for every $n \in \mathbb{N}$, there exist two homogeneous real valued polynomials of degree n, P_n and Q_n, such that:
$$z^n = P_n(x, y) + iQ_n(x, y).$$
(These polynomials are closely related to the Tchebytcheff polynomials.)

2. Let X and Y be two centered, reduced, independent Gaussian variables. Show that, for every $n \in \mathbb{N}^*$,
$$A_n = \frac{P_n(X, Y)}{(X^2 + Y^2)^{\frac{n-1}{2}}} \quad \text{and} \quad B_n = \frac{Q_n(X, Y)}{(X^2 + Y^2)^{\frac{n-1}{2}}}$$
are two independent, centered, reduced Gaussian variables.

3. Prove that the double sequence
$$\mathcal{S} = \{A_n, B_m; n, m \in \mathbb{N}\}$$
consists of orthogonal variables, i.e.
if C and D are two distinct elements of $\mathcal{S}$, then: $E[CD] = 0$.

4. Let $n \neq m$. Compute the complex cross moments of (A_n, B_n, A_m, B_m), that is $E[Z_n^a (\overline{Z_m})^b]$, for any integers a, b, where $Z_n = A_n + iB_n$ and $Z_m = A_m + iB_m$. Is the vector (A_n, B_n, A_m, B_m) Gaussian?

* 3.7 Isotropy property of multidimensional Gaussian laws

Let X and Y be two r.v.s taking their values in $\mathbb{R}^k$. Let, moreover, $\mathbf{G}_k$ be a Gaussian r.v. with values in $\mathbb{R}^k$, which is centered, and has I_k for covariance matrix, and let U be an r.v. which is uniformly distributed on S_{k-1}, the unit sphere of $\mathbb{R}^k$.

We assume moreover that $\mathbf{G}_k$, U, X and Y are independent.

Prove that the following properties are equivalent:

(i) $\|X\| \overset{(\text{law})}{=} \|Y\|$; (ii) $\langle \mathbf{G}_k, X \rangle \overset{(\text{law})}{=} \langle \mathbf{G}_k, Y \rangle$; (iii) $\langle U, X \rangle \overset{(\text{law})}{=} \langle U, Y \rangle$, where $\langle x, y \rangle$ is the scalar product of x and y, two generic elements of $\mathbb{R}^k$, and $\|x\| = (\langle x, x \rangle)^{1/2}$ is the Euclidian norm of $x \in \mathbb{R}^k$.

** 3.8 The Gaussian distribution and matrix transposition

1. Let $\mathbf{G}_n$ be an n-sample of the Gaussian, centered distribution with variance 1 on $\mathbb{R}$. Prove that, for every $n \times n$ matrix A with real coefficients, one has:

$$\|A\mathbf{G}_n\| \overset{(\text{law})}{=} \|A^*\mathbf{G}_n\|,$$

where A^* is the transpose of A.

2. Let X be an r.v. with values in $\mathbb{R}$, which has a finite second moment. Prove that if, for every $n \in \mathbb{N}$, and every $n \times n$ matrix A, one has:

$$\|A\mathbf{X}_n\| \overset{(\text{law})}{=} \|A^*\mathbf{X}_n\|,$$

where $\mathbf{X}_n$ is an n-sample of the law of X, then X is a centered Gaussian variable.

Comments. The solution we propose for question 1 uses systematically the Gauss transform, as defined in Exercise **4.18**. Another proof can be given using the fact that AA^* and A^*A have the same eigenvalues, with the same order of multiplicity: write $AA^* = U^*DU$ and $A^*A = V^*DV$, where U and V are orthogonal matrices and D is diagonal, then use:

$$\mathbf{G}_n \overset{(\text{law})}{=} V\mathbf{G}_n \overset{(\text{law})}{=} U\mathbf{G}_n.$$

* 3.9 A law whose n-samples are preserved by every orthogonal transformation is Gaussian

Let X be an r.v. with values in $\mathbb{R}$, and, for every $n \in \mathbb{N}$, let $\mathbf{X}_n$ denote an n-sample of (the law of) X.

1. Prove that if X is Gaussian and centered, then for every $n \in \mathbb{N}$, and every orthogonal $n \times n$ matrix R, one has:

$$\mathbf{X}_n \overset{(\text{law})}{=} R\mathbf{X}_n.$$

2. Prove that, conversely, if the above property holds for every $n \in \mathbb{N}$, and every orthogonal $n \times n$ matrix R, then X is Gaussian and centered.

Comments and references about Exercises **3.7**, **3.8** *and* **3.9**:

This characterization of the normal distribution has a long history going back to the 1930s and early 1940s with H. Cramér and S. Bernstein, the main reference being:

M. KAC: On a characterization of the normal distribution. *Amer. J. Math.*, **61**, 726–728 (1939).

Amongst more recent references, we suggest:

C. DONATI-MARTIN, S. SONG AND M. YOR: On symmetric stable random variables and matrix transposition. *Annales de l'I.H.P.*, **30** (3), 397–413 (1994)

G. LETAC AND X. MILHAUD: Une suite stationnaire et isotrope est sphérique. *Zeit. für Wahr.*, **49**, 33–36 (1979)

G. LETAC: Isotropy and sphericity: some characterizations of the normal distribution. *The Annals of Statistics.*, **9** (2), 408–417 (1981)

S. BERMAN: Stationarity, isotropy and sphericity in ℓ_p. *Zeit für Wahr.*, **54**, 21–23 (1980)

and the book by S. Janson [41].

** 3.10 Non-canonical representation of Gaussian random walks

Consider $X_1, \ldots, X_n, \ldots$ a sequence of centered, independent Gaussian r.v.s, each with variance 1.

Define for $n \geq 1$, $S_n = \sum_{i=1}^{n} X_i$ and $\Sigma_n = S_n - \sum_{j=1}^{n} \left(\frac{S_j}{j} \right)$.

We define the σ-fields:

$$\mathcal{F}_n = \sigma\{X_1, \ldots, X_n\} \text{ and } \mathcal{G}_n = \sigma\{X_i - X_j; 1 \leq i, j \leq n\}.$$

1. Prove that: $\mathcal{G}_n = \sigma\{\Sigma_1, \Sigma_2, \ldots, \Sigma_n\} = \sigma\left\{ \frac{S_i}{i} - \frac{S_j}{j}; 1 \leq i, j \leq n \right\}$.

2. Fix n and let $\{\hat{S}_k, k \leq n\} = \{S_k - \frac{k}{n} S_n, k \leq n\}$. Prove that $\{\hat{S}_k, k \leq n\}$ is independent of S_n and is distributed as $\{S_k, k \leq n\}$ conditioned by $S_n = 0$. Prove that $\mathcal{G}_n = \sigma\{\hat{S}_k, k \leq n\}$.

3. Define $Y_n = X_n - \frac{S_n}{n}$.

Show that, for any $n \geq 1$, the equality: $\mathcal{G}_n = \sigma\{Y_1, Y_2, \ldots, Y_n\}$ holds, and that, moreover, the variables $(Y_n, n \geq 1)$ are independent. Hence, the sequence $\Sigma_n = \sum_{j=1}^n Y_j, n = 1, 2, \ldots$ has independent increments.

Let $(\alpha_k, k = 1, 2, \ldots)$ be a sequence of reals, such that $\alpha_k \neq 0$, for all k. Prove that $\Sigma_n^{(\alpha)} \stackrel{(\text{def})}{=} S_n - \sum_{k=1}^n \alpha_k S_k$ has independent increments if and only if $\alpha_k = \frac{1}{k}$, for every k.

4. Prove the equality: $\dfrac{S_n}{n} - \dfrac{S_{n+1}}{n+1} = -\dfrac{Y_{n+1}}{n}$.

Deduce therefrom an expression for $E\left[\dfrac{S_n}{n}\Big|\mathcal{G}_{n+k}\right]$ $(k \in \mathbb{N})$ as a linear combination, to be computed, of $(Y_{n+1}, Y_{n+2}, \ldots, Y_{n+k})$.

5. Prove that S_n is measurable with respect to $\sigma\{Y_{n+1}, Y_{n+2}, \ldots\}$.

6. Prove that: $\mathcal{F}_\infty = \mathcal{G}_\infty$, where: $\mathcal{F}_\infty = \lim_k \uparrow \mathcal{F}_k$, and $\mathcal{G}_\infty = \lim_k \uparrow \mathcal{G}_k$.

Comments and references. This is the discrete version of the following "non-canonical" representation of Brownian motion, which is given by:

$$\Sigma_t = S_t - \int_0^t \frac{ds}{s} S_s \equiv \int_0^t \left(1 - \log\frac{t}{u}\right) dS_u, \tag{3.10.1}$$

where we use the (unusual!) notation $(S_t, t \geq 0)$ for a one-dimensional Brownian motion. Then, $(\Sigma_t, t \geq 0)$ is another Brownian motion, and (3.10.1) is a non canonical representation of Brownian motion, in the sense that $(S_u; u \leq t)$ cannot be reconstructed from $(\Sigma_u, u \leq t)$; in fact, just as in question 2, the random variable S_t is independent of $\sigma\{\Sigma_u; u \leq t\}$; nonetheless, just as in question 6, the two σ-fields $\sigma\{S_u; u \geq 0\}$ and $\sigma\{\Sigma_u; u \geq 0\}$ are equal up to negligible sets.

For details, and references, see, e.g., Chapter 1 in M. Yor [101]. Another example, due to Lévy is:

$$\Sigma'_t = S_t - \frac{3}{t}\int_0^t S_u \, du = \int_0^t \left(3\frac{u}{t} - 2\right) dS_u.$$

$(\Sigma'_t, t \geq 0)$ is a Brownian motion and $(S_u, u \leq t)$ cannot be constructed from $(\Sigma'_u, u \leq t)$ since $\int_0^t u \, dS_u = tS_t - \int_0^t S_u \, du$ is independent from $(\Sigma'_u, u \leq t)$. For a more extended discussion, see:

Y. CHIU: From an example of Lévy's. *Séminaire de Probabilités XXIX*, 162–165, *Lecture Notes in Mathematics*, **1613**, Springer, Berlin, 1995.

A recent contribution to this topic is:

Y. HIBINO AND M. HITSUDA: Canonical property of representations of Gaussian processes with singular Volterra kernels. *Infin. Dimens. Anal. Quantum Probab. Relat. Top.*, **5** (2), 293–296 (2002).

* 3.11 Concentration inequality for Gaussian vectors

Let X and Y be two independent centered d-dimensional Gaussian vectors with covariance matrix I_d.

1. Prove that for any $f, g \in \mathcal{C}_b^2(\mathbb{R}^d)$,

$$\text{Cov}(f(X), g(X)) = \int_0^1 E\left[\langle \nabla f(X), \nabla g(\alpha X + \sqrt{1-\alpha^2}Y)\rangle\right] d\alpha, \quad (3.11.1)$$

where $\nabla f(x) - \left(\frac{\partial f}{\partial x_i}(x)\right)$.

Hint: First check (3.11.1) for $f(x) = e^{i\langle t, x\rangle}$ and $g(x) = e^{i\langle s, x\rangle}$, $s, t, x \in \mathbb{R}^d$.

2. Let μ_α be the Gaussian measure in $\mathbb{R}^{2d}$ which is the distribution of $(X, \alpha X + \sqrt{1-\alpha^2}Y)$ and denote by μ the probability measure $\int_0^1 \mu_\alpha \, d\alpha$. Let Z be a random vector in $\mathbb{R}^d$ such that the $2d$-dimensional random vector (X, Z) has law μ.

   Prove that for any Lipschitz function f, such that $\|f\|_{\text{Lip}} \leq 1$ and $E[f(X)] = 0$, the inequality

$$E[f(X)e^{tf(X)}] \leq tE[e^{tf(Z)}], \quad (3.11.2)$$

holds for all $t \geq 0$ and deduce that

$$E[e^{tf(X)}] \leq e^{\frac{t^2}{2}}. \quad (3.11.3)$$

3. Prove that for every Lipschitz function f on $\mathbb{R}^d$ with $\|f\|_{\text{Lip}} \leq 1$, the inequality

$$P(f(X) - E[f(X)] \geq \lambda) \leq e^{-\frac{\lambda^2}{2}}. \quad (3.11.4)$$

holds for any $\lambda \geq 0$.

Comments and references. The inequality (3.11.4) is known as the concentration property of the Gaussian law. Many developments based on stochastic calculus may be found in:

M. LEDOUX: Concentration of measure and logarithmic Sobolev inequalities. *Séminaire de Probabilités XXXIII*, 120–216, *Lecture Notes in Mathematics*, **1709**, Springer, Berlin, 1999.

M. LEDOUX: *The Concentration Measure Phenomenon. Mathematical Surveys and Monographs*, **89**. American Mathematical Society, Providence, RI, 2001.

See also Theorem 2.6 in:

W.V. LI AND Q.M. SHAO: Gaussian processes: inequalities, small ball probabilities and applications. In: *Stochastic Processes: Theory and Methods*, 533–597, *Handbook of Statistics*, **19**, North-Holland, Amsterdam, 2001.

This inequality has many applications in large deviations theory, see Exercise **5.6**; the present exercise was suggested by C. Houdré, who gave us the next reference:

C. HOUDRÉ: Comparison and deviation from a representation formula. In: *Stochastic Processes and Related Topics*, 207–218, Trends in Mathematics, Birkhäuser Boston, Boston, MA, 1998.

R. AZENCOTT: Grandes déviations et applications. *Eighth Saint Flour Probability Summer School–1978* (Saint Flour, 1978), 1–176, *Lecture Notes in Mathematics*, **774**, Springer, Berlin, 1980.

The inequality (3.11.4) is somehow related to the so-called Chernov's inequality, i.e.

$$E[(f(X) - E[f(X)])^2] \leq E[(f'(X))^2],$$

for any $f \in \mathcal{C}^1(\mathbb{R}^d)$. Note that this inequality was re-discovered by Chernov in 1981; it had been already proved much earlier by Nash in 1958. A proof of it may be found in Exercises 4.29 and 4.30, pp. 126 and 127 in Letac [52]. See also Exercise 3.13, Chapter 5 in Revuz and Yor [75] for a proof based on stochastic calculus.

* 3.12 Determining a jointly Gaussian distribution from its conditional marginals

Give a necessary and sufficient condition on the six-tuple $(\alpha, \beta, \gamma, \delta, a, b) \in \mathbb{R}^6$, for the existence of a two-dimensional Gaussian variable (X, Y) (not necessarily centered) with both X and Y non-degenerate satisfying:

$$\mathcal{L}(Y \mid X = x) = \mathcal{N}(\alpha + \beta x, a^2) \tag{3.12.1}$$

$$\mathcal{L}(X \mid Y = y) = \mathcal{N}(\gamma + \delta y, b^2). \tag{3.12.2}$$

* 3.13 Gaussian integration by parts

We say that a function $f : \mathbb{R}^n \to \mathbb{R}$ is polynomially bounded if there exist integers $k = (k_1, \ldots, k_n)$ and a real $a > 0$ such that $|f(x)| \leq |x_1|^{k_1} \ldots |x_n|^{k_n}$, for all $x = (x_1, \ldots, x_n)$, such that $|x| \geq a$.

1. Prove that if G is real valued Gaussian and centered, then for any polynomially bounded and continuously differentiable function Φ, we have:

$$E[G\Phi(G)] = E[G^2]E[\Phi'(G)]. \tag{3.13.1}$$

2. Prove that if the $(n+1)$-dimensional vector $(G, G_1, G_2, \ldots, G_n)$ is Gaussian, centered, then for any polynomially bounded and continuously differentiable function $\Phi : \mathbb{R}^n \to \mathbb{R}$, one has

$$E[G\Phi(G_1, \ldots, G_n)] = \sum_{l \leq n} E[GG_l]E\left[\frac{\partial \Phi}{\partial x_l}(G_1, \ldots, G_n)\right]. \qquad (3.13.2)$$

Comments and references. In fact property (3.13.1) characterizes the centered Gaussian variables; it is the first step in Stein's method. See formula (14), p. 21, in:

C. STEIN: *Approximate Computation of Expectations. Institute of Mathematical Statistics Lecture Notes—Monograph Series*, **7**, Institute of Mathematical Statistics, Hayward, CA, 1986.

The multidimensional extension of (3.13.2) is found in:

M. TALAGRAND: *Mean Field Models for Spin Glasses. Volume I. Basic examples*, *Ergebnisse der Mathematik und ihrer Grenzgebiete*, **54**, Springer-Verlag, Berlin, 2011.

It is formula A.17, p. 440, there.

* 3.14 Correlation polynomials

Let q belong to $[-1, +1]$. For the benefit of the reader, we introduce the following slightly unusual notation: (G_q, G) denotes a Gaussian vector consisting of two centered, reduced Gaussian variables, such that $E[G_q G] = q$.

1. (i) Note that the law of (G_q, G) is symmetric. Hence, we shall also denote the vector (G_q, G) by (G, G_q).

   (ii) Prove that the cross moment $E[G_q^n G^m]$ is 0, if $n - m$ $(\in \mathbb{Z})$ is odd.

2. We introduce the notation: $Q_m^{(l)}(q) = E[G_q^{m+2l} G^m]$, $m, l \in \mathbb{N}$, thus considering all remaining "interesting" cross moments. In fact, for now, we consider only the two sequences:

$$Q_m(q) = E[(G_q G)^m] \quad \text{and} \quad Q_m^{(-)}(q) = E[G_q^m G^{m-2}], \quad m \geq 2.$$

Show that the sequences (Q_m) and $(Q_m^{(-)})$ satisfy the relations:

$$Q_m(q) = (m-1)Q_m^{(-)}(q) + mqQ_{m-1}(q) \qquad (3.14.1)$$
$$Q_m^{(-)}(q) = (m-2)qQ_{m-1}^{(-)}(q) + (m-1)Q_{m-2}(q). \qquad (3.14.2)$$

3. Show that (3.14.1) and (3.14.2) are equivalent to the following recurrence relations:

$$Q_m(q) = (2m-1)qQ_{m-1}(q) + (m-1)^2(1-q^2)Q_{m-2}(q) \qquad (3.14.3)$$

$$Q_m^{(-)}(q) = (2m-3)qQ_{m-1}^{(-)}(q) + (m-3)(m-1)(1-q^2)Q_{m-2}^{(-)}(q). \qquad (3.14.4)$$

Note that (3.14.3) only involves the sequence (Q_m) and that (3.14.4) only involves the sequence $(Q_m^{(-)})$.

4. (a) Under the same hypothesis, show that, for all λ, μ, reals,

$$E\left[\exp\left(\lambda G_q - \lambda^2/2\right)\exp\left(\mu G - \mu^2/2\right)\right] = \exp(\lambda\mu q). \qquad (3.14.5)$$

(b) Deduce that for all $m, n \in \mathbb{N}$,

$$E[h_n(G_q)h_m(G)] = \delta_{m,n}(q^n)(n!),$$

where $(h_n(x))$ is the sequence of Hermite polynomials that are defined by the relation:

$$\exp\left(\lambda x - \lambda^2/2\right) = \sum_{n=0}^{\infty} \frac{\lambda^n}{n!}h_n(x).$$

Comments and references: (i) The sequence of polynomials $(Q_m(q))$ and $(Q_m^{(-)}(q))$ enjoy a number of interesting properties which are discussed in detail in:

L. CHAUMONT AND M. YOR: On correlation polynomials associated with 2-dimensional centered Gaussian vectors. Work in progress, March 2012.

(ii) We learnt for the first time about Q_m and $Q_m^{(-)}$ when reading M. Talagrand's book, as mentioned in Exercise **3.13**.

(iii) Our proof of (3.14.1) with the help of two q-correlated Brownian motions was inspired by J. Neveu's celebrated proof of the hypercontractivity property of the Ornstein–Uhlenbeck semigroup.

J. NEVEU: Sur l'espérance conditionnelle par rapport à un mouvement brownien. *Ann. Inst. H. Poincaré Sect. B*, **12**, no. 2, 105–109 (1976).

Solutions for Chapter 3

Solution to Exercise 3.1

1. Let f and g be two bounded measurable functions. We prove the result by writing

$$
\begin{aligned}
E[f(\varepsilon X)g(Y)] &= E[f(X)g(Y)\mathbb{1}_{\{\varepsilon=1\}}] + E[f(-X)g(Y)\mathbb{1}_{\{\varepsilon=-1\}}] \\
&= E[f(X)]\left(E[g(Y)\mathbb{1}_{\{\varepsilon=1\}}] + E[g(Y)\mathbb{1}_{\{\varepsilon=-1\}}]\right) \\
&= E[f(X)]E[g(Y)].
\end{aligned}
$$

2. For any $a \in \mathbb{R}$, let $\varepsilon_a^Y = \left(\mathbb{1}_{\{Y \le a\}} - \mathbb{1}_{\{Y > a\}}\right)$, then $Z_a = \varepsilon_a^Y X$, is a Gaussian variable that is independent of Y and $\sigma(X, Y) = \sigma(X, Z_a : a \in \mathbb{R})$. Since, from question 1, each Z_a is independent of Y, then the latter identity proves that $\sigma(X, Y) = \Sigma$.

3. Let X be a Gaussian variable which is independent of $\mathbb{H}$. We want to prove that the sub-σ-field of $\sigma(X, \mathbb{H})$ which is generated by the r.v.s satisfying:

(i) Z is Gaussian and (ii) Z is independent of $\mathbb{H}$

is the σ-field $\sigma(X, \mathbb{H})$ itself.

The proof is essentially the same as in the previous question. To any $Y \in \mathbb{H}$, we associate the family $(\varepsilon_a^Y : a \in \mathbb{R})$ as in question 2. Moreover, we can check as in question 1. that for fixed $a \in \mathbb{R}$ and $Y \in \mathbb{H}$, each $\varepsilon_a^Y X$, is a Gaussian variable which is independent of $\mathbb{H}$. Finally, it is clear that $\sigma(\mathbb{H}) = \sigma(\varepsilon_a^Y : a \in \mathbb{R}, Y \in \mathbb{H})$ and $\sigma(X, \mathbb{H}) = \sigma(X, \varepsilon_a^Y X : a \in \mathbb{R}, Y \in \mathbb{H})$.

Solution to Exercise 3.2

1. Let $\mathbb{H}$ be the closed vector space generated by Z and $\mathbb{G}$, and let $\mathbb{F}$ be the Gaussian subspace of $\mathbb{H}$ such that $\mathbb{H} = \mathbb{F} \oplus \mathbb{G}$. Z can be decomposed as: $Z = Z_1 + Z_2$, with $Z_1 \in \mathbb{F}$ and $Z_2 \in \mathbb{G}$. It follows from general properties of Gaussian spaces that Z_1 is

independent of $\mathbb{G}$; hence $E[Z \,|\, \sigma(\mathbb{G})] = Z_2$. But Z is $\sigma(\mathbb{G})$-measurable, so $Z \,(= Z_2)$ belongs to $\mathbb{G}$.

2. Let Y and $\mathbb{G}$ be such that $\mathbb{G} = \{\alpha Y : \alpha \in \mathbb{R}\}$. Pick a real $a > 0$ and define $Z = -Y\mathbb{1}_{\{|Y|\leq a\}} + Y\mathbb{1}_{\{|Y|\geq a\}}$. We easily check that Z is Gaussian, centered and is $\sigma(\mathbb{G})$-measurable although the vector (Y, Z) is not Gaussian. Indeed, $Y + Z = 2Y\mathbb{1}_{\{|Y|\geq a\}}$, thus $P(Y + Z = 0) = P(|Y| \leq a)$ belongs to the open interval $(0, 1)$; the closed vector space generated by Z and $\mathbb{G}$ cannot be Gaussian.

Comments on the solution. In the case dim($\mathbb{G}$)=2, any variable Z constructed in question 2 of Exercise **3.1** constitutes an example of an r.v. which is $\sigma(\mathbb{G})$-measurable, Gaussian, centered, and such that (γ) is not satisfied.

Solution to Exercise 3.3

1 (a) It suffices to decompose orthogonally X_2 and X_3 with respect to X_1, thus "creating" the vector (X_1', X_1''). These two variables belong to the Gaussian space generated by (X_1, X_2, X_3). Since X_1' is orthogonal to X_1 and X_1'' is orthogonal to X_1, the pair (X_1', X_1'') is independent from X_1.

(b) From $\rho = E[X_1' X_1'']$, it follows that

$$\sqrt{1 - a^2}\sqrt{1 - b^2}\rho = E[(X_2 - aX_1)(X_3 - bX_1)]$$
$$= E[X_2 X_3] - aE[X_1 X_3] - bE[X_1 X_2] + ab$$
$$= E[X_2 X_3] - ab.$$

Finally, we obtain

$$\rho = \frac{c - ab}{\sqrt{1 - a^2}\sqrt{1 - b^2}}.$$

2. Assume that X_2 and X_3 are independent and let us consider the Gaussian vector (X_1', X_1'') of question 1(a). Since $c = 0$, it follows from question 1(b) that

$$\sqrt{1 - a^2}\sqrt{1 - b^2}\rho = -ab. \tag{3.3.a}$$

Then set $A = a/\sqrt{1 - a^2}$. An elementary computation shows that the unique solution of equation (3.3.a) such that $b \neq 0$ and $|b| < 1$ is

$$b = -\rho\frac{\mathrm{sgn}(a)}{\sqrt{A^2 + \rho^2}}.$$

Solution to Exercise 3.4

1. A classical computation shows that $|Z|^2 = X^2 + Y^2$ is exponentially distributed with parameter $1/2$. The result follows.

2. Set $\xi = (\xi_1, \xi_2) = Z' - A^*Z$. From (a), the vector (X, Y, ξ_1, ξ_2) is Gaussian, so to prove that ξ and Z are independent, it is enough to show that their covariance matrix equals 0. But by definition, this matrix is the difference between A and the covariance matrix of Z and A^*Z, which is precisely A.

The covariance matrix of A^*Z is A^*A and the covariance matrix of Z' is I_2, so ξ has covariance matrix: $I_2 - A^*A$ and the result follows.

3. (i) We know from question 1 that if $p < 2$, then the conditional expectation $E\left[\frac{1}{|Z'|^p} \mid Z\right]$ is integrable, hence a.s. finite. On the other hand, we deduce from the independence between Z and ξ that

$$E\left[\frac{1}{|Z'|^p} \mid Z\right] = E\left[\frac{1}{|A^*Z + \xi|^p} \mid Z\right]$$

$$= E\left[\frac{1}{|\xi + x|^p}\right]_{x = A^*Z}. \tag{3.4.a}$$

Now let us check that the function $x \mapsto E\left[\frac{1}{|\xi + x|^p}\right]$ is defined and bounded on $\mathbb{R}^2$. Since the covariance matrix $I_2 - A^*A$ of ξ is invertible, then the law of ξ has a density and moreover, there exists a matrix M such that $M\xi$ is centered and has covariance matrix I_2. Let $|M|$ be the norm of the linear operator represented by M, then from the inequality $|M\xi + Mx| \leq |M||\xi + x|$, we can write

$$E\left[\frac{1}{|\xi + x|^p}\right] \leq |M|^p E\left[\frac{1}{|M\xi + Mx|^p}\right] = \frac{|M|^p}{2\pi} \int_{\mathbb{R}^2} \frac{e^{-[(y_1 - x_1')^2 + (y_2 - x_2')^2]/2}}{(y_1^2 + y_2^2)^{p/2}} \, dy_1 \, dy_2,$$

where $Mx = (x_1', x_2')$. The latter expression is clearly bounded in $(x_1', x_2') \in \mathbb{R}^2$. Hence, the conclusion follows from equality (3.4.a) above.

(ii) From (i), we have

$$E\left[\frac{1}{|Z|^p} \frac{1}{|Z'|^p}\right] = E\left[\frac{1}{|Z|^p} E\left[\frac{1}{|Z'|^p} \mid Z\right]\right] \leq CE\left[\frac{1}{|Z|^p}\right] < +\infty.$$

Solution to Exercise 3.5

1. Since β_i is centered and has covariance matrix $c_{i-1}^{-1} I_d$, then for any orthogonal matrix U, $U\beta_i \overset{\text{(law)}}{=} \beta_i$. Moreover, from the independence between the β_is and since for every $k \in \mathbb{N}$, U^k is an orthogonal matrix, then we have

$$(\beta_0, \ldots, \beta_{p+1}) \overset{\text{(law)}}{=} (\beta_0, U\beta_1, U^2\beta_2, \ldots, U^{p+1}\beta_{p+1}).$$

This identity in law implies

$$\sum_{i=0}^{p} <\beta_i, \beta_{i+1}> \overset{\text{(law)}}{=} \sum_{i=0}^{p} \left\langle U^i \beta_i, U^{i+1} \beta_{i+1} \right\rangle.$$

We conclude by noticing that for any i, $\langle U^i \beta_i, U^{i+1} \beta_{i+1} \rangle = \langle \beta_i, U \beta_{i+1} \rangle$.

2. First, for $n = p$, we have

$$
\begin{aligned}
E[e^{ixS_p} \mid \beta_p = m] &= E[e^{i\langle x\beta_p, \beta_{p+1} \rangle} \mid \beta_p = m] \\
&= E[e^{i\langle xm, \beta_{p+1} \rangle}] \\
&= e^{-\frac{x^2 |m|^2}{2c_p}},
\end{aligned}
$$

and the latter expression has the required form. Now, suppose that for a given index $n \leq p$, $E[e^{ixS_n} \mid \beta_n = m]$ has the form which is given in the statement. Using this expression, we get

$$
\begin{aligned}
E[e^{ixS_{n-1}} \mid \beta_{n-1} = m] &= E[e^{i\langle x\beta_{n-1}, \beta_n \rangle} e^{ixS_n} \mid \beta_{n-1} = m] = E[e^{i\langle xm, \beta_n \rangle} e^{ixS_n}] \\
&= E[e^{i\langle xm, \beta_n \rangle} E[e^{ixS_n} \mid \beta_n]] = E[e^{i\langle xm, \beta_n \rangle} h_n(x) e^{-\frac{|\beta_n|^2}{2} k_n(x)}] \\
&= h_n(x) \prod_{j=1}^{2} \sqrt{\frac{c_{n-1}}{2\pi}} \int_{-\infty}^{+\infty} dy\, e^{ixm_j y} e^{-k_n(x) y^2/2} e^{-c_{n-1} y^2/2} \\
&= h_n(x) \prod_{j=1}^{2} \sqrt{\frac{c_{n-1}}{2\pi}} \frac{1}{\sqrt{k_n(x) + c_{n-1}}} \int_{-\infty}^{+\infty} dz\, e^{\frac{ixm_j z}{\sqrt{k_n(x) + c_{n-1}}}} e^{-z^2/2} \\
&= h_n(x) \frac{c_{n-1}}{k_n(x) + c_{n-1}} e^{-\frac{x^2 |m|^2}{2(k_n(x) + c_{n-1})}}.
\end{aligned}
$$

From the above identity, we deduce the following recurrence relations:

$$
h_{n-1}(x) = \frac{h_n(x) c_{n-1}}{k_n(x) + c_{n-1}} \quad \text{and} \quad k_{n-1}(x) = \frac{x^2}{k_n(x) + c_{n-1}}, \qquad (3.5.a)
$$

which implies (3.5.2).

Expression (3.5.1) yields

$$
\begin{aligned}
E[\exp ixS_0] &= h_0(x) E[e^{-|\beta_0|^2 k_0(x)/2}] \\
&= h_0(x) \left(\sqrt{\frac{c_{-1}}{2\pi}} \int_{-\infty}^{+\infty} dy\, e^{-(k_0(x) + c_{-1}) y^2/2} \right)^2 \\
&= h_0(x) \left(\sqrt{\frac{c_{-1}}{2\pi (k_0(x) + c_{-1})}} \int_{-\infty}^{+\infty} dz\, e^{-z^2/2} \right)^2 = \frac{h_0(x) c_{-1}}{k_0(x) + c_{-1}}.
\end{aligned}
$$

Combining this identity with the above recurrence relation, we obtain

$$
\begin{aligned}
E[\exp ixS_0] &= h_0(x) c_{-1} \frac{k_{-1}(x)}{x^2} = \frac{h_1(x) c_0}{k_1(x) + c_0} c_{-1} \frac{k_{-1}(x)}{x^2} \\
&= h_1(x) c_0 \frac{h_0(x)}{x^2} c_{-1} \frac{k_{-1}(x)}{x^2} = \cdots = \prod_{n=-1}^{p} \left(\frac{c_n k_n(x)}{x^2} \right).
\end{aligned}
$$

3. Ag being given by

$$Ag = (a_0g_1, a_0g_0 + a_1g_2, a_1g_1 + a_2g_3, \ldots, a_{p-1}g_{p-1} + a_pg_{p+1}, a_pg_p),$$

we obtain

$$\langle g, Ag \rangle = (2a_0g_0g_1 + 2a_1g_1g_2 + 2a_2g_2g_3 + \cdots + 2a_pg_pg_{p+1}),$$

which allows us to conclude that

$$
\begin{aligned}
S_0 &= \beta_0\beta_1 + \cdots + \beta_p\beta_{p+1} \\
&\overset{\text{(law)}}{=} \frac{1}{\sqrt{c_{-1}c_0}}g_0g_1 + \frac{1}{\sqrt{c_0c_1}}g_1g_2 + \cdots + \frac{1}{\sqrt{c_{p-1}c_p}}g_pg_{p+1} \\
&= 2a_0g_0g_1 + \cdots + 2a_pg_pg_{p+1}.
\end{aligned}
$$

A may be diagonalized as $A = U^{-1}DU$, where U is an orthogonal matrix. Since $Ug \overset{\text{(law)}}{=} g$, we have $\langle g, U^{-1}DUg \rangle \overset{\text{(law)}}{=} \langle g, Dg \rangle = \sum_{n=1}^{p+1} d_ng_n^2$.

4. When $d = 2$, S_0 can be developed as follows:

$$S_0 = \beta_{0,1}\beta_{1,1} + \beta_{0,2}\beta_{1,2} + \beta_{1,1}\beta_{2,1} + \beta_{1,2}\beta_{2,2} + \cdots,$$

so it clearly appears that S_0 is equal in law to the sum of two independent copies of the sum $\sum_{i=0}^{p} \beta_i\beta_{i+1}$ in the case $d = 1$. Now let us compute $E[e^{ixS_0}]$ for $d = 1$:

$$
\begin{aligned}
E[e^{ixS_0}] &= E\left[e^{ix\sum_{n=0}^{p+1} d_ng_n^2}\right] = \prod_{n=0}^{p+1} \int_{-\infty}^{+\infty} dy \frac{1}{\sqrt{2\pi}} e^{-(1-2xid_n)y^2/2} \\
&= \prod_{n=0}^{p+1} \frac{1}{\sqrt{1 - 2xid_n}}.
\end{aligned}
$$

So, when $d = 2$, taking account of the above remark, one has:

$$E[e^{ixS_0}] = \prod_{n=0}^{p+1} \frac{1}{1 - 2xid_n} = \left(\frac{i}{2x}\right)^{p+1} \frac{1}{\det\left(A + \frac{i}{2x}I\right)}.$$

5. Developing the determinant of $A + \lambda I_d$ with respect to the first column, we obtain:

$$D_{a_0,\ldots,a_p}(\lambda) = \lambda D_{a_1,\ldots,a_p}(\lambda) - a_0\det(M),$$

where

$$
M = \begin{pmatrix}
a_0 & a_1 & 0 & 0 & \ldots & 0 \\
0 & \lambda & a_2 & 0 & \ldots & 0 \\
0 & a_2 & \lambda & a_3 & \ldots & 0 \\
& \ddots & \ddots & \ddots & \ddots & \\
0 & \ldots & 0 & a_{p-1} & \lambda & a_p \\
0 & \ldots & \ldots & 0 & a_p & \lambda
\end{pmatrix}
$$

and $\det(M) = a_0 D_{a_2,\ldots,a_p}(\lambda)$, so that

$$D_{a_0,\ldots,a_p}(\lambda) = \lambda D_{a_1,\ldots,a_p}(\lambda) - a_0^2 D_{a_2,\ldots,a_p}(\lambda).$$

This recurrence relation is a classical way to compute the characteristic polynomial of A. Henceforth formulae (b) and (c) combined together yield another manner to obtain this determinant.

6. Set $X_n = \sum_{i=0}^{n} \alpha_i g_i g_{i+1}$, then for any n and m such that $n \leq m$ one has

$$E[(X_m - X_n)^2] = \sum_{i=n+1}^{m} \alpha_i^2,$$

so (X_n) is a Cauchy sequence in L^2 (which is equivalent to the convergence of (X_n) in L^2) if and only if $\sum_{i=0}^{\infty} \alpha_i^2 < \infty$.

Solution to Exercise 3.6

1. We prove the result by induction. For $n = 0$ and $n = 1$, the statement is clearly true. Let us suppose that for a fixed $n \in \mathbb{N}$, there exist two homogeneous polynomials of degree n, P_n and Q_n, such that:

$$z^n = P_n(x,y) + iQ_n(x,y).$$

We can write z^{n+1} as:

$$z^{n+1} = xP_n(x,y) - yQ_n(x,y) + i[yP_n(x,y) + xQ_n(x,y)].$$

The polynomials $P_{n+1}(x,y) = xP_n(x,y) - yQ_n(x,y)$ and $Q_{n+1} = yP_n(x,y) + xQ_n(x,y)$ are homogeneous with degree $n+1$, so the result is true for z^{n+1}.

2. Set $Z = X + iY$. It is well known that Z may be written as $Z = R\exp(i\Theta)$, where $R = (X^2 + Y^2)^{1/2}$ and Θ is a uniformly distributed r.v. over $[0, 2\pi)$ which is independent of R. On the one hand, note that $\frac{Z^n}{|Z|^{n-1}} = R\exp(in\Theta)$ has the same law as Z. On the other hand, from question 1, we have

$$\frac{Z^n}{|Z|^{n-1}} = A_n + iB_n.$$

So the complex r.v. $A_n + iB_n$ has the same law as Z, which proves that A_n and B_n are two independent, centered Gaussian variables.

3. First, from question 2, we know that $E[A_n B_n] = 0$ for any $n \in \mathbb{N}$. Let $n \neq m$. With the notations introduced above, we have

$$E[(A_n + iB_n)(A_m + iB_m)] = E[R^2 \exp i(n+m)\Theta] = E[R^2]E[\exp i(n+m)\Theta] = 0,$$

and

$$E[(A_n + iB_n)(A_m - iB_m)] = E[R^2 \exp i(n-m)\Theta] = E[R^2]E[\exp i(n-m)\Theta] = 0.$$

Developing the left hand side of the above expressions, we see that the terms $E[A_nA_m] + E[B_nB_m]$, $E[A_nA_m] - E[B_nB_m]$, $E[A_mB_n] + E[B_mA_n]$, and $E[A_mB_n] - E[B_mA_n]$ vanish, from which we deduce that $E[A_nA_m] = E[B_nB_m] = E[A_mB_n] = 0$.

4. (A_n, B_n) and (A_m, B_m) are pairs of independent, centered Gaussian r.v.s such that $Z_n = A_n + iB_n = R\exp(in\Theta)$ and $Z_m = A_m + iB_m = R\exp(im\Theta)$. So, for integers a, b:

$$\begin{aligned} E[Z_n^a(\overline{Z_m})^b] &= E[R^{a+b}\exp(i(na - mb)\Theta)] \\ &= E[R^{a+b}]E[\exp(i(na - mb)\Theta)]. \end{aligned}$$

From question 1 of Exercise **3.4**, $R^2 \overset{\text{(law)}}{=} 2\varepsilon$, where ε is exponentially distributed with parameter 1, thus $E[R^{a+b}] = 2^{\frac{a+b}{2}}E[\varepsilon^{\frac{a+b}{2}}] = 2^{\frac{a+b}{2}}\Gamma\left(\frac{a+b}{2} + 1\right)$. Moreover, $E[\exp(i(na - mb)\Theta)] = \frac{\exp(2i\pi(na-mb)) - 1}{2i\pi(na-mb)}$, if $na \neq mb$, and 1 if $na = mb$. Hence

$$\begin{aligned} E[Z_n^a(\overline{Z_m})^b] &= 2^{\frac{a+b-2}{2}}\Gamma\left(\frac{a+b}{2} + 1\right)\frac{(\exp(2i\pi(na - mb)) - 1)}{i\pi(na - mb)}, \quad \text{if } na \neq mb, \\ &= 1, \quad \text{if } na = mb. \end{aligned}$$

If the vector (A_n, B_n, A_m, B_m) were Gaussian, then from question 3, the pairs (A_n, B_n) and (A_m, B_m) would be independent, hence Z_n and Z_m would be independent. But this is impossible since $|Z_n| = |Z_m| = R$ is not deterministic. Note that the vector (A_n, B_n) is not Gaussian either. This can easily be verified for $n = 2$.

Solution to Exercise 3.7

(i)$\Longleftrightarrow$(ii): The identity in law $\langle X, G\rangle \overset{\text{(law)}}{=} \langle Y, G\rangle$ is equivalent to $E[e^{i\lambda\langle X,G\rangle}] = E[e^{i\lambda\langle Y,G\rangle}]$ for every $\lambda \in \mathbb{R}$. Computing each member of the equality, we get $E[e^{-\frac{\lambda^2}{2}\|X\|^2}] = E[e^{-\frac{\lambda^2}{2}\|Y\|^2}]$ for every $\lambda \in \mathbb{R}$, hence, the Laplace transforms of the (non-negative) r.v.s $\|X\|^2$ and $\|Y\|^2$ are the same, which is equivalent to $\|X\| \overset{\text{(law)}}{=} \|Y\|$.

(ii)$\Longleftrightarrow$(iii): First observe that since the law of G is preserved by any rotation in $\mathbb{R}^k$, $\|G\|U \overset{\text{(law)}}{=} G$, hence (ii)$\Longleftrightarrow$ $\|G\|\langle U, X\rangle \overset{\text{(law)}}{=} \|G\|\langle U, Y\rangle$. Now, it is easy to check that $\|G\|$ is a simplifiable r.v., so from Exercise **1.13** (or use Exercise **1.14**) $\|G\|\langle U, X\rangle \overset{\text{(law)}}{=} \|G\|\langle U, Y\rangle \Longleftrightarrow \langle U, X\rangle \overset{\text{(law)}}{=} \langle U, Y\rangle$.

Comments on the solution. It appears, from above, that these equivalences also hold if we replace G by any other isotropic random variable (i.e. whose law is invariant under orthogonal transformations) whose norm is simplifiable.

Solution to Exercise 3.8

1. Let $\mathbf{G}'_n$ be an independent copy of $\mathbf{G}_n$, then for any real λ, we have:

$$E[e^{i\lambda\langle \mathbf{G}'_n, A\mathbf{G}_n\rangle}] = E[e^{i\lambda\langle A^*\mathbf{G}'_n, \mathbf{G}_n\rangle}].$$

By conditioning, on the left hand side by $\mathbf{G}_n$, and using the characteristic function of $\mathbf{G}'_n$ (and similarly for the right hand side, with the noles of $\mathbf{G}_m$ and $\mathbf{G}'_n$ exchanged), the previous identity can be written as:

$$E[e^{-\frac{\lambda^2}{2}\|A\mathbf{G}_n\|^2}] = E[e^{-\frac{\lambda^2}{2}\|A^*\mathbf{G}'_n\|^2}].$$

This proves the identity in law: $\|A\mathbf{G}_n\| \overset{(\text{law})}{=} \|A^*\mathbf{G}_n\|$, by the uniqueness property of the Laplace transform.

2. Let $n \in \mathbb{N}^*$ and denote by $X^{(i)}$, $i = 1, \ldots, n$ the coordinates of $\mathbf{X}_n$, then by considering the equality in law of the statement for

$$A = \frac{1}{\sqrt{n}}\begin{pmatrix} 1 & \cdots & 1 \\ 0 & \cdots & 0 \\ \cdot & \cdots & \cdot \\ 0 & \cdots & 0 \end{pmatrix},$$

we get the following identity in law

$$\frac{1}{\sqrt{n}}\Big|\sum_{i=1}^{n} X^{(i)}\Big| \overset{(\text{law})}{=} |X|. \tag{3.8.a}$$

Since X is square integrable, we can apply the Central Limit Theorem, so, letting $n \to \infty$ in (3.8.a), we see that $|X|$ is distributed as the absolute value of a centered Gaussian variable.

It remains for us to show that X is symmetric. By considering the identity in law in the statement of question 2 for

$$A = \begin{pmatrix} 1 & 1 \\ -1 & -1 \end{pmatrix}$$

we obtain

$$|X^{(1)} + X^{(2)}| \overset{(\text{law})}{=} |X^{(1)} - X^{(2)}|.$$

From Exercise **3.7**, this identity in law is equivalent to

$$\varepsilon(X^{(1)} + X^{(2)}) \overset{(\text{law})}{=} \varepsilon(X^{(1)} - X^{(2)}), \tag{3.8.b}$$

where ε is a symmetric Bernoulli r.v. which is independent of both $X^{(1)} + X^{(2)}$ and $X^{(1)} - X^{(2)}$. Let φ be the characteristic function of X. The identity (3.8.b) may be expressed in terms of φ as: $\frac{1}{2}(\varphi^2(\lambda) + \overline{\varphi^2(\lambda)}) = \varphi(\lambda)\overline{\varphi(\lambda)}$, or equivalently $(\varphi(\lambda) - \overline{\varphi(\lambda)})^2 = 0$. Hence, $\varphi(\lambda)$ takes only real values, that is: X is symmetric.

Solution to Exercise 3.9

1. It suffices to note that $R\boldsymbol{X}_n$ is a centered Gaussian vector with covariance matrix $RR^* = I_d$.

2. Let $n = 2$ and $\boldsymbol{X}_2 = (X^{(1)}, X^{(2)})$, then by taking

$$A = \begin{pmatrix} \frac{1}{\sqrt{2}} & \frac{-1}{\sqrt{2}} \\ \frac{1}{\sqrt{2}} & \frac{1}{\sqrt{2}} \end{pmatrix},$$

we get

$$X^{(1)} + X^{(2)} \overset{\text{(law)}}{=} X^{(1)} - X^{(2)} \overset{\text{(law)}}{=} \sqrt{2}\, X^{(1)}. \tag{3.9.a}$$

Writing this identity in terms of the characteristic function φ of X, we obtain the equation: $\varphi(t)^2 = \varphi(t)\overline{\varphi}(t) = \varphi(\sqrt{2}t)$, for any $t \in \mathbb{R}$. Therefore, φ is real and non-negative. Since moreover, $\varphi(0) = 1$, then from the above equation and by the continuity of φ, we deduce that it is strictly positive on $\mathbb{R}$. We may then define $\psi = \log\varphi$ on $\mathbb{R}$. From the equation $2\psi(t) = \psi(\sqrt{2}t)$, which holds for all $t \in \mathbb{R}$, and from the continuity of ψ, we deduce by classical arguments that ψ has the form $\psi(t) = at^2$, for some $a \in \mathbb{R}$. So φ has the form $\varphi(t) = e^{at^2}$. Finally, differentiating each member of the equation $\varphi(t)^2 = \varphi(\sqrt{2}t)$, we obtain $a = -\text{var}(X)/2$.

Assuming the variance of X is finite, a more probabilistic proof consists in writing:

$$\varphi(t) = \left(\varphi\left(\frac{1}{\sqrt{2}}t\right)\right)^2 = \cdots = \left(\varphi\left(\frac{1}{(\sqrt{2})^n}t\right)\right)^{2^n},$$

from which we deduce:

$$X \overset{\text{(law)}}{=} \frac{1}{2^{\frac{n}{2}}} \sum_{i=1}^{2^n} X_i.$$

Since from (3.9.a) X is centered, letting $n \to \infty$, we see from the central limit theorem that X is a Gaussian variable.

Solution to Exercise 3.10

1. First note that $\mathcal{G}_n = \sigma\{X_{i+1} - X_i; 1 \le i \le n-1\}$ and $\sigma\{\Sigma_1, \Sigma_2, \ldots, \Sigma_n\} = \sigma\{\Sigma_{i+1} - \Sigma_i; 1 \le i \le n-1\}$, so the first equality comes from the relation:

$$X_{i+1} - X_i = \frac{i+1}{i}(\Sigma_{i+1} - \Sigma_i) + \Sigma_{i-1} - \Sigma_i, \quad 1 \le i \le n-1.$$

Note also that

$$\sigma\left\{\frac{S_i}{i} - \frac{S_j}{j}; 1 \le i, j \le n\right\} = \sigma\left\{\frac{S_{i+1}}{i+1} - \frac{S_i}{i}; 1 \le i \le n-1\right\},$$

so that the second equality is due to:

$$\frac{S_{i+1}}{i+1} - \frac{S_i}{i} = \frac{1}{i}\left(X_{i+1} - \frac{S_{i+1}}{i+1}\right) \tag{3.10.a}$$

$$= \frac{1}{i}(\Sigma_{i+1} - \Sigma_i), \quad 1 \le i \le n-1.$$

2. The vector $(\hat{S}_1, \ldots, \hat{S}_n, S_n)$ is Gaussian and for each $k \le n$, $\mathrm{Cov}(\hat{S}_k, S_n) = 0$, hence, S_n is independent of $\{\hat{S}_k, k \le n\}$. Writing $\{S_k, k \le n\} = \{\hat{S}_k + \frac{k}{n} S_n, k \le n\}$, we deduce from this independence that $\{\hat{S}_k, k \le n\}$ has the same law as $\{S_k, k \le n\}$ conditioned by $S_n = 0$. Finally, we deduce from question 1 and the equality, $\frac{1}{i}\hat{S}_i - \frac{1}{j}\hat{S}_j = \frac{1}{i}S_i - \frac{1}{j}S_j$, that $\mathcal{G}_n = \sigma\{\hat{S}_k, k \le n\}$.

3. The equality $\mathcal{G}_n = \sigma\{Y_1, \ldots, Y_n\}$ is due to (3.10.a). For each $n \ge 1$, $(Y_1, \ldots, Y_n)$ is a Gaussian vector whose covariance matrix is diagonal, hence, $Y_1, \ldots, Y_n$ are independent.

Since the vector $\Sigma_n^{(\alpha)}$, $n = 1, 2, \ldots$ is Gaussian, its increments are independent if and only if

$$E[(X_k - \alpha_k S_k)(X_n - \alpha_n S_n)] = 0,$$

for any $k < n$. Under the condition $\alpha_k \ne 0$, this is verified if and only if: $\alpha_k = \frac{1}{k}$.

4. This equality has been established in (3.10.a).

By successive iterations of equality (3.10.a), we obtain:

$$\frac{S_n}{n} = \frac{S_{n+k}}{n+k} - \sum_{j=1}^{k} \frac{Y_{n+j}}{n+j-1}. \tag{3.10.b}$$

Note that from question 2, S_{n+k} is independent of $\mathcal{G}_{n+k}$. Since moreover S_{n+k} is centered and $(Y_{n+1}, \ldots, Y_{n+k})$ is $\mathcal{G}_{n+k}$-measurable, we have:

$$E\left[\frac{S_n}{n} \mid \mathcal{G}_{n+k}\right] = E\left[\frac{S_{n+k}}{n+k}\right] - \sum_{j=1}^{k} E\left[\frac{Y_{n+j}}{n+j-1} \mid \mathcal{G}_{n+k}\right]$$

$$= -\sum_{j=1}^{k} \frac{Y_{n+j}}{n+j-1}.$$

5. Letting $k \to \infty$ in (3.10.b), we obtain:

a) $\displaystyle\lim_{k\to\infty} \frac{S_{n+k}}{n+k} = 0$, a.s. and in L^2, by the law of large numbers,

b) $\displaystyle\frac{S_n}{n} = -\sum_{j=1}^{\infty} \frac{Y_{n+j}}{n+j-1}$, a.s.,

the convergence of the series holds in L^2 and a.s. In particular, note that a.s. $\lim_{k\to\infty} \frac{S_{n+k}}{n+k} = 0$ by the law of large numbers.

6. It is clear, from the definition of $\mathcal{F}_n$ and $\mathcal{G}_n$ that for every n, $\mathcal{F}_n = \mathcal{G}_n \vee \sigma(X_1)$, which implies $\mathcal{F}_\infty = \mathcal{G}_\infty \vee \sigma(X_1)$. Moreover, from question 3, $\mathcal{G}_\infty = \sigma\{Y_1, Y_2, \ldots\}$ and from question 5, $\sigma(X_1) \subset \sigma\{Y_1, Y_2, \ldots\}$. The conclusion follows.

Solution to Exercise 3.11

1. By standard approximation arguments, it is enough to show identity (3.11.1) for $f(x) = e^{i\langle t, x \rangle}$ and $g(x) = e^{i\langle s, x \rangle}$, with $s, t, x \in \mathbb{R}^d$. Let us denote by φ the characteristic function of X, that is $E[e^{i\langle t, X \rangle}] = \varphi(t) = e^{\frac{-|t|^2}{2}}$, $t \in \mathbb{R}^d$. First, it follows from the definition of the covariance that $\mathrm{Cov}(f(X), g(X)) = \varphi(s+t) - \varphi(s)\varphi(t)$. On the other hand, the computation of the integral in (3.11.1) gives:

$$\int_0^1 E[\langle \nabla f(X), \nabla g(\alpha X + \sqrt{1-\alpha^2} Y) \rangle] \, d\alpha$$
$$= \int_0^1 d\alpha \, E\left[\left\langle \left(it_j \exp(i \langle t, X \rangle) \right)_j , \left(is_j \exp(i \langle s, \alpha X + \sqrt{1-\alpha^2} Y \rangle) \right)_j \right\rangle \right]$$
$$= -\int_0^1 d\alpha \sum_j s_j t_j E[e^{i\langle t + \alpha s, X \rangle}] E[e^{i\langle s\sqrt{1-\alpha^2}, Y \rangle}]$$
$$= -\int_0^1 d\alpha \, \langle s, t \rangle \exp -\frac{1}{2}(\|t\|^2 + 2\alpha \langle s, t \rangle + \|s\|^2)$$
$$= -\varphi(s)\varphi(t)\frac{1 - e^{-\langle s, t \rangle}}{2} = \frac{1}{2}(\varphi(s+t) - \varphi(s)\varphi(t)).$$

2. Let f be any Lipschitz function such that $\|f\|_{\mathrm{Lip}} \leq 1$ and $E[f(X)] = 0$, and define $g_t \stackrel{\mathrm{def}}{=} e^{tf}$, for $t \geq 0$. Then applying (3.11.1) to f and g_t gives

$$E[f(X)g_t(X)] = E[\langle \nabla f(X), \nabla g_t(Z) \rangle]$$
$$= tE[\langle \nabla f(X), \nabla f(Z) \rangle e^{tf(Z)}]$$
$$\leq tE[e^{tf(Z)}],$$

thanks to the condition $\|f\|_{\mathrm{Lip}} \leq 1$. Now let the function u be defined via $E[e^{tf(X)}] = e^{u(t)}$. Then $E[f(X)e^{tf(X)}] = u'(t)e^{u(t)}$, and from the above inequality: $u'(t) \leq t$. Since $u(0) = 0$, we conclude that $u(t) \leq \frac{t^2}{2}$, that is $E[e^{tf(X)}] \leq e^{\frac{t^2}{2}}$.

3. By symmetry, one can also apply (3.11.3) to $-f$, hence (3.11.3) holds for all $t \in \mathbb{R}$ and for every Lipschitz function f on $\mathbb{R}^d$ with $\|f\|_{\mathrm{Lip}} \leq 1$ and $E[f(X)] = 0$. Then (3.11.4) follows from Chebychev's inequality, since:

$$e^{t\lambda} P(f(X) - E[f(X)] > \lambda) \leq e^{\frac{t^2}{2}},$$

for every t and λ,

$$P(f(X) - E[f(X)] > \lambda) \leq \exp\inf_{t>0}\left(\frac{t^2}{2} - \lambda t \right) = \exp\left(-\frac{\lambda^2}{2} \right).$$

Solution to Exercise 3.12 (Given by A. Cherny)

Let (X, Y) be a Gaussian vector with the mean vector (m_X, m_Y) and with the covariance matrix

$$\begin{pmatrix} c_{XX}, & c_{XY} \\ c_{XY}, & c_{YY} \end{pmatrix}.$$

By the "normal correlation" theorem (see for example, Theorem 2, Chapter II, § 13 in Shiryaev [82], but prove the result directly...),

$$\mathcal{L}(Y \mid X = x) = \mathcal{N}\left(m_Y + \frac{c_{XY}}{c_{XX}}(x - m_X), c_{YY} - \frac{c_{XY}^2}{c_{XX}}\right),$$

$$\mathcal{L}(X \mid Y = y) = \mathcal{N}\left(m_X + \frac{c_{XY}}{c_{YY}}(y - m_Y), c_{XX} - \frac{c_{XY}^2}{c_{YY}}\right).$$

In order for $(\alpha, \beta, \gamma, \delta, a, b)$ to satisfy the desired property, we should have

$$\alpha = m_Y - m_X \frac{c_{XY}}{c_{XX}}, \quad \beta = \frac{c_{XY}}{c_{XX}}, \quad a^2 = \frac{c_{XX} c_{YY} - c_{XY}^2}{c_{XX}},$$

$$\gamma = m_X - m_Y \frac{c_{XY}}{c_{YY}}, \quad \delta = \frac{c_{XY}}{c_{YY}}, \quad b^2 = \frac{c_{XX} c_{YY} - c_{XY}^2}{c_{YY}}$$

for some $(m_X, m_Y, c_{XX}, c_{XY}, c_{YY})$.

Owing to the inequality $c_{XY}^2 \leq c_{XX}^2 c_{YY}^2$, we have $0 \leq \beta\delta \leq 1$. Let us first consider the case when $\beta\delta = 0$. Then $c_{XY} = 0$, which yields $\beta = 0$, $\delta = 0$, $a^2 > 0$, and $b^2 > 0$. On the other hand, for any $(\alpha, \beta, \gamma, \delta, a, b)$ with these properties, we can find corresponding $(m_X, m_Y, c_{XX}, c_{XY}, c_{YY})$.

Let us now consider the case when $0 < \beta\delta < 1$. Then $\beta b^2 = \delta a^2$, $a^2 > 0$, and $b^2 > 0$. On the other hand, for any $(\alpha, \beta, \gamma, \delta, a, b)$ with these properties, we can find corresponding $(m_X, m_Y, c_{XX}, c_{XY}, c_{YY})$.

Let us finally consider the case when $\beta\delta = 1$. Then $c_{XY}^2 = c_{XX} c_{YY}$, which yields $\alpha = -\beta\gamma$, $a^2 = 0$, and $b^2 = 0$. On the other hand, for any $(\alpha, \beta, \gamma, \delta, a, b)$ with these properties, we can find corresponding $(m_X, m_Y, c_{XX}, c_{XY}, c_{YY})$.

As a result, the desired necessary and sufficient condition is:

$$(\alpha, \beta, \gamma, \delta, a, b) \in \{\beta = 0, \, \delta = 0, \, a^2 > 0, \, b^2 > 0\}$$

$$\cup \{0 < \beta\delta < 1, \, \beta b^2 = \delta a^2, \, a^2 > 0, \, b^2 > 0\}$$

$$\cup \{\beta\delta = 1, \, \alpha = -\beta\gamma, \, a^2 = 0, \, b^2 = 0\}.$$

Solution to Exercise 3.13

1. Set $\sigma^2 = E[G^2]$, then we have:

$$E[G\Phi(G)] = \frac{1}{\sqrt{2\pi}\sigma} \int_{\mathbb{R}} x e^{-\frac{x^2}{2\sigma^2}} \Phi(x) \, dx$$

$$= \frac{\sigma^2}{\sqrt{2\pi}\sigma} \int_{\mathbb{R}} e^{-\frac{x^2}{2\sigma^2}} \Phi'(x) \, dx$$

$$= E[G^2] E[\Phi'(G)],$$

where the existence of the integrals clearly follows from the assumption on the function Φ.

2. Write $G_i = c_i G + \sqrt{1 - c_i^2} G_i'$, where $(G_1', \ldots, G_n')$ is a Gaussian vector independent from G (we may assume that $G_1, \ldots, G_n$ are all normal Gaussian variables). Then (3.13.2) is obtained by conditioning the left hand side of this identity on $(G_1', \ldots, G_n')$ and by applying (3.13.1).

Solution to Exercise 3.14

1. (i) The law of (G_q, G) is determined by the covariance matrix of this vector and this matrix is symmetric.

(ii) From the identity in law $(G_q, G) \overset{\text{(law)}}{=} (-G_q, -G)$, we derive that $E[G_q^n G^m] = (-1)^{n+m} E[G_q^n G^m] = -E[G_q^n G^m]$, if $m + n$ is odd. It then follows that $E[G_q^n G^m] = 0$.

2. Let B_1 and B_2 be standard correlated Brownian motions with covariance $\langle B_1, B_2 \rangle_t = qt$. Then on the one hand, from the scaling property of (B_1, B_2), one has

$$E[B_1(t)^m B_2(t)^m] = t^m Q_m(q).$$

On the other hand, from Itô's formula, we may write

$$
\begin{aligned}
E[B_1(t)^m B_2(t)^m] &= E\left[2\int_0^t B_2(s)^m \, dB_1(s)^m\right] + E\left[m^2 \int_0^t B_1(s)^{m-1} B_2(s)^{m-1} q \, ds\right] \\
&= E\left[2\int_0^m B_2(s)^m \frac{m(m-1)}{2} B_1(s)^{m-2} \, ds\right] \\
&\quad + m^2 q \int_0^t ds E\left[(G_q G)^{m-1} s^{m-1}\right] \\
&= m(m-1) \int_0^t ds\, s^{\frac{2m-2}{2}} Q_m^-(q) + mq Q_{m-1}(q) t^m \\
&= (m-1) t^m Q_m^-(q) + mq t^m Q_{m-1}(q).
\end{aligned}
$$

This proves (3.14.1).

The Gaussian integration by parts formula (3.13.2) in Exercise 3.13 provides another proof of this identity. Indeed, applying this formula to the function $\Phi(x, y) = x^{m-1} y^m$, one obtains

$$\frac{\partial \Phi}{\partial x}(x, y) = (m-1) x^{m-2} y^m \quad \text{and} \quad \frac{\partial \Phi}{\partial y}(x, y) = m x^{m-1} y^{m-1},$$

so that

$$
\begin{aligned}
E[G_q^m G^m] &= E[G_q \Phi(G_q, G)] \\
&= E[G_q^2] E[(m-1) G_q^{m-2} G^m] + E[G_q G] E[m G_q^{m-1} G^{m-1}].
\end{aligned}
$$

The identity (3.14.2) may be proved similarly by considering the function $\Phi(x, y) = x^{m-1} y^{m-2}$. Then we obtain:

$$E[G_q^m G^{m-2}] = E[G_q \Phi(G_q, G)]$$
$$= E[G_q^2] E[(m-1) G_q^{m-2} G^{m-2}] + E[G_q G] E[(m-2) G_q^{m-1} G^{m-3}].$$

3. From (3.14.1) it follows $(m-1) Q_m^{(-)}(q) = Q_m(q) - mq Q_{m-1}(q)$. Plugging this into (3.14.2) yields

$$(m-1) Q_m^{(-)}(q) = (m-1)(m-2) q Q_{m-1}^{(-)}(q) + (m-1)^2 Q_{m-2}(q).$$

Then we deduce from this identity that

$$Q_m(q) - mq Q_{m-1}(q) = q(m-1)[Q_{m-1}(q) - (m-1) q Q_{m-2}(q)] + (m-1)^2 Q_{m-2}(q).$$

Finally, we have

$$Q_m(q) = (2m-1) q Q_{m-1}(q) - (m-1)^2 q^2 Q_{m-2}(q) + (m-1)^2 Q_{m-2}(q)$$
$$= (2m-1) q Q_{m-1}(q) + (m-1)^2 (1 - q^2) Q_{m-2}(q).$$

The other identity is obtained through the same arguments.

4. (a) This follows directly from the expression of the Laplace transform of Gaussian vectors

$$E\left[\exp\left(\lambda G_q + \mu G\right)\right] = \exp(\lambda^2/2 + \mu^2/2 + \lambda\mu q).$$

4. (b) From the definition of Hermite polynomials and the previous question, we obtain

$$E\left[\left(\sum_{n=0}^{\infty} \frac{\lambda^n}{n!} h_n(G_q)\right)\left(\sum_{m=0}^{\infty} \frac{\mu^m}{m!} h_m(G)\right)\right] = \sum_{m,n\geq 0} \frac{\lambda^n}{n!} \frac{\mu^m}{m!} E\left[h_n(G_q) h_m(G)\right]$$
$$= E[\exp\left(\lambda G_q - \lambda^2/2\right) \exp\left(\mu G - \mu^2/2\right)]$$
$$= \exp(\lambda\mu q)$$
$$= \sum_{j=0}^{\infty} \frac{(\lambda\mu)^j}{j!} q^j.$$

Then the desired identity follows by identifying the terms of both series

$$\sum_{m,n\geq 0} \frac{\lambda^n}{n!} \frac{\mu^m}{m!} E\left[h_n(G_q) h_m(G)\right] \quad \text{and} \quad \sum_{j=0}^{\infty} \frac{(\lambda\mu)^j}{j!} q^j,$$

as functions of λ and μ.

Chapter 4

Distributional computations

Contents of this chapter

Probabilists often deal with the following two topics.

(i) Given a random vector $Y = (X_1, \ldots, X_n)$ whose distribution on $\mathbb{R}^n$ is known, e.g. through its density, compute the law of $\varphi(X_1, \ldots, X_n)$, where $\varphi : \mathbb{R}^n \to \mathbb{R}$ is a certain Borel function. This involves essentially changing variables in multiple integrals, hence computing Jacobians, etc.

(ii) The distribution of $Y = (X_1, \ldots, X_n)$ may not be so easy to express explicitly. Then, one resorts to finding expressions for some transform of this distribution which characterizes it, e.g. its characteristic function $\Phi_Y(y) = E[\exp i < y, Y >]$, $y \in \mathbb{R}^n$, or, if $Y \equiv X_1$ takes values in $\mathbb{R}_+$, its Laplace transform $\lambda \to E[e^{-\lambda Y}]$, $\lambda \geq 0$, or its Mellin transform: $m \mapsto E[Y^m]$, $m \geq 0$, or $m \in \mathbb{C}$, or its Stieltjes transform: $s \mapsto E\left[\frac{1}{1+sY}\right]$, $s \geq 0$ (or variants of this function). For this topic, we refer, for instance, to the books by Lukacs [55] and Widder [96].

In this chapter, we have focussed mainly on the families of beta and gamma variables, as well as on stable(α) unilateral variables. Most of the identities on these stable variables are found in the books by Zolotarev [103] and Uchaikin and Zolotarev [94]. For some proofs (by analysis) of identities involving the gamma and beta functions, we refer the reader to Andrews, Askey and Roy [1] or Lebedev [51].

---

* 4.1 Hermite polynomials and Gaussian variables

Preliminaries. Let (x, t) denote the generic point in $\mathbb{R}^2$.

There exists a sequence of polynomials in two variables (the Hermite polynomials) such that, for every $a \in \mathbb{C}$:

$$e_a(x, t) \equiv \exp\left(ax - \frac{a^2 t}{2}\right) = \sum_{n=0}^{\infty} \frac{a^n}{n!} h_n(x, t).$$

In the following, we write simply $e_a(x)$ for $e_a(x, 1)$, resp. $h_n(x)$ for $h_n(x, 1)$. Throughout the exercise, X and Y denote two centered, independent, Gaussian variables, with variance 1.

Part A.

1. Compute $E\left[e_a(X)e_b(X)\right] (a, b \in \mathbb{C})$ and deduce: $E\left[h_n(X)h_m(X)\right] (n, m \in \mathbb{N})$.

2. Let $a \in \mathbb{C}$, and $c \in \mathbb{R}$.

   Compute $E\left[\exp a(X + cY) \mid X\right]$ in terms of the function of two variables e_a, as well as of X and c.

   Compute $E\left[(X + cY)^k \mid X\right]$, for $k \in \mathbb{N}$, in terms of the polynomial of two variables h_k, and of X and c.

Part B. Define $T = \frac{1}{2}(X^2 + Y^2), g = \dfrac{X^2}{X^2 + Y^2}$, and $\varepsilon = \operatorname{sgn}(X)$.

1. Compute explicitly the joint law of the triple (T, g, ε). What can be said about these three variables?

   What is the distribution of $\sqrt{2T}$?

2. Define $\mu = \varepsilon\sqrt{g} = \dfrac{X}{\sqrt{X^2+Y^2}}$.

   Show that, for any $a \in \mathbb{C}$, the following identities hold.

   (i) $E\left[\exp(aX) \mid \mu\right] = \varphi(a\mu)$, where: $\varphi(x) = 1 + x \exp\left(\frac{x^2}{2}\right) F(x)$, and
   $F(x) = \int\limits_{-\infty}^{x} dy\, e^{-y^2/2}$.

   (ii) $E\left[\sinh(aX) \mid \mu\right] = \sqrt{\frac{\pi}{2}} a\mu \exp\left(\frac{a^2\mu^2}{2}\right)$.

   (iii) $E\left[\cosh(aX) \mid \mu\right] = 1 + a^2\mu^2 \int\limits_{0}^{1} dy \exp\left(\frac{a^2\mu^2}{2}(1 - y^2)\right)$.

3. Show that, for any $p \in \mathbb{N}$:

$$E[h_{2p+1}(X) \mid \mu] = c_{2p+1}\mu(\mu^2 - 1)^p$$

$$E[h_{2p}(X) \mid \mu] = c_{2p} + c'_{2p}\mu^2 \int_0^1 dy\,(\mu^2(1 - y^2) - 1)^{p-1}\,(p \geq 1),$$

where c_{2p}, c_{2p+1}, and c'_{2p} are constants to be determined.

4. Define $\nu = \dfrac{Y}{\sqrt{2T}}$. Are the variables μ and ν independent?

Compute, for $a, b \in \mathbb{C}$, the conditional expectation:

$$E[\exp(aX + bY) \mid \mu, \nu]$$

with the help of the function φ introduced in question 2 of part B.

Comments and references. This exercise originates from the article:

J. AZÉMA AND M. YOR: Étude d'une martingale remarquable. *Séminaire de Probabilités XXIII*, 88–130, *Lecture Notes in Mathematics*, **1372**, Springer, Berlin, 1989.

We now explain how the present exercise has been constructed. The remarkable martingale which is studied in this article is the so-called Azéma martingale: $\mu_t = \mathrm{sgn}(B_t)\sqrt{t - \gamma_t}$, where (B_t) is a one-dimensional Brownian motion and $\gamma_t = \sup\{s \leq t : B_s = 0\}$. Let us take $t = 1$, and denote $\mu = \mu_1$, $\varepsilon = \mathrm{sgn}(B_1)$, $g = 1 - \gamma_1$, $X = B_1$. Then the triplet (X^2, ε, g) is distributed as $\left(X^2, \varepsilon, \frac{X^2}{X^2+Y^2}\right)$ in the present exercise (Brownian enthusiasts can check this assertion!) so that the (Brownian) quantity $E[\varphi(B_1) \mid \mu_1 = z]$ found in the above article is equal to $E\left[\varphi(X) \mid \varepsilon\sqrt{\frac{X^2}{X^2+Y^2}} = z\right]$.

** 4.2 The beta–gamma algebra and Poincaré's Lemma

Let $a, b > 0$. Throughout the exercise, Z_a is a gamma variable with parameter a, and $Z_{a,b}$ a beta variable with parameters a and b, that is:

$$P(Z_a \in dt) = t^{a-1}e^{-t}dt/\Gamma(a),\ (t \geq 0)$$

and

$$P(Z_{a,b} \in dt) = \frac{t^{a-1}(1-t)^{b-1}dt}{\beta(a,b)},\ (t \in [0,1]).$$

1. Show the identity in law between the two-dimensional variables:

$$(Z_a, Z_b) \overset{(\text{law})}{=} (Z_{a,b}Z_{a+b}, (1 - Z_{a,b})Z_{a+b}), \tag{4.2.1}$$

where, on the right hand side, the r.v.s $Z_{a,b}$ and Z_{a+b} are assumed to be independent, and on the left hand side, the r.v.s Z_a and Z_b are independent.

2. Show the identity:

$$Z_{a,b+c} \overset{(\text{law})}{=} Z_{a,b} Z_{a+b,c} \qquad (4.2.2)$$

with the same independence assumption on the right hand side, as in the preceding question.

Hint. Use Exercise **1.13** about simplifiable variables.

3. Let $k \in \mathbb{N}$, and $N_1, \ldots, N_2, \ldots, N_k$ be k independent centered Gaussian variables, with variance 1.

Show that, if we denote: $R_k = \left(\sum\limits_{i=1}^{k} N_i^2 \right)^{1/2}$, then

$$R_k^2 \overset{(\text{law})}{=} 2 Z_{k/2} \qquad (4.2.3)$$

and $\quad \underline{x}_k = (x_1, x_2, \ldots, x_k) \equiv \frac{1}{R_k}(N_1, N_2, \ldots, N_k)$ is uniformly distributed on the unit sphere of $\mathbb{R}^k$, and is independent of R_k.

4. Prove *Poincaré's Lemma.*
   Fix $k > 0, k \in \mathbb{N}$. Consider, for every $n \in \mathbb{N}$, $\mathbf{x}_n = (x_1^{(n)}, \ldots, x_n^{(n)})$ a uniformly distributed r.v. on the unit sphere of $\mathbb{R}^n$.

   Prove that: $\sqrt{n}(x_1^{(n)}, \ldots, x_k^{(n)})$ converges in law, as $n \to \infty$, towards $(N_1, \ldots, N_k)$, where the r.v.s $(N_i; 1 \le i \le k)$ are Gaussian, centered with variance 1.

5. Let $\underline{x}_k = (x_1, x_2, \ldots, x_k)$ be uniformly distributed on the unit sphere of $\mathbb{R}^k$, and let $p \in \mathbb{N}$, $1 \le p < k$.

   Show that:

$$\sum_{i=1}^{p} x_i^2 \overset{(\text{law})}{=} Z_{\frac{p}{2}, \frac{k-p}{2}} . \qquad (4.2.4)$$

6. Let $k \ge 3$. Assume that $\underline{x}_k$ is distributed as in the previous question.

   Show that: $\left(\sum\limits_{i=1}^{k-2} x_i^2 \right)^{\left(\frac{k}{2}-1\right)}$ is uniformly distributed on $[0, 1]$.

7. We shall say that a positive r.v. T is (s) distributed if:

$$P(T \in dt) = \frac{dt}{\sqrt{2\pi t^3}} \exp\left(-\frac{1}{2t}\right) .$$

(a) Show that if T is (s)-distributed, then

$$T \stackrel{\text{(law)}}{=} 1/N^2 \,, \tag{4.2.5}$$

where N is a Gaussian r.v., centered, with variance 1. Deduce from this identity that

$$E\left[\exp\left(-\frac{\alpha^2 T}{2}\right)\right] = \exp(-\alpha), \quad \alpha \geq 0 \,.$$

(b) Show that if $(p_i)_{1 \leq i \leq k}$ is a probability distribution, and if $T_1, T_2, \ldots, T_k$ are k independent, (s)-distributed r.v.s, then $\sum_{i=1}^{k} p_i^2 T_i$ is (s) distributed.

8. (a) Show that, if $\underline{x}_k = (x_1, x_2, \ldots, x_k)$ is uniformly distributed on the unit sphere of $\mathbb{R}^k$, and if $(p_i)_{1 \leq i \leq k}$ is a probability distribution, then the distribution of: $\left(\sum_{i=1}^{k} \frac{p_i^2}{x_i^2}\right)$ does not depend on $(p_i)_{1 \leq i \leq k}$.

(b) Compute this distribution.

9. (This question presents a multi-dimensional extension of question 1; it is independent of the other ones.)

Let $(Z_{a_i}; i \leq k)$ be k independent gamma variables, with respective parameters a_i.

Show that $\sum_{i=1}^{k} Z_{a_i}$ is independent of the vector: $\left(Z_{a_i} / \sum_{j=1}^{k} Z_{a_j}; i \leq k\right)$, and identify the law of this vector, which takes its values in the simplex

$$\sum_k = \left\{(x_1, \ldots, x_k); x_i \geq 0; \sum_{i=1}^{k} x_i = 1\right\}.$$

Comments and references:

(a) There is extensive literature about beta and gamma variables. The identities in law (4.2.1) and (4.2.2) are used very often, and it is convenient to refer to them and some of their variants as "beta–gamma algebra"; see e.g.:

D. DUFRESNE: Algebraic properties of beta and gamma distributions, and applications. *Adv. in Appl. Math.*, **20**, no. 3, 285–299 (1998).

JF. CHAMAYOU AND G. LETAC: Additive properties of the Dufresne laws and their multivariate extensions. *J. Theoret. Probab.*, **12**, no. 4, 1045–1066 (1999).

(b) As outlined in:

P. DIACONIS AND D. FREEDMAN: A dozen of de Finetti results in search of a theory. *Ann. Inst. H. Poincaré*, **23**, Supp. to no. 2, 397–423 (1987),

Poincaré's Lemma is misattributed. It goes back at least to Mehler. Some precise references are given in

D. W. STROOCK: *Probability Theory: An Analytic View.* Cambridge University Press, 1993. See, in particular, pp. 78 and 96.

G. LETAC: *Exercises and Solutions Manual for Integration and Probability.* Springer-Verlag, New York, 1995. Second French edition: Masson, 1997.

(c) The distribution of T introduced in question 7 is the unilateral stable distribution with parameter $1/2$. The unilateral stable laws will be studied more deeply in Exercises **4.19**, **4.20** and **4.21**.

** 4.3 An identity in law between reciprocals of gamma variables

We keep the same notations as in Exercise **4.2**.

Let $\alpha, \mu > 0$, with $\alpha < \mu$; the aim of the present exercise is to prove the identity in law:

$$\frac{1}{Z_\alpha} \stackrel{\text{(law)}}{=} \frac{1}{Z_\mu} + \left(\frac{1}{Z_\alpha} + \frac{1}{Z_\mu}\right) \frac{Z_A}{Z_B}, \qquad (4.3.1)$$

where $A = (\mu - \alpha)/2$, $B = (\mu + \alpha)/2$ and on the right hand side of (4.3.1), the four gamma variables are independent.

1. Prove that the identity (4.3.1) is equivalent to:

$$\frac{1}{Z_{a+2b}} + \left(\frac{1}{Z_a} + \frac{1}{Z_{a+2b}}\right) \frac{Z_b}{Z_{a+b}} \stackrel{\text{(law)}}{=} \frac{1}{Z_a}, \qquad a, b > 0. \qquad (4.3.2)$$

2. Prove that (4.3.2) follows from the identity in law

$$\frac{1}{Z_{a+2b}} + \left(\frac{1}{C} + \frac{1}{Z_{a+2b}}\right) \frac{Z_b}{Z_{a+b}} \stackrel{\text{(law)}}{=} \frac{1}{Z_{a+b}} \left(1 + \frac{Z_b}{C}\right), \qquad (4.3.3)$$

where C is distributed as Z_a and, in the above identity, C, Z_b, Z_{a+b}, Z_{a+2b} are independent. (For the moment, take (4.3.3) for granted.)

Hint. Use the fundamental beta–gamma algebra identity (4.2.1) in Exercise **4.2**.

3. We shall now show that the identity (4.3.3) holds for any random variable C assumed independent of the rest of the r.v.s in (4.3.3). We denote this general identity as $(4.3.3)_{\text{gen}}$. Prove that $(4.3.3)_{\text{gen}}$ holds if and only if there is the identity in law between the two pairs:

$$\left(\frac{1}{Z_{a+2b}} + \frac{1}{Z_{a+2b}} \frac{Z_b}{Z_{a+b}}, \frac{Z_b}{Z_{a+b}}\right) \stackrel{\text{(law)}}{=} \left(\frac{1}{Z_{a+b}}, \frac{Z_b}{Z_{a+b}}\right). \qquad (4.3.4)$$

4. Finally, deduce (4.3.4) from the fundamental beta–gamma algebra identity (4.2.1).

Comments. The identity in law (4.3.1) has been found as the probabilistic translation of a complicated relation between McDonald functions K_ν (see Exercise **4.21** for their definition). The direct proof proposed here is due to D. Dufresne.

* 4.4 The Gamma process and its associated Dirichlet processes

This exercise is a continuation, at process level, of the discussion of the gamma–beta algebra in Exercise **4.2**, especially question 9.

A subordinator is a continuous time process with increasing paths whose increments are independent and time homogeneous. Let $(\gamma_t,\ t \geq 0)$ be a gamma process, i.e. a subordinator whose law at time t is the gamma distribution with parameter t:

$$P(\gamma_t \in du) = \frac{u^{t-1}e^{-u}}{\Gamma(t)}\, du\,.$$

1. Prove that for $t_0 > 0$ fixed, the Dirichlet process of duration $t_0 > 0$, which we define as

$$\left(\frac{\gamma_t}{\gamma_{t_0}},\ t \leq t_0\right) \tag{4.4.1}$$

is independent from γ_{t_0}, hence from $(\gamma_u,\ u \geq t_0)$.

2. We call the process $\left(\frac{\gamma_t}{\gamma_1},\ t \leq 1\right)$, the standard Dirichlet process $(D_t, t \leq 1)$.

   Prove that the random vector $(D_{t_1}, D_{t_2} - D_{t_1}, \ldots, D_{t_k} - D_{t_{k-1}}, D_1 - D_{t_k})$, for $t_1 < t_2 < \cdots < t_k \leq 1$, follows the Dirichlet law with parameters $(t_1, t_2 - t_1, \ldots, 1 - t_k)$, i.e. with density

$$\frac{u_1^{t_1-1} u_2^{t_2-t_1} \ldots u_k^{t_k-t_{k-1}}(1 - (u_1 + \ldots + u_k))^{-t_k}}{\Gamma(t_1)\Gamma(t_2 - t_1)\ldots\Gamma(1 - t_k)}\,,$$

   with respect to the Lebesgue measure $du_1 du_2 \ldots du_k$, on the simplex

$$\textstyle\sum_{k+1} = \{(u_i)_{1 \leq i \leq k+1} : \sum_i u_i = 1,\ u_i \geq 0\}\,.$$

3. Prove the following relation for any Borel, bounded function $f : [0,1] \to \mathbb{R}_+$

$$E\left[\frac{1}{1 + \int_0^1 f(u)\, dD_u}\right] = E\left[\exp - \int_0^1 f(u)\, d\gamma_u\right]. \tag{4.4.2}$$

Comments and references. In the following papers

D.M. CIFARELLI AND E. REGAZZINI: Distribution functions of means of a Dirichlet process. *Ann. Statist.*, **18**, no. 1, 429–442 (1990)

P. DIACONIS AND J. KEMPERMAN: Some new tools for Dirichlet priors. *Bayesian Statistics 5* (Alicante, 1994), 97–106, Oxford Sci. Publ., Oxford Univ. Press, New York, 1996

D.M. CIFARELLI AND E. MELILLI: Some new results for Dirichlet priors. *Ann. Statist.*, **28**, no. 5, 1390–1413 (2000)

the authors have used the formula (4.4.2) together with the inversion of the Stieltjes transform to obtain remarkably explicit formulae for the densities of $\int_0^1 f(u)\, dD_u$. As an example,

$$\int_0^1 u\, dD(u) \stackrel{\text{(law)}}{=} \int_0^1 D(u)\, du$$

has density: $\frac{1}{\pi} \sin(\pi x) \frac{1}{x^x (1-x)^{1-x}}$ on $[0,1]$.

** 4.5 Gamma variables and Gauss multiplication formulae

Preliminary. The classical gamma function

$$\Gamma(z) = \int_0^\infty dt\, t^{z-1} e^{-t}$$

satisfies *Gauss's multiplication formula:* for any $n \in \mathbb{N}$, $n \geq 1$,

$$\Gamma(nz) = \frac{1}{(2\pi)^{\frac{n-1}{2}}} n^{nz - \frac{1}{2}} \prod_{k=0}^{n-1} \Gamma\left(z + \frac{k}{n}\right). \tag{4.5.1}$$

(See e.g. Section 1.5 of [1].) In particular, for $n = 2$, formula (4.5.1) is known as the *duplication formula* (see e.g. Lebedev [51])

$$\Gamma(2z) = \frac{1}{\sqrt{2\pi}} 2^{2z - \frac{1}{2}} \Gamma(z) \Gamma\left(z + \frac{1}{2}\right), \tag{4.5.2}$$

and, for $n = 3$, as the *triplication formula:*

$$\Gamma(3z) = \frac{1}{2\pi} 3^{3z - \frac{1}{2}} \Gamma(z) \Gamma\left(z + \frac{1}{3}\right) \Gamma\left(z + \frac{2}{3}\right). \tag{4.5.3}$$

1. Prove that if, for $a > 0$, Z_a denotes a gamma variable with parameter a, then the identity in law:

$$(Z_{na})^n \stackrel{\text{(law)}}{=} n^n Z_a Z_{a+\frac{1}{n}} \ldots Z_{a+\frac{n-1}{n}} \tag{4.5.4}$$

holds, where the right hand side of (4.5.4) features n independent gamma variables, with respective parameters $a + \frac{k}{n}$ ($0 \leq k \leq n - 1$).

2. We denote by $Z_{a,b}$ a beta variable with parameters (a, b). Prove that the identity in law:

$$(Z_{na,nb})^n \overset{\text{(law)}}{=} Z_{a,b} Z_{a+\frac{1}{n},b} \cdots Z_{a+\frac{n-1}{n},b} \tag{4.5.5}$$

holds, where the right hand side of (4.5.5) features n independent beta variables with respective parameters $\left(a + \frac{k}{n}, b\right)$ for $0 \le k \le n - 1$.

Hint. Use the identity (4.2.1) in Exercise **4.2**.

Comments and references:

(a) The identity in law (4.5.4) may be considered as a translation in probabilistic terms of Gauss' multiplication formula (4.5.1). For some interesting consequences of (4.5.4) in terms of one-sided stable random variables, see Exercises **4.19**, **4.20** and **4.21** hereafter and the classical reference of V.M. Zolotarev [103].

(b) In this exercise, some identities in law for beta and gamma variables are deduced from the fundamental properties of the gamma function. The inverse attitude is taken by

L. GORDON: A stochastic approach to the Gamma function. *Amer. Math. Monthly*, **101**, 858–865 (1994).

See also:

A. FUCHS AND G. LETTA: Un résultat élémentaire de fiabilité. Application à la formule de Weierstrass sur la fonction gamma. *Séminaire de Probabilités XXV*, 316–323, *Lecture Notes in Mathematics*, **1485**, Springer, Berlin, 1991.

** 4.6 The beta–gamma algebra and convergence in law

1. Let $a > 0$, and $n \in \mathbb{N}$, $n \ge 1$.

   Deduce from the identity (4.2.1) that

   $$Z_a \overset{\text{(law)}}{=} \left(Z_{a,1} Z_{a+1,1} \cdots Z_{a+n-1,1}\right) Z_{a+n} \tag{4.6.1}$$

   where, on the right hand side, the $(n + 1)$ random variables which appear are independent.

2. Prove that, as $n \to \infty$, the sequence $\left(\dfrac{Z_{a+n}}{n}, n \to \infty\right)$ converges in law, hence in probability, towards a constant; compute this constant.

3. Let $U_1, U_2, \dots, U_n, \dots$ be a sequence of independent random variables, each of which is uniformly distributed on $[0, 1]$. Let $a > 0$.

Prove that:

$$nU_1^{1/a}U_2^{\frac{1}{a+1}}\dots U_n^{\frac{1}{a+(n-1)}} \xrightarrow[n\to\infty]{\text{(law)}} Z_a.$$

4. Let $a > 0$ and $r > 0$. With the help of an appropriate extension of the two first questions above, prove that:

$$nZ_{a,r}Z_{a+r,r}\dots Z_{a+(n-1)r,r} \xrightarrow[n\to\infty]{\text{(law)}} Z_a,$$

where, on the left hand side, the n beta variables are assumed to be independent.

* 4.7 Beta–gamma variables and changes of probability measures

(We keep the same notation as in Exercises **4.2**, **4.5**, and **4.6**.)

Consider, on a probability space $(\Omega, \mathcal{A}, P)$, a pair (A, L) of random variables such that A takes its values in $[0, 1]$, and L in $\mathbb{R}_+$, and, moreover:

$$\left(\frac{A}{L}, \frac{1-A}{L}\right) \overset{\text{(law)}}{=} \left(\frac{1}{Z_b}, \frac{1}{Z_a}\right),$$

where Z_a and Z_b are independent (gamma) variables.

We write $M = \dfrac{L}{A(1-A)}$.

1. Prove the identity in law:

$$(A, M) \overset{\text{(law)}}{=} (Z_{a,b}, Z_{a+b}),$$

where, on the right hand side (hence, also on the left hand side), the two variables are independent.

2. Let $\beta > 0$. Define a new probability measure Q_β on $(\Omega, \mathcal{A})$ by the formula:

$$Q_\beta = c_\beta L^\beta \cdot P.$$

(a) Compute the normalizing constant c_β in terms of a, b and β.

(b) Prove that, under Q_β, the variables A and M are independent and that A is now distributed as $Z_{a+\beta,b+\beta}$, whereas M is distributed as $Z_{a+b+\beta}$.

(c) Compute the joint law under Q_β of $\dfrac{A}{L}$ and $\dfrac{1-A}{L}$, and prove that they are no longer independent.

* 4.8 Exponential variables and powers of Gaussian variables

1. Let Z be an exponential variable with parameter 1, i.e:

$$P(Z \in dt) = e^{-t} dt \qquad (t > 0).$$

Prove that:

$$Z \stackrel{\text{(law)}}{=} \sqrt{2Z} |N|, \tag{4.8.1}$$

where, on the right hand side, N denotes a centered Gaussian variable, with variance 1, which is independent of Z.

Hint: Use the identity (4.5.4) for $n = 2$ and $a = \frac{1}{2}$.

2. Let $p \in \mathbb{N}$, $p > 0$. Prove that:

$$Z \stackrel{\text{(law)}}{=} Z^{\frac{1}{2^p}} 2^{\frac{1}{2} + \cdots + \frac{1}{2^p}} \left(|N_1| \, |N_2|^{\frac{1}{2}} \ldots |N_p|^{\frac{1}{2^{p-1}}} \right) \tag{4.8.2}$$

where, on the right hand side of (4.8.2), $N_1, N_2, \ldots, N_p$ are independent, centered Gaussian variables with variance 1, which are also independent of Z.

3. Let $(N_p, p \in \mathbb{N})$ be a sequence of independent centered Gaussian variables with variance 1.

Prove that:

$$\prod_{p=0}^{n} |N_p|^{\frac{1}{2^p}} \xrightarrow[n \to \infty]{\text{(law)}} \frac{1}{2} Z \ .$$

* 4.9 Mixtures of exponential distributions

Let λ_1 and λ_2 be two positive reals such that: $0 < \lambda_1 < \lambda_2 < \infty$. Let T_1 and T_2 be two independent r.v.s such that for every $t \geq 0$, $P(T_i > t) = \exp(-\lambda_i t)$ $(i = 1, 2)$.

1. Show that there exist two constants α and β such that, for every Borel set B of $\mathbb{R}_+$,

$$P(T_1 + T_2 \in B) = \alpha P(T_1 \in B) + \beta P(T_2 \in B).$$

Compute explicitly α and β.

2. Consider a third variable T_3, which is independent of the pair (T_1, T_2), and which satisfies:

for every $t \geq 0$, $P(T_3 \geq t) = \exp -(\lambda_2 - \lambda_1)t$.

   (a) Compute $P(T_1 > T_3)$.

   (b) Compare the law of the pair (T_1, T_3), conditionally on $(T_1 > T_3)$, and the law of the pair $(T_1 + T_2, T_2)$.

   (c) Deduce therefrom a second proof of the formula stated in question 1.

Hint. Write $E[f(T_1)] = E\left[f(T_1)1_{(T_1 > T_3)}\right] + E\left[f(T_1)1_{(T_1 \leq T_3)}\right]$ for $f \geq 0$, Borel.

3. (This question is independent of question 2.)
   Let $\lambda_1, \lambda_2, \ldots, \lambda_n$ be n positive reals, which are distinct and let $T_1, T_2, \ldots, T_n$ be n independent r.v.s such that:
   for every $t \geq 0$, $P(T_i > t) = \exp(-\lambda_i t)$ $(i = 1, 2, \ldots, n)$.

   Show that there exist n constants $\alpha_i^{(n)}$ such that:
   for every Borel set B in $\mathbb{R}_+$,

   $$P\left(\sum_{i=1}^n T_i \in B\right) = \sum_{i=1}^n \alpha_i^{(n)} P(T_i \in B). \qquad (4.9.1)$$

   Prove the formula

   $$\alpha_i^{(n)} = \prod_{j \neq i} \left(1 - \frac{\lambda_i}{\lambda_j}\right)^{-1}. \qquad (4.9.2)$$

Comments and references. Sums of independent exponential variables play an important role in:

PH. BIANE, J. PITMAN AND M. YOR: Probability laws related to the Jacobi theta and Riemann zeta functions, and Brownian excursions. *Bull. (N.S.) Amer. Math. Soc.*, **38**, no. 4, 435–465 (2001).

** 4.10 Some computations related to the lack of memory property of the exponential law

Let T and T' be two independent, exponentially distributed r.v.s, i.e.

$$P(T \in dt) = P(T' \in dt) = e^{-t} dt \ (t > 0).$$

1. Compute the joint law of $\left(\frac{T}{T+T'}, T + T'\right)$.
   What can be said of this pair of r.v.s?

2. Let U be a uniformly distributed r.v. on $[-1, +1]$, which is independent of the pair (T, T').
   Define: $X = \log \frac{1+U}{1-U}$, and $Y = U(T + T')$.

   Prove that the pairs (X, Y) and $\left(\log \frac{T}{T'}, T - T'\right)$ have the same law.

   Hint. It may be convenient to consider $\tilde{U} = \frac{1}{2}(1 + U)$.

3. Compute, for all $\mu, \nu \in \mathbb{R}$.

   $$E[\exp i(\nu X + \mu Y)].$$

Hints.

(a)   It is a rational fraction of: $\dfrac{\pi\nu}{\sinh(\pi\nu)}$, $(1+i\mu)^{1-i\nu}$, $(1-i\mu)^{1+i\nu}$.

(b) The formula of complements: $\Gamma(x)\Gamma(1-x) = \dfrac{\pi}{\sin(\pi x)}$ for the classical Γ function may be used.

4. Prove that, for any $\mu \in \mathbb{R}$, $E[Y \exp i\mu(Y{-}X)] = 0$.

Compute $E[Y \mid Y{-}X]$.

5. Prove that the conditional law of (T, T'), given $(T > T')$, is identical to the law of $\left(T + \frac{T'}{2}, \frac{T'}{2}\right)$; in short:

$$((T,T') \mid (T > T')) \overset{\text{(law)}}{=} \left(T + \frac{T'}{2}; \frac{T'}{2}\right).$$

Deduce that the law of $(|X|, |Y|)$ is identical to that of: $\left(\log\left(1 + \frac{2T'}{T}\right), T'\right)$.

Comments and references. The origin of questions 4 and 5 is found in the study of certain Brownian functionals in

PH. BIANE AND M. YOR: Valeurs principales associées aux temps locaux Browniens. *Bull. Sci. Math.*, (2) **111**, no. 1, 23–101 (1987).

*4.11 Some identities in law between Gaussian and exponential variables

1. Let N_1, N_2, N_3 be three independent Gaussian variables, centered, with variance 1. Define: $R = \left(\sum\limits_{i=1}^{3} N_i^2\right)^{1/2}$, and compute its law.

2. Let X be an r.v. such that $P(X \in dx \mid R = r) = \frac{1}{r}\mathbf{1}_{]0,r]}(x)dx$. Define: $U = \dfrac{X}{R}$; what is the law of U? What can be said of the pair (R, U)?

3. Let $Y = R - X$; what is the law of Y given R?

Compute the law of the pair (X, Y).

4. Define $G = X - Y$ and $T = 2XY$.

What is the law of the pair (G, T)?

What is the law of G? of T? What can be said of these two variables?

5. Define $V = 2U - 1$.

Remark that the following identities hold:

$$G = RV \quad \text{and} \quad T = R^2 \left(\frac{1 - V^2}{2} \right),$$

and, therefore:

$$G^2 = 2T \left(\frac{V^2}{1 - V^2} \right). \tag{4.11.1}$$

What is the law of $\dfrac{V^2}{1 - V^2}$ given T?

(6) Let A be an r.v. which is independent of T, and which has the arcsine distribution, that is:

$$P(A \in da) = \frac{da}{\pi \sqrt{a(1 - a)}} \qquad (a \in]0, 1[) .$$

Show that:

$$G^2 \overset{\text{(law)}}{=} 2TA. \tag{4.11.2}$$

Note and comment upon the difference with the above representation (4.11.1) of G^2.

* 4.12 Some functions which preserve the Cauchy law

1. Let Θ be an r.v. which is uniformly distributed on $[0, 2\pi)$. Give a simple argument to show that $\tan(2\Theta)$ and $\tan(\Theta)$ have the same law.

2. Let N and N' be two centered, independent Gaussian variables with variance 1.

   (i) Prove that $C \overset{\text{def}}{=} \dfrac{N}{N'}$ and $\tan(\Theta)$ have the same distribution.

   (ii) Show that C is a Cauchy variable, i.e.

$$P(C \in dx) = \frac{dx}{\pi(1 + x^2)}.$$

3. Prove, only using the two preceding questions, that if C is a Cauchy variable, then so are:

$$\frac{1}{2} \left(C - \frac{1}{C} \right) \quad \text{and} \quad \frac{1 + C}{1 - C} . \tag{4.12.1}$$

Comments and references. Equation (4.12.1) is an instance of a much more general result which, roughly speaking, says that the complex Cauchy distribution is preserved by a large class of meromorphic functions of the complex upper half plane.

E.J.G. PITMAN AND E.J. WILLIAMS: Cauchy-distributed functions of Cauchy variates. *Ann. Math. Statist.*, **8**, 916–918 (1967).

F.B. KNIGHT AND P.A. MEYER: Une caractérisation de la loi de Cauchy. *Z. Wahrscheinlichkeitstheorie und Verw. Gebiete*, **34**, no. 2, 129–134 (1976).

G. LETAC: Which functions preserve Cauchy laws? *Proc. Amer. Math. Soc.*, **67**, no. 2, 277–286 (1977).

One may find some other related references in:

J. PITMAN AND M. YOR: Some properties of the arcsine law related to its invariance under a family of rational maps. In: *A Festschrift for Herman Rubin*, IMS-AMS series, ed: A. DasGupta, **45**, 126–137 (2004).

* 4.13 Uniform laws on the circle

Let U and V be two random variables which are independent and uniformly distributed on $[0, 1)$. Let m, n, m', n' be four integers, all different from 0, that is elements of $\mathbb{Z} \setminus \{0\}$.

What is the law of $\{mU + nV\}$, where $\{x\}$ denotes the fractional part of x. Give some criterion which ensures that the two variables $\{mU + nV\}$ and $\{m'U + n'V\}$ are independent, where $\{x\}$ denotes the fractional part of x.

Hint. Consider $E\left[\exp 2i\pi \left(p\{mU + nV\} + q\{m'U + n'V\}\right)\right]$ for $p, q \in \mathbb{Z}$.

* 4.14 Fractional parts of random variables and the uniform law

Recall that $\{x\}$ denotes the fractional part of the real x.

1. Let U be uniform on $[0, 1[$ and let Y be an r.v. independent from U. Prove that $\{U + Y\}$ is uniform.

2. Prove that any random variable V, with values in $[0, 1[$, such that, for every r.v. Y independent from V,

$$\{V + Y\} \overset{(\text{law})}{=} V$$

is uniformly distributed.

Comments and references. (a) This characteristic property of the uniform law plays an important role in the study of the already cited Tsirel'son's equation, see Exercise **2.5**.

J. AKAHORI, C. UENISHI AND K. YANO: Stochastic equations on compact groups in discrete negative time. *Probab. Theory Related Fields*, **140**, no. 3–4, 569–593 (2008).

K. YANO AND M. YOR: Around Tsirelson's equation, or: The evolution process may not explain everything. Preprint (2010), `arXiv:0906.3442`, to appear in *Ergodic Theory and Dynamical Systems*.

(b) In fact, more generally, on a compact group, it is the Haar probability measure that may be characterized as the unique probability measure which is left (and/or right) invariant. For thorough discussions, see

W. RUDIN: *Real and Complex Analysis*. Third edition. McGraw-Hill Book Co., New York, 1987.

Y. KATZNELSON: *An Introduction to Harmonic Analysis*. Third edition. Cambridge Mathematical Library, Cambridge University Press, Cambridge, 2004.

* 4.15 Non-infinite divisibility and signed Lévy–Khintchine representation

In this exercise, U denotes a uniformly distributed random variable over $[0,1]$. Moreover, for $a > 0$, e_a denotes an exponential variable with parameter a, U_a is its fractional part, and N_a its integer part ($e_a = U_a + N_a$, with $N_a = \lfloor e_a \rfloor$).

1. Show that a non constant r.v. which is uniformly bounded is not infinitely divisible.

2. Prove that nevertheless, U_a admits the following Lévy–Khintchine type representation with respect to a signed "Lévy" measure μ_a:

$$E[\exp(-\lambda U_a)] = \exp\left(-\int d\mu_a(x)(1 - e^{-\lambda x})\right), \quad \lambda > 0.$$

3. Can we find V_1 and V_2 i.i.d. such that:

$$U \overset{(\text{law})}{=} V_1 + V_2 \, ?$$

4. Let $k \in \mathbb{N}$. Give an example of an r.v. X such that:

   (i) there exist $X_1, \ldots, X_k$ i.i.d. such that

$$X \overset{(\text{law})}{=} X_1 + \ldots + X_k,$$

(ii) there do not exist $Y_1, \ldots Y_{k+1}$ i.i.d. such that:

$$X \stackrel{(\text{law})}{=} Y_1 + \ldots + Y_{k+1} .$$

5. Let $U_1, \ldots, U_n, \ldots$ be a sequence of r.v.s which are uniformly distributed and independent. Give a recurrence formula which relates f_{k+1} to f_k, where f_k is the density function of $(U_1 + \ldots + U_k)/k$.

○ 6. Characterize uniformly bounded r.v.s $V \geq 0$ which admit a signed Lévy–Khintchine representation, i.e.

$$E\left[\exp(-\lambda V)\right] = \exp\left(-\int d\nu(x) \left(1 - e^{-\lambda x}\right)\right), \quad \lambda > 0,$$

where ν is a signed measure, such that $\int |\nu(dx)|(x \wedge 1) < \infty$.

*4.16 Trigonometric formulae and probability

1. Let Θ be an r.v. which is uniformly distributed on $[0, \pi)$. Compute the law of $X = \cos(\Theta)$.

2. Prove that if Θ and Θ' are independent and uniformly distributed on $[0, 2\pi)$, then:
$$\cos(\Theta) + \cos(\Theta') \stackrel{(\text{law})}{=} \cos(\Theta + \Theta') + \cos(\Theta - \Theta').$$

   Hint: Use Exercise **4.13**.

3. Prove that, if X and Y are independent, and have both the distribution of $\cos(\Theta)$ [which was computed in question 1], then:

$$\frac{1}{2}(X + Y) \stackrel{(\text{law})}{=} XY.$$

*4.17 A multidimensional version of the Cauchy distribution

Let N be a centered Gaussian r.v. with variance 1.

1. Prove that if $T \stackrel{\text{def}}{=} 1/N^2$, then, one has:

$$P(T \in dt) = \frac{dt}{\sqrt{2\pi t^3}} \exp\left(-\frac{1}{2t}\right) \qquad (t > 0).$$

2. Prove that if: C is a standard Cauchy variable, that is: its distribution is given by $P(C \in dx) = \dfrac{dx}{\pi(1 + x^2)}$ $(x \in \mathbb{R})$, then its characteristic function is

$$E\left[\exp(i\lambda C)\right] = \exp(-|\lambda|) \qquad (\lambda \in \mathbb{R}).$$

3. Deduce from the preceding question that the variable T, defined in question 1, satisfies:

$$E\left[\exp\left(-\frac{\lambda^2}{2}T\right)\right] = \exp(-|\lambda|) \qquad (\lambda \in \mathbb{R}).$$

Hint. Use the representation of a Cauchy variable given in Exercise **4.12**, question 2.

4. Let $T_1, T_2, \ldots, T_n, \ldots$ be a sequence of independent r.v.s with common law that of T.

Compare the laws of $\frac{1}{n^2} \sum\limits_{i=1}^{n} T_i$ and T.

How does this result compare with the law of large numbers?

5. Let $N_0, N_1, \ldots, N_n$ be $(n+1)$ independent, centered Gaussian variables, with variance 1.

Compute explicitly the law of the random vector:

$$\left(\frac{N_1}{N_0}, \frac{N_2}{N_0}, \ldots, \frac{N_n}{N_0}\right), \qquad (4.17.1)$$

and compute also its characteristic function:

$$E\left[\exp\left(i\sum_{j=1}^{n}\lambda_j\frac{N_j}{N_0}\right)\right], \qquad \text{where } (\lambda_j)_{j\leq n} \in \mathbb{R}^n.$$

6. Prove that this distribution may be characterized as the unique law of X valued in $\mathbb{R}^n$, such that $\langle \theta, X \rangle$ is a standard Cauchy r.v. for any $\theta \in \mathbb{R}^n$, with $|\theta| = 1$.

Comments and references:

(a) The random vector (4.17.1) appears as the limit in law, for $h \to 0$, of :

$$\frac{1}{(\beta_{t+h}^0 - \beta_t^0)}\left(\sum_{j=0}^{n}\int_t^{t+h} H_s^j \, d\beta_s^j\right) \qquad (4.17.2)$$

where $\beta^0, \beta^1, \ldots, \beta^n$ are $(n+1)$ independent Brownian motions, and $\{H^j; j = 0, 1 \ldots, n\}$ are $(n+1)$ continuous adapted processes. The limit in law, as $h \to 0$, for (4.17.2) may be represented as

$$H_t^0 + \sum_{j=1}^{n} H_t^j\left(\frac{N_j}{N_0}\right)$$

with $(H_t^0, \ldots H_t^n)$ independent of the random vector (4.17.1). For more details, we refer to:

C. YOEURP: Sur la dérivation des intégrales stochastiques. *Séminaire de Probabilités XIV* (Paris, 1978/1979), 249–253, *Lecture Notes in Mathematics,* **784**, Springer, Berlin, 1980

D. ISAACSON: Stochastic integrals and derivatives. *Ann. Math. Statist.,* **40**, 1610–1616 (1969).

(b) Student's laws $\left(\text{of which a particular case is } N_0/\left(\sum_{i=1}^{n} N_i^2\right)^{1/2}\right)$ are classically considered when testing mean and variances in Gaussian statistics. See e.g. Chapters 26–27 in N.L. Johnson, S. Kotz and N. Balakrishnan [42], and Sections 7.2, 7.4, 7.5 in C. J. Stone: *A Course in Probability and Statistics.* Duxbury Press, 1996.

** 4.18 Some properties of the Gauss transform

Let A be a strictly positive random variable, which is independent of N, a Gaussian, centered r.v. with variance 1. Let X be such that:

$$X \stackrel{\text{(law)}}{=} \sqrt{A} N.$$

We shall say that the law of X is the Gauss transform of the law of A.

0. *Preliminaries.* This question is independent of the sequel of the exercise.

(i) Prove that X admits a density which is given by:

$$E\left[\frac{1}{\sqrt{2\pi A}} \exp\left(-\frac{x^2}{2A}\right)\right].$$

(ii) Prove that for any $\lambda \in \mathbb{R}$, $E[\exp(i\lambda X)] = E\left[\exp\left(-\frac{\lambda^2}{2}A\right)\right]$. Hence the Gauss transform is injective on the set of probability measures on $\mathbb{R}_+$.

1. Prove that, for every $\alpha \geq 0$, one has

$$E\left[\frac{1}{|X|^\alpha}\right] = E\left[\frac{1}{A^{\alpha/2}}\right] E\left[\frac{1}{|N|^\alpha}\right]. \qquad (4.18.1)$$

Give a criterion on A and α, which ensures that: $E\left[\frac{1}{|X|^\alpha}\right] < \infty$.

2. Prove that, for every $\alpha > 0$, the quantity $E\left[\frac{1}{A^{\alpha/2}}\right]$ is also equal to:

$$E\left[\frac{1}{A^{\alpha/2}}\right] = \frac{1}{\Gamma\left(\frac{\alpha}{2}\right) 2^{\frac{\alpha}{2}-1}} \int_0^\infty dx\, x^{\alpha-1} E\left[\cos(\alpha X)\right]. \qquad (4.18.2)$$

Hint. Use the elementary formula: $\dfrac{1}{r^{\alpha/2}} = \dfrac{1}{\Gamma\left(\frac{\alpha}{2}\right)} \int\limits_0^\infty dt \, t^{\frac{\alpha}{2}-1} e^{-tr}$.

3. The aim of this question is to compare formulae (4.18.1) and (4.18.2) in the case $E\left(\dfrac{1}{|X|^\alpha}\right) < \infty$.

   3.1. Preliminaries.

      (i) Recall (see, e.g. Lebedev [51]) the *duplication formula* for the gamma function:
      $$\Gamma(2z) = \frac{1}{\sqrt{2\pi}} 2^{2z-\frac{1}{2}} \Gamma(z)\Gamma\left(z+\frac{1}{2}\right)$$
      and the *formula of complements* $\Gamma(z)\Gamma(1-z) = \dfrac{\pi}{\sin(\pi z)}$.

      (ii) For $0 < \alpha < 1$, $a_\alpha \stackrel{\text{def}}{=} \lim\limits_{t\uparrow\infty} \int\limits_0^t dx \, x^{\alpha-1} \cos(x)$ exists in $\mathbb{R}$.

      (iii) Compute $E\left(\dfrac{1}{|N|^\alpha}\right)$, when this quantity is finite, in terms of the gamma function.

   3.2. Prove that, if $E\left(\dfrac{1}{|X|^\alpha}\right) < \infty$, then:
      $$E\left[\frac{1}{A^{\alpha/2}}\right] = \frac{a_\alpha}{\Gamma\left(\frac{\alpha}{2}\right) 2^{\frac{\alpha}{2}-1}} E\left(\frac{1}{|X|^\alpha}\right). \qquad (4.18.3)$$

      Hint. Use formula (4.18.2) and dominated convergence.

   3.3. Comparing (4.18.1) and (4.18.3) and using the result in 3.1 (iii), prove that:
      $$a_\alpha = \Gamma(\alpha) \cos\left(\frac{\pi\alpha}{2}\right) \qquad (0 < \alpha < 1). \qquad (4.18.4)$$

   3.4. Give a direct proof of formula (4.18.4).

4. [A random Cauchy transform.]

   Develop a similar discussion as for the Gauss transform, starting with an identity in law:
   $$Y \stackrel{\text{(law)}}{=} BC ,$$
   where, on the right hand side, B and C are independent, $B > 0$ P a.s. and C is a Cauchy variable with parameter 1.

   In particular, show that: $E\left(\dfrac{1}{|C|^\alpha}\right) = E\left(|C|^\alpha\right)$, and compute this quantity, when it is finite, in terms of the gamma function.
   Hint. Use question 2 in Exercise **4.12**.

Comments and references.

(a) One of the advantages of using the Gauss transform for certain variables A with complicated distributions is that $X \overset{\text{(law)}}{=} N\sqrt{A}$ may have a simple distribution. A noteworthy example is the Kolmogorov–Smirnov statistics: $A = \left(\sup_{s\leq 1}|b(s)|\right)^2$, where $(b(s),\ s \leq 1)$ denotes the one-dimensional Brownian bridge; A satisfies

$$P\left(\left(\sup_{s\leq 1}|b(s)|\right)^2 \leq x\right) = \sum_{n=-\infty}^{\infty}(-1)^n \exp(-2nx^2)\,,$$

whereas there is the simpler expression:

$$P\left(|N|\sup_{s\leq 1}|b(s)| \leq a\right) = \tanh(a)\,, \quad a \geq 0\,.$$

For a number of other examples, see the paper by Biane–Pitman–Yor referred to in Exercise **4.9**.

(b) The Gauss transform appears also quite naturally when subordinating an increasing process $(\tau_t,\ t \geq 0)$ to an independent Brownian motion $(\beta_u,\ u \geq 0)$ following Bochner (1955), that is defining: $X_t = \beta_{\tau_t}$. Then, for each t, one has: $X_t \overset{\text{(law)}}{=} \sqrt{\tau_t}N$.

* 4.19 Unilateral stable distributions (1)

In this exercise, Z_a denotes a gamma variable with parameter $a > 0$, and, for $0 < \mu < 1$, T_μ denotes a stable unilateral random variable with exponent μ, i.e. the Laplace transform of T_μ is given by:

$$E\left[\exp(-\lambda T_\mu)\right] = \exp(-\lambda^\mu) \qquad (\lambda \geq 0)\,, \tag{4.19.1}$$

1. Prove that, for any $\gamma > 0$, and for any Borel function $f : \mathbb{R}_+ \to \mathbb{R}_+$, the following equality holds:

$$E\left[f\left((Z_{\frac{\gamma}{\mu}})^{\frac{1}{\mu}}\right)\right] = E\left[f\left(\frac{Z_\gamma}{T_\mu}\right)\frac{c}{(T_\mu)^\gamma}\right]\,, \tag{4.19.2}$$

where $c = \mu\frac{\Gamma(\gamma)}{\Gamma(\frac{\gamma}{\mu})}$, and, on the right hand side, Z_γ and T_μ are assumed to be independent. In particular, in the case $\mu = \gamma$, one has:

$$E\left[f(Z^{\frac{1}{\mu}})\right] = E\left[f\left(\frac{Z_\mu}{T_\mu}\right)\frac{\Gamma(\mu+1)}{(T_\mu)^\mu}\right]\,. \tag{4.19.3}$$

where Z denotes a standard exponential variable (with expectation 1).

2. Prove that for $s \in \mathbb{R}$, $s < 1$, one has

$$E\left[(T_\mu)^{\mu s}\right] = \frac{\Gamma(1-s)}{\Gamma(1-\mu s)} . \tag{4.19.4}$$

Comments and references:

(a) Formula (4.19.2) may be found in Theorem 1 of:

D.N. SHANBHAG AND M. SREEHARI: An extension of Goldie's result and further results in infinite divisibility. *Z. für Wahr*, **47**, 19–25 (1979).

Some extensions have been established in:

W. JEDIDI: Stable processes: mixing and distributional properties. *Prépublication du Laboratoire de Probabilités et Modèles Aléatoires* (2000).

(b) The (s) variable considered in Exercise **4.2** satisfies $E[\exp(-\lambda T)] = \exp(-\sqrt{2\lambda})$. Hence, we have: $\frac{1}{2}T \overset{\text{(law)}}{=} T_{\frac{1}{2}}$, with the above notations.

(c) Except for the case $\mu = \frac{1}{2}$ (see the discussion in Exercise **4.20**), there is no simple expression for the density of T_μ. However, see:

H. POLLARD: The representation of e^{-x^λ} as a Laplace integral. *Bull. Amer. Math. Soc.*, **52**, 908–910 (1946).

V.M. ZOLOTAREV: On the representation of the densities of stable laws by special functions. (In Russian.) *Teor. Veroyatnost. i Primenen.*, **39**, no. 2, 429–437 (1994); translation in *Theory Probab. Appl.* **39**, no. 2, 354–362 (1994).

* 4.20 Unilateral stable distributions (2)

(We keep the notation of Exercise **4.19**.)

0. *Preliminary.* The aim of this exercise is to study the following relation between the laws of two random variables X and Y taking values in $(0, \infty)$: for every Borel function $f : (0, \infty) \to \mathbb{R}_+$,

$$E[f(X)] = E\left[f\left(\frac{1}{Y}\right)\frac{c}{Y^\mu}\right] \tag{4.20.1}$$

where $\mu \geq 0$, and c is the constant such that: $E\left[\frac{c}{Y^\mu}\right] = 1$.

In the case $\mu = 0$, one has obviously $c = 1$, and (4.20.1) means that $X \overset{\text{(law)}}{=} \frac{1}{Y}$; moreover, in the general case $\mu > 0$, replacing X and Y by some appropriate powers, we may assume that $0 < \mu < 1$.

1. Prove that the following properties are equivalent:

    (i) X and Y satisfy (4.20.1);

    (ii) $P(Z_\mu X \in dt) = \dfrac{c}{\Gamma(\mu)} t^{\mu-1} \varphi(t) dt$,

    where Z_μ and X are assumed to be independent, and $\varphi(t) = E[\exp(-tY)]$;

    (iii) $P\left(Z^{\frac{1}{\mu}} X \in du\right) = c\mu u^{\mu-1} \varphi_\mu(u) du$,

    where $\varphi_\mu(u) = E\left[\exp -(uY)^\mu\right]$, and X and Z are assumed to be independent.

2. Explain the equivalence between (ii) and (iii) using the result of Exercise **4.19**.

Comments and references. The scaling property of Brownian motion yields many illustrations of formula (4.20.1). For $\mu = 0$, a classical example is the following identity in law between B_1, the Brownian motion taken at time 1 and T_1^B, the first hitting time by this process of the level 1:

$$T_1^B \overset{\text{(law)}}{=} \frac{1}{B_1^2}.$$

For $\mu = \frac{1}{2}$ an example is given by the relationship between the law of the local time for the Brownian bridge and the inverse local time of Brownian motion. We refer to:

PH. BIANE, J.F. LE GALL AND M. YOR: Un processus qui ressemble au pont brownien. *Séminaire de Probabilités XXI*, 270–275, *Lecture Notes in Mathematics*, **1247**, Springer, Berlin, 1987.

J. PITMAN AND M. YOR: Arcsine laws and interval partitions derived from a stable subordinator. *Proc. London Math. Soc.*, (3) **65**, no. 2, 326–356 (1992).

** 4.21 Unilateral stable distributions (3)

(We keep the notation of Exercise **4.19**.)

1. Let $0 < \mu \leq 1$ and $0 < \nu \leq 1$. Prove the double identity in law

$$\frac{Z^{\frac{1}{\mu}}}{T_\nu} \overset{\text{(law)}}{=} \frac{Z^{\frac{1}{\nu}}}{T_\mu} \overset{\text{(law)}}{=} \frac{Z}{T_\nu T_\mu} \tag{4.21.1}$$

where the variables Z, T_μ and T_ν are assumed to be independent, and we make the convention that $T_1 = 1$ a.s. Prove that (4.21.1) is equivalent to

$$Z^{\frac{1}{\mu}} \overset{\text{(law)}}{=} \frac{Z}{T_\mu}. \tag{4.21.2}$$

2. Recall the identity in law (4.5.4) obtained in Exercise **4.5**:

$$(Z_{na})^n \overset{\text{(law)}}{=} n^n Z_a Z_{a+\frac{1}{n}} \cdots Z_{a+\frac{n-1}{n}} .$$

Combining this identity in law for an appropriate choice of a, with (4.21.2), prove that for $n \in \mathbb{N}^*$, one has:

$$T_{\frac{1}{n}} \overset{\text{(law)}}{=} \left(n^n (Z_{\frac{1}{n}} \cdots Z_{\frac{n-1}{n}}) \right)^{-1} . \tag{4.21.3}$$

3. Let $Z_{a,b}$ denote a beta variable with parameters a and b, as defined in Exercise **4.2**, and recall the identity in law (4.2.1):

$$Z_a \overset{\text{(law)}}{=} Z_{a,b} Z_{a+b} ,$$

obtained in this exercise (on the right hand side, Z_{a+b} and $Z_{a,b}$ are independent). Deduce from this identity that for all $n, m \in \mathbb{N}$, such that $m < n$, we have (with the convention, for $m = 1$, that $\prod_{k=1}^0 = 1$):

$$\left(\frac{m}{T_{\frac{m}{n}}} \right)^m \overset{\text{(law)}}{=} n^n \left(\prod_{k=1}^{m-1} Z_{\frac{k}{n}, k(\frac{1}{m} - \frac{1}{n})} \right) \left(\prod_{k=m}^{n-1} Z_{\frac{k}{n}} \right) , \tag{4.21.4}$$

where the r.v.s on the right hand side are independent.

4. Using identity (4.8.2) in Exercise **4.8**, prove that for any $p \geq 1$:

$$T_{2^{-p}} \overset{\text{(law)}}{=} \left(2^{2^p - 1} (N_1^2 N_2^{2^2} \cdots N_p^{2^p}) \right)^{-1} , \tag{4.21.5}$$

where $N_1, \ldots, N_p$ are independent centered Gaussian variables with variance 1.

5. Applications:

(a) Deduce from above the formulae:

$$P(T_{\frac{1}{2}} \in dt) = \frac{dt\, e^{-\frac{1}{4t}}}{\sqrt{4\pi t^3}} , \tag{4.21.6}$$

$$P((T_{\frac{1}{3}})^{-1} \in dt) = \frac{dt}{\Gamma(\frac{1}{3})\Gamma(\frac{2}{3})} \left(\frac{2}{3^{\frac{2}{3}} \sqrt{t}} \right) K_{\frac{1}{3}} \left(\frac{2}{3} \sqrt{\frac{t}{3}} \right) , \tag{4.21.7}$$

where K_ν is the MacDonald function with index ν, which admits the integral representation (see e.g. Lebedev [51]):

$$K_\nu(x) = \frac{1}{2} \left(\frac{x}{2} \right)^\nu \int_0^\infty \frac{dt}{t^{\nu+1}} \exp - \left(t + \frac{x^2}{4t} \right) .$$

(b) From the representation (4.21.5), deduce the formula:

$$P \left(\frac{1}{4(T_{\frac{1}{4}})^{\frac{1}{3}}} > y \right) = \frac{1}{\pi} \int_0^1 \frac{da}{\sqrt{a(1-a)}} \exp \left(-\frac{y}{a^{\frac{1}{3}} (1-a)^{\frac{2}{3}}} \right) . \tag{4.21.8}$$

(c) Set $T = \frac{4}{27}\left(T_{\frac{2}{3}}\right)^{-2}$, prove that

$$P(T \in dx) = \frac{C\,dx}{x^{\frac{1}{3}}} \int_0^1 dz\, \frac{\exp\left(-\frac{x}{z}\right)}{z^{\frac{4}{3}}(1-z)^{\frac{5}{6}}}, \qquad (4.21.9)$$

with $C = \left(\Gamma\left(\frac{2}{3}\right) B\left(\frac{1}{3}, \frac{1}{6}\right)\right)^{-1}$.

Comments and references:

(a) Formula (4.21.2) is found in:

D.N. SHANBHAG AND M. SREEHARI: On certain self-decomposable distributions. *Z. für Wahr*, **38**, 217–222 (1977).

These authors later conveniently extended this formula to gamma variables, as is presented in Exercise **4.19**.

As an interpretation of formula (4.21.2) in terms of stochastic processes, we mention:

$$l_S^{(\alpha)} \overset{\text{(law)}}{=} S^\alpha \frac{1}{(T_\alpha)^\alpha},$$

where $l_S^{(\alpha)}$ is the local time of the Bessel process with dimension $2(1 - \alpha)$, considered at the independent exponential time S. This relation is due to the scaling property of Bessel processes; in fact, one has:

$$l_1^{(\alpha)} \overset{\text{(law)}}{=} \frac{1}{(T_\alpha)^\alpha}.$$

Here, $l_1^{(\alpha)}$ is the local time taken at time 1. Its distribution is known as the Mittag–Leffler distribution with parameter α, whose Laplace transform is given by:

$$E[\exp(x l_1^{(\alpha)})] = E[\exp(x(T_\alpha)^{-\alpha})] = \sum_{n=0}^\infty \frac{x^n}{\Gamma(\alpha n + 1)}.$$

For more details about Mittag–Leffler distributions, we refer to

S.A. MOLCHANOV AND E. OSTROVSKII: Symmetric stable processes as traces of degenerate diffusion processes. *Teor. Verojatnost. i Primenen.*, **14**, 127–130 (1969)

D.A. DARLING AND M. KAC: On occupation times for Markoff processes. *Trans. Amer. Math. Soc.*, **84**, 444–458 (1957)

R. N. PILLAI: On Mittag–Leffler functions and related distributions. *Ann. Inst. Statist. Math.*, **42**, no. 1, 157–161 (1990).

(b) Formulae (4.21.3), (4.21.5) and their applications, (4.21.7) and (4.21.8) are found in V.M. Zolotarev [103].

* 4.22 A probabilistic translation of Selberg's integral formulae

Preliminary. A particular case of the celebrated Selberg's integral formulae (see below for references) is the following:

$$\text{for every } \gamma > 0, \quad E[|\Delta_n(X_1,\ldots,X_n)|^{2\gamma}] = \prod_{j=2}^{n} \frac{\Gamma(1+j\gamma)}{\Gamma(1+\gamma)}, \qquad (4.22.1)$$

where $X_1,\ldots,X_n$ are independent Gaussian variables, centered, with variance 1, and $\Delta_n(x_1,\ldots,x_n) = \prod_{j<l\leq n}(x_j - x_l)$ is the Vandermonde determinant of the vector $(x_i)_{i\leq n}$. See Exercise 3, Chapter 8, p. 440 of [1].
   Prove that

$$\text{(i)} \quad (\Delta_n(X_1,\ldots,X_n))^2 \stackrel{\text{(law)}}{=} \prod_{j=2}^{n} \frac{1}{T_{\frac{1}{j}}} \quad \text{and} \quad \text{(ii)} \quad \prod_{j=2}^{n} \frac{1}{T_{\frac{1}{j}}} \stackrel{\text{(law)}}{=} \prod_{j=2}^{n}(j^j) \prod_{1\leq i<j} Z_{\frac{i}{j}},$$
$$(4.22.2)$$

where the variables $T_{\frac{1}{j}}$, featured in Exercise **4.21** are independent, and the $(Z_{\frac{i}{j}})$, $1 \leq i < j \leq n$ are independent gamma variables with respective parameters $\frac{i}{j}$.

Hint. To prove (4.22.2) (i), resp. (4.22.2) (ii), use (4.21.2), resp. (4.21.3).

Comments and references:

(a) A discussion of Selberg integral formulae is found in Chapter 17 of

   M. MEHTA: *Random Matrices and the Statistical Theory of Energy Levels.* Second Edition. Academic Press, 1991.

(b) Some extensions of formulas (4.22.1) and (4.22.2) can be found in

   I. LU AND D. RICHARDS: Random discriminants. *Ann. Statist.*, **21**, no. 4, 1982–2000 (1993).

(c) It would be most interesting to be able to obtain an a.s. identity in (i) above, instead of an identity in law, with the T_k's constructed from $X_1,\ldots,X_n$.

** 4.23 Mellin and Stieltjes transforms of stable variables

Let $0 < \mu < 1$ and let T_μ and T'_μ be two independent, unilateral, stable variables with exponent μ, as defined in (4.19.1). Define $X = \dfrac{T_\mu}{T'_\mu}$.

1. Prove the formula:

$$E\left[\frac{1}{1+sX}\right] = \frac{1}{1+s^\mu} \qquad (s \geq 0).$$ (4.23.1)

2. Prove the formula:

$$E[X^s] = \frac{\sin \pi s}{\mu \sin \frac{\pi s}{\mu}}, \quad 0 < s < \mu.$$ (4.23.2)

(X^s indicates, simply, X raised to the power s.)

3. Show that the density of X^μ is given by:

$$P(X^\mu \in dy) = \frac{\sin(\pi\mu)}{\pi\mu} \frac{dy}{y^2 + 2y\cos(\pi\mu) + 1} \qquad (y \geq 0),$$ (4.23.3)

or equivalently,

$$\left(\frac{T_\mu}{T'_\mu}\right)^\mu \overset{\text{(law)}}{=} (C_\mu \,|\, C_\mu > 0),$$ (4.23.4)

where C denotes a standard Cauchy variable and $C_\mu = \sin(\pi\mu)C - \cos(\pi\mu) \overset{\text{(law)}}{=} \frac{\sin(\pi\mu-\Theta)}{\sin(\Theta)}$, and Θ is uniform on $[0, 2\pi)$.

Hint. Using residue calculus, show that $\frac{1}{\pi} \int_{\mathbb{R}} e^{isx} \frac{\sin \pi\mu}{\cosh x + \cos \pi\mu} \, dx = \frac{\sinh \pi\mu s}{\sinh \pi s}$.

Comments and references:

(a) Formula (4.23.3) is very simple and striking, since it shows that the density of the ratio of two independent, unilateral, stable variables can be made explicit, whereas there is no such explicit formula for the density of T_μ, except in the particular cases which are considered in Exercise **4.21**.

When $\mu = \frac{1}{2}$, the identity

$$\frac{T_{\frac{1}{2}}}{T_{\frac{1}{2}} + T'_{\frac{1}{2}}} \overset{\text{(law)}}{=} \frac{1}{1+C^2} \overset{\text{(law)}}{=} \cos^2(\Theta),$$

where $T_{\frac{1}{2}}$ and $T'_{\frac{1}{2}}$ are independent, C is a Cauchy r.v. and Θ is uniformly distributed on $[0, 2\pi)$, plays some role in the derivation by:

P. LÉVY: Sur certains processus stochastiques homogènes. *Compositio Math.*, **7**, 283–339 (1939)

of the arcsine law for $\int_0^1 ds \, \mathbb{I}_{\{B_s > 0\}}$, where B is the standard Brownian motion. His results have been extended to some multidimensional cases in:

M. BARLOW, J. PITMAN AND M. YOR: Une extension multidimensionnelle de la loi de l'arcsinus. *Séminaire de Probabilités XXIII*, 294–312, *Lecture Notes in Mathematics*, **1372**, Springer (1989).

(b) Formula (4.23.3) is found in both:

V.M. ZOLOTAREV: Mellin–Stieltjes transforms in probability theory. *Teor. Veroyatnost. i Primenen.*, **2**, 444–469 (1957)

and

J. LAMPERTI: An occupation time theorem for a class of stochastic processes. *Trans. Amer. Math. Soc.*, **88**, 380–387 (1958).

(c) The same law occurs in free probability, with the free convolution of two free stable variables. See Biane's appendix (Proposition A4.4) in:

H. BERCOVICI AND V. PATA: Stable laws and domains of attraction in free probability theory. With an appendix by Philippe Biane. *Ann. of Math.*, (2) **149**, no. 3, 1023–1060 (1999).

(d) See Exercise **5.11** for some discussion of the asymptotic distribution for T_μ, as $\mu \to 0$, using partly (4.23.3).

** 4.24 Solving certain moment problems via simplification

Part A. Let $(\mu(n),\ n \in \mathbb{N})$ be the sequence of moments of a positive random variable V, which is moments determinate.

Let $(\nu(n),\ n \in \mathbb{N})$ be the sequence of moments of a positive random variable W, which is simplifiable.

Assume further that $V \overset{\text{(law)}}{=} WR$, for R a positive r.v., independent of W.

Prove that there exists only one moments determinate distribution $\theta(dx)$ on $\mathbb{R}_+$, such that:
$$\int_0^\infty x^n\, \theta(dx) = \frac{\mu(n)}{\nu(n)}, \qquad n \in \mathbb{N}, \tag{4.24.1}$$
and that θ is the distribution of R.

Part B. (Examples.) Identify R when:

(i) $\mu(n) = (n!)^2$ and $\nu(n) = \Gamma(n+\alpha)\Gamma(n+\beta)$, $\alpha, \beta > 1$;

(ii) $\mu(n) = (2n)!$ and $\nu(n) = n!$;

(iii) $\mu(n) = (3n)!$ and $\nu(n) = (2n)!$.

Comments and references:

(a) This exercise combines some of our previous discussions involving moment problems, simplifications, beta–gamma algebra, etc.

(b) This exercise has been strongly motivated from a discussion with K. Penson, in particular about the paper:

J.R. KLAUDER, K.A. PENSON AND J.M. SIXDENIERS: Constructing coherent states through solutions of Stieltjes and Hausdorff moment problems. *Phys. Review A.*, **64**, 1–18 (2001).

Solutions for Chapter 4

Solution to Exercise 4.1

Part A.

1. On the one hand, we have

$$E[e_a(X)e_b(X)] = E[\exp{(a+b)X}]\exp{-\frac{1}{2}(a^2+b^2)} = \exp{ab} = \sum_{n\geq 0}\frac{(ab)^n}{n!}\,, \quad a,b \in \mathbb{C}\,.$$

On the other hand,

$$E[e_a(X)e_b(X)] = \sum_{n,m}\frac{a^n b^m}{n!m!}E[h_n(X)h_m(X)]\,, \quad a,b \in \mathbb{C}\,.$$

This shows that $E[h_n(X)h_m(X)] = 0$, if $n \neq m$ and $E[h_n(X)h_m(X)] = n!$, if $n = m$.

2. From the expression of the Laplace transform of Y, we deduce

$$E[\exp(a(X+cY))\,|\,X] = \exp(aX)E[\exp acY] = e_a(X,-c^2)\,.$$

On the other hand, we have

$$E[\exp(a(X+cY))\,|\,X] = \sum_{n=0}^{\infty}\frac{a^n}{n!}E[(X+cY)^n\,|\,X]\,,$$

so $E[(X+cY)^k\,|\,X] = \frac{\partial^k}{\partial a^k}e_a(X,-c^2)|_{a=0} = h_k(X,-c^2)$.

Part B.

1. First, note that $\varepsilon = \mathrm{sgn}(X)$ is a symmetric Bernoulli r.v. which is independent of (X^2, Y), therefore it is independent of (T,g), and we only need to compute the law of this pair of variables. Let f be a bounded measurable function on $\mathbb{R}_+ \times [0,1]$, then

$$E[f(T,g)] = \frac{2}{\pi}\int_{\mathbb{R}_+^2} f\left(\frac{1}{2}(x^2+y^2), \frac{x^2}{x^2+y^2}\right)e^{-(x^2+y^2)/2}\,dx\,dy\,.$$

The change of variables:

$$u = (x^2 + y^2)/2 \qquad x = \sqrt{2uv}$$
$$v = x^2/(x^2 + y^2) \qquad y = \sqrt{2u - 2uv},$$

gives

$$E[f(T, g)] = \int_{\mathbb{R}_+ \times [0,1]} f(u, v) \frac{e^{-u}}{\pi} \frac{\mathbb{I}_{\{v \in]0,1[\}}}{\sqrt{v(1-v)}} \, du \, dv.$$

Therefore, the density of the law of (T, g) on $\mathbb{R}^2$ is given by

$$e^{-u} \mathbb{I}_{\{u \in [0,\infty]\}} \frac{1}{\pi \sqrt{v(1-v)}} \mathbb{I}_{\{v \in]0,1[\}}, \qquad u, v \in \mathbb{R}.$$

This expression shows that T and g are independent and since ε is independent of the pair (T, g), it follows that ε, T and g are independent. From above, the law of $\sqrt{2T}$ has density $t \exp -\frac{1}{2}t^2 \mathbb{I}_{\{t \in [0,\infty[\}}$ (so T is exponentially distributed), and g is arcsine distributed.

2. (i) Observing that $X = \mu\sqrt{2T}$, and using the previous question, we can write

$$E[\exp(aX) \,|\, \mu] = E[\exp(a\mu\sqrt{2T}) \,|\, \mu]$$
$$= \int_0^\infty \exp(at\mu) t \exp\left(-\frac{1}{2}t^2\right) dt$$
$$= \varphi(a\mu),$$

where φ is given by $\varphi(z) = E[\exp(z\sqrt{2T})]$, for any $z \in \mathbb{C}$. For any real x, we have

$$\varphi(x) = \int_0^\infty t e^{xt - \frac{t^2}{2}} \, dt = 1 + x \int_0^\infty e^{xt - \frac{t^2}{2}} \, dt$$
$$= 1 + x \exp\left(\frac{x^2}{2}\right) \int_0^\infty e^{-\frac{1}{2}(t-x)^2} \, dt$$
$$= 1 + x \exp\left(\frac{x^2}{2}\right) \int_{-\infty}^x e^{-\frac{y^2}{2}} \, dy,$$

which gives the expression of φ on $\mathbb{C}$, by analytical continuation.

The formulas given in (ii) and (iii) simply follow from the development of $\cosh(aX) = \frac{1}{2}\left(e^{aX} + e^{-aX}\right)$.

3. Since sinh is an odd function, from the definition of $e_a(X)$, we can write

$$\exp\left(-\frac{a^2}{2}\right) E[\sinh(aX) \,|\, \mu] = \sum_{n \geq 0} \frac{a^{2n+1}}{(2n+1)!} E[h_{2n+1}(X) \,|\, \mu].$$

On the other hand, from question 2, we have

$$\exp\left(-\frac{a^2}{2}\right) E[\sinh(aX)\,|\,\mu] = \sqrt{\frac{\pi}{2}}\,a\mu\,\exp\left(\frac{a^2(\mu^2-1)}{2}\right)$$

$$= \sqrt{\frac{\pi}{2}}\,\mu \sum_{n\geq 0} \frac{1}{2^n n!} a^{2n+1}(\mu^2-1)^n\,.$$

Equating the two above expressions, we obtain

$$E[h_{2p+1}(X)\,|\,\mu] = \sqrt{\frac{\pi}{2}}\frac{(2p+1)!}{2^p p!}\mu(\mu^2-1)^p\,.$$

From similar arguments applied to $\cosh(aX)$, we obtain

$$\sum_{n\geq 0} \frac{a^{2n}}{(2n)!} E[h_{2n}(X)\,|\,\mu]$$

$$= \exp\left(-\frac{a^2}{2}\right)\left[1 + a^2\mu^2 \int_0^1 dy\, \exp\left(\frac{a^2\mu^2}{2}(1-y^2)\right)\right]$$

$$= 1 + \sum_{n\geq 1} \frac{(-1)^n a^{2n}}{2^n n!} + \frac{\mu^2}{2^{n-1}(n-1)!} \int_0^1 dy\,(\mu^2(1-y^2)-1)^{n-1} a^{2n}\,,$$

which gives, for $p \geq 1$,

$$E[h_{2p}(X)\,|\,\mu] = \frac{(-1)^p(2p)!}{2^p p!} + \frac{(2p)!\mu^2}{2^{p-1}(p-1)!} \int_0^1 dy\,(\mu^2(1-y^2)-1)^{p-1}\,.$$

4. The r.v.s μ and ν cannot be independent since they are related by $\mu^2 + \nu^2 = 1$.

Set $\varepsilon' = \mathrm{sgn}(Y)$ then it is obvious that T, g, ε and ε' are independent. Moreover note that $\nu = \varepsilon'\sqrt{g}$, so $\sqrt{2T}$ is independent of the pair of variables (μ,ν), and

$$E[\exp(aX+bY)\,|\,\mu,\nu] = E[\exp(a\mu\sqrt{2T}+b\nu\sqrt{2T})\,|\,\mu,\nu]$$

$$= \int_0^\infty \exp\left(at\mu+bt\nu\right) t \exp\left(-\frac{1}{2}t^2\right) dt$$

$$= \varphi(a\mu+b\nu)\,.$$

Solution to Exercise 4.2

1. We obtain this identity in law by writing for any bounded Borel function f defined on $\mathbb{R}_+^2$,

$$E[f(Z_{a,b}Z_{a+b}, (1-Z_{a,b})Z_{a+b})]$$

$$= \int_0^1 \int_0^\infty dx\,dy\, f(xy, (1-\dot{x})y) \frac{1}{\beta(a,b)} x^{a-1}(1-x)^{b-1} \frac{1}{\Gamma(a+b)} y^{a+b-1} e^{-y}$$

$$= \frac{1}{\Gamma(a+b)\beta(a,b)} \int_0^\infty \int_0^\infty dz\,dt\, f(z,t) z^{a-1} e^{-z} t^{b-1} e^{-t}\,,$$

where the second equality has been obtained with an obvious change of variables. Note that the above calculation (with $f \equiv 1$) allows us to deduce the formula

$$\Gamma(a)\Gamma(b) = \Gamma(a+b)\beta(a,b).$$

2. Let Z_a, Z_{a+b}, Z_{a+b+c}, $Z_{a,b}$, $Z_{a,b+c}$, and $Z_{a+b,c}$ be independent variables. Since Z_{a+b+c} satisfies the hypothesis of Exercise **1.13**, the identity in law we want to prove is equivalent to

$$Z_{a+b+c}Z_{a,b+c} \overset{\text{(law)}}{=} Z_{a+b+c}Z_{a,b}Z_{a+b,c}.$$

Applying question 1 to the products $Z_{a+b+c}Z_{a,b+c}$ and $Z_{a+b+c}Z_{a+b,c}$, we see that the above identity also is equivalent to

$$Z_a \overset{\text{(law)}}{=} Z_{a+b}Z_{a,b},$$

which has been proved in question 1.

3. A classical computation shows that for any i, $N_i^2 \overset{\text{(law)}}{=} 2Z_{1/2}$. The identity in law $R_k^2 \overset{\text{(law)}}{=} 2Z_{k/2}$ follows from the expression of the Laplace transform of the gamma variable with parameter a, i.e. $E[\exp -\lambda Z_a] = \left(\frac{1}{\lambda+1}\right)^a$, $\lambda > 0$.

The law of a uniformly distributed vector $\underline{x}_k$ on the unit sphere S_{k-1} of $\mathbb{R}^k$ is characterized by the identities $|\underline{x}_k| = 1$ and $\underline{x}_k \overset{\text{(law)}}{=} \rho\underline{x}_k$ for every rotation matrix ρ in $\mathbb{R}^k$. Since such matrices are orthogonal, from Exercise **3.8** it holds that $(N_1,\ldots,N_k) \overset{\text{(law)}}{=} \rho(N_1,\ldots,N_k)$. Hence for every rotation matrix ρ, one has

$$\left(\frac{1}{R_k}(N_1,\ldots,N_k), R_k\right) \overset{\text{(law)}}{=} \left(\rho\frac{1}{R_k}(N_1,\ldots,N_k), R_k\right)$$

and since $|\frac{1}{R_k}(N_1,\ldots,N_k)| = 1$, we conclude that $\frac{1}{R_k}(N_1,\ldots,N_k)$ is uniformly distributed on S_{k-1} and independent of R_k.

4. Let $(N_i)_{i\geq 1}$ be an infinite sequence of independent centered Gaussian r.v.s with variance 1. It follows from question 3 that for any $k \leq n$,

$$\sqrt{n}(x_1^{(n)},\ldots,x_k^{(n)}) \overset{\text{(law)}}{=} \frac{\sqrt{n}}{R_n}(N_1,\ldots,N_k),$$

where R_n is the norm of $(N_1,\ldots,N_k,\ldots,N_n)$. From question 3 (i) and the law of large numbers, $\frac{\sqrt{n}}{R_n}$ converges almost surely to $\sqrt{1/E[2Z_{1/2}]} = 1$, which proves the result.

5. From question 3 (ii), $R_k^2 \sum_{i=1}^{p} x_i^2 = \sum_{i=1}^{p} N_i^2$ and R_k is independent of $\sum_{i=1}^{p} x_i^2$. Moreover, we know from question 3 (i) that $\sum_{i=1}^{p} N_i^2$ is distributed as $2Z_{\frac{p}{2}}$ and that $\frac{1}{R_k^2}$ is distributed as $\frac{1}{2}Z_{\frac{k}{2}}^{-1}$. So we have $Z_{\frac{p}{2}} \overset{\text{(law)}}{=} \sum_{i=1}^{p} x_i^2 Z_{\frac{k}{2}}$. Applying the result of

question 2 we also obtain $Z_{\frac{p}{2}} \overset{\text{(law)}}{=} Z_{\frac{p}{2}, \frac{k-p}{2}} Z_{\frac{k}{2}} \overset{\text{(law)}}{=} \sum_{i=1}^{p} x_i^2 Z_{\frac{k}{2}}$. And we conclude thanks to the fact that $Z_{\frac{k}{2}}$ is simplifiable.

6. From above, we have $(\sum_{i=1}^{k-2} x_i^2)^{k/2-1} \overset{\text{(law)}}{=} (Z_{k/2-1,1})^{k/2-1}$ and we easily check from the above definition of beta variables that for any $a > 0$, $Z_{a,1}^a$ is uniformly distributed on $[0,1]$.

7. (a) By the change of variables $t = 1/x^2$, for any bounded Borel function f, we have

$$E\left[f\left(\frac{1}{N^2}\right)\right] = \sqrt{\frac{2}{\pi}} \int_0^\infty f\left(\frac{1}{x^2}\right) e^{-\frac{x^2}{2}} dx = \frac{1}{\sqrt{2\pi}} \int_0^\infty f(t) e^{-\frac{1}{2t}} \frac{dt}{\sqrt{t^3}} = E[f(T)].$$

To obtain the other identity in law, set $F(\alpha) = E[\exp(-\frac{\alpha^2}{2T})]$. From above, $F(\alpha) = \sqrt{\frac{2}{\pi}} \int_0^\infty e^{-\frac{\alpha^2}{2x^2}} e^{-\frac{x^2}{2}} dx$, hence one has $F'(\alpha) = -\alpha \sqrt{\frac{2}{\pi}} \int_0^\infty e^{-\frac{\alpha^2}{2x^2}} e^{-\frac{x^2}{2}} \frac{dx}{x^2}$. The change of variables $u = \frac{\alpha}{x}$ gives $F'(\alpha) = -\sqrt{\frac{2}{\pi}} \int_0^\infty e^{-\frac{u^2}{2}} e^{-\frac{\alpha^2}{2u^2}} du = -F(\alpha)$. Therefore, $F(\alpha) = ce^{-\alpha}$, where c is some constant. Since $F(0) = 1$, we conclude that $F(\alpha) = e^{-\alpha}$.

(b) This simply follows from the expression of the Laplace transform of the (s) distribution: $E\left[\exp\left(-\frac{1}{2}\alpha^2 \sum_{i=1}^k p_i^2 T_i\right)\right] = \prod_{i=1}^k \exp(-p_i \alpha) = \exp(-\alpha)$.

8. (a) Applying the result of question 7, we obtain that for any $k \geq 1$:

$$T \overset{\text{(law)}}{=} \sum_{i=1}^k \frac{p_i^2}{N_i^2} \overset{\text{(law)}}{=} \frac{1}{R_k^2} \sum_{i=1}^k \frac{p_i^2}{x_i^2},$$

where, on the right hand side, R_k^2 is independent from $(x_1, \ldots, x_k)$. Then we have

$$\frac{1}{R_k^2} \sum_{i=1}^k \frac{p_i^2}{x_i^2} \overset{\text{(law)}}{=} \frac{1}{R_k^2} \frac{1}{x_1^2}.$$

Hence, since $\frac{1}{R_k^2}$ is a simplifiable variable (use Exercise **1.13**), we obtain:

$$\sum_{i=1}^k \frac{p_i^2}{x_i^2} \overset{\text{(law)}}{=} \frac{1}{x_1^2} \overset{\text{(law)}}{=} \frac{1}{x_j^2},$$

for any $j \leq k$.

(b) Now we give a more computational proof which also yields the density of $\sum_{i=1}^k \frac{p_i^2}{x_i^2}$. The calculation of the characteristic functions of $\log \frac{1}{R_k^2}$ and $\log T$ gives

$$E\left[\exp\left(i\lambda \log \frac{1}{R_k^2}\right)\right] = E[(2Z_{k/2})^{-i\lambda}] = 2^{-i\lambda} \frac{\Gamma(k/2 - i\lambda)}{\Gamma(k/2)} \quad \text{and}$$

$$E[e^{i\lambda \log T}] = E[(N^2)^{-i\lambda}] = 2 \int_0^\infty (x^2)^{-i\lambda} \frac{e^{-x^2/2}}{\sqrt{2\pi}} dx = \frac{2^{-i\lambda}}{\sqrt{\pi}} \int_0^\infty s^{-i\lambda-1/2} e^{-s} ds$$

$$= 2^{-i\lambda} \Gamma(1/2 - i\lambda).$$

So that for all $\lambda \in \mathbb{R}$,

$$E\Big[\exp i\lambda \log \sum_{j=1}^{k} \frac{p_j^2}{x_j^2}\Big] = \frac{\Gamma(\frac{1}{2} - i\lambda)\Gamma(\frac{k}{2})}{\Gamma(\frac{k}{2} - i\lambda)} = \frac{\Gamma(k/2)}{\Gamma(k/2 - 1/2)} \beta(\tfrac{1}{2} - i\lambda, \tfrac{k}{2} - \tfrac{1}{2})$$

$$= \frac{\Gamma(k/2)}{\Gamma(k/2 - 1/2)} \int_0^1 \exp\Big(i\lambda \log \frac{1}{t}\Big) \frac{(1-t)^{\frac{k-3}{2}}}{t^{1/2}} \, dt$$

$$= \frac{k-2}{2} \int_1^\infty \exp(i\lambda \log x) \frac{(x-1)^{\frac{k-3}{2}}}{x^{k/2}} \, dx \, .$$

In conclusion, the law of the r.v. $\sum_{i=1}^{k} \frac{p_i^2}{x_i^2}$ has density $\frac{k-2}{2} \frac{(x-1)^{\frac{k-3}{2}}}{x^{k/2}} \mathbb{1}_{[1,\infty)}(x)$.

9. Since $\sum_{i=1}^{k} \big(Z_{a_i} / \sum_{j=1}^{k} Z_{a_j} \big) = 1$, it suffices to show the independence between $\sum_{i=1}^{k} Z_{a_i}$ and the vector $\Big(Z_{a_i} / \sum_{j=1}^{k} Z_{a_j} : i \le k-1 \Big)$.

Let $f : \mathbb{R}_+ \mapsto \mathbb{R}$ and $g : \mathbb{R}_+^{k-1} \mapsto \mathbb{R}$ be Borel and bounded. From the change of variables,

$$\begin{cases} y_1 = \frac{x_1}{y_k} \\ \cdot \\ \cdot \\ \cdot \\ y_{k-1} = \frac{x_{k-1}}{y_k} \\ y_k = x_1 + \cdots + x_k \end{cases} \quad \text{or equivalently} \quad \begin{cases} x_1 = y_1 y_k \\ \cdot \\ \cdot \\ \cdot \\ x_{k-1} = y_{k-1} y_k \\ x_k = y_k - \sum_{i=1}^{k-1} y_i y_k \end{cases}$$

and putting $Z'_{a_i} = Z_{a_i} / \sum_{j=1}^{k} Z_{a_j}$, we obtain:

$$E[f(Z_{a_1} + \cdots + Z_{a_k}) g(Z'_{a_1}, \ldots, Z'_{a_{k-1}})]$$

$$= \int_{\mathbb{R}_+^k} f(x_1 + \cdots + x_k) g\Big(\frac{x_1}{x_1 + \cdots + x_k}, \ldots, \frac{x_{k-1}}{x_1 + \cdots + x_k} \Big)$$

$$\times \frac{x_1^{a_1 - 1} \ldots x_k^{a_k - 1}}{\Gamma(a_1) \ldots \Gamma(a_k)} e^{-(x_1 + \cdots + x_k)} \, dx_1 \ldots dx_k$$

$$= \int_{\mathbb{R}_+^k} f(y_k) g(y_1, \ldots, y_{k-1}) (y_1 y_k)^{a_1 - 1} \ldots (y_{k-1} y_k)^{a_{k-1} - 1}$$

$$\times \Big(y_k - \sum_{i=1}^{k-1} y_i y_k \Big)^{a_k - 1} \frac{e^{-y_k} y_k^{k-1}}{\Gamma(a_1) \ldots \Gamma(a_k)} \mathbb{1}_{\{(y_1, \ldots, y_{k-1}) \in [0,1]^{k-1}\}} \, dy_1 \ldots dy_k$$

$$= \int_{\mathbb{R}_+} f(y_k) \frac{y_k^{a_1 + \cdots + a_{k-1} - 1} e^{-y_k}}{\Gamma(a_1 + \cdots + a_k)} \, dy_k \int_{\mathbb{R}_+^{k-1}} g(y_1, \ldots, y_{k-1}) (y_1)^{a_1 - 1} \ldots (y_{k-1})^{a_{k-1} - 1}$$

$$\times \Big(1 - \sum_{i=1}^{k-1} y_i \Big)^{a_k - 1} \frac{\Gamma(a_1 + \cdots + a_k)}{\Gamma(a_1) \ldots \Gamma(a_k)} \mathbb{1}_{\{(y_1, \ldots, y_{k-1}) \in [0,1]^{k-1}\}} \, dy_1 \ldots dy_{k-1} \, ,$$

which shows the required independence. The above identity also gives the density of the law of the vector: $\left(Z_{a_i} / \sum_{j=1}^{k} Z_{a_j} : i \le k - 1 \right)$ on $\mathbb{R}^{k-1}$:

$$\frac{\Gamma(a_1 + \cdots + a_k)}{\Gamma(a_1) \ldots \Gamma(a_k)} (y_1)^{a_1 - 1} \ldots (y_{k-1})^{a_{k-1}-1} \Big(1 - \sum_{i=1}^{k-1} y_i\Big)^{a_k - 1} \mathbb{1}_{\{(y_1, \ldots, y_{k-1}) \in [0,1]^{k-1}\}} ,$$

and this law entirely characterizes the law of the vector $\left(Z_{a_i} / \sum_{j=1}^{k} Z_{a_j} : i \le k \right)$ on the simplex $\sum_k$.

Solution to Exercise 4.3

1. We prove the equivalence only by taking $a = \alpha$ and $b = A$.

2. It suffices to show that the right hand sides of (4.3.2) and (4.3.3) are identical in law, that is:

$$\frac{1}{Z_a} \overset{\text{(law)}}{=} \frac{1}{Z_{a+b}} \left(1 + \frac{Z_b}{Z_a}\right) .$$

But the above identity follows directly from the fundamental beta–gamma identity (4.2.1).

3. We obtain (4.3.3), by adding the first coordinate to the second one multiplied by $\frac{1}{C}$, in each member of (4.3.4).

4. This follows (as in question 2) from the equivalent form of (4.2.1):

$$(Z_{a+b}, Z_b) \overset{\text{(law)}}{=} (Z_{a+b,b} Z_{a+2b}, (1 - Z_{a+b,b}) Z_{a+2b})$$

by trivial algebraic manipulations.

Solution to Exercise 4.4

Both questions 1 and 2 are solved by considering the finite-dimensional variables $(\gamma_{t_1}, \gamma_{t_2}, \ldots, \gamma_{t_k})$, for $t_1, t_2, \ldots, t_k$, and using question 9 of Exercise **4.2**.

3. The independence between $(D_u,\, u \le 1)$ and γ_1 allows us to write:

$$E\left[\exp\left(-\gamma_1 \int_0^1 f(u)\, dD_u\right)\right] = \int_0^\infty dt\, e^{-t} E\left[\exp -t \int_0^1 f(u)\, dD_u\right]$$

$$= E\left[\frac{1}{1 + \int_0^1 f(u)\, dD_u}\right] .$$

Solution to Exercise 4.5

1. It suffices to identify the Mellin transforms of both sides of (4.5.4). (See comment (b) in Exercise **1.14**.)

Observe that for all $a > 0$ and $k \geq 0$, $E[(Z_a)^k] = \frac{\Gamma(k+a)}{\Gamma(a)}$. Therefore, for every $k \geq 0$,

$$E[(n^n Z_a Z_{a+\frac{1}{n}} \ldots Z_{a+\frac{n-1}{n}})^k] = n^{nk} \Pi_{j=0}^{n-1} E\left[\left(Z_{a+\frac{i}{n}}\right)^k\right] = n^{nk} \Pi_{j=0}^{n-1} \frac{\Gamma(k + a + \frac{j}{n})}{\Gamma(a + \frac{j}{n})}$$

$$= \frac{\Gamma(na + nk)}{\Gamma(na)} = E[(Z_{na})^{nk}],$$

where the second to last equality follows from (4.5.1).

2. From question 1 of Exercise **4.2**, we deduce that for all $n \geq 1$ and $a > 0$,

$$(Z_{na})^n \overset{\text{(law)}}{=} (Z_{na+nb})^n (Z_{na,nb})^n.$$

Applying the result of the previous question to Z_{na} and $Z_{n(a+b)}$ gives:

$$Z_a Z_{a+\frac{1}{n}} \ldots Z_{a+\frac{n-1}{n}} \overset{\text{(law)}}{=} Z_{a+b} Z_{a+b+\frac{1}{n}} \ldots Z_{a+b+\frac{n-1}{n}} (Z_{na,nb})^n.$$

Now, from question 1 of Exercise **4.2**, we have $Z_{a+\frac{i}{n}} = Z_{a+\frac{i}{n}+b} Z_{a+\frac{i}{n},b}$, so that

$$\left(Z_{a+b} Z_{a+b+\frac{1}{n}} \ldots Z_{a+b+\frac{n-1}{n}}\right) \left(Z_{a,b} Z_{a+\frac{1}{n},b} \ldots Z_{a+\frac{n-1}{n},b}\right)$$

$$\overset{\text{(law)}}{=} \left(Z_{a+b} Z_{a+b+\frac{1}{n}} \ldots Z_{a+b+\frac{n-1}{n}}\right) (Z_{na,nb})^n.$$

The variable $Z_{a+b} Z_{a+b+\frac{1}{n}} \ldots Z_{a+b+\frac{n-1}{n}}$ being simplifiable, we can apply the result of Exercise **1.13** to get the conclusion. (We might also invoke the injectivity of the Mellin transform.)

Solution to Exercise 4.6

1. From question 1 of Exercise **4.2**, we have:

$$Z_a \overset{\text{(law)}}{=} Z_{a,1} Z_{a+1}, \quad Z_a \overset{\text{(law)}}{=} Z_{a,1} Z_{a+1,1} Z_{a+2}, \ldots,$$

which allows to obtain (4.6.1) by successive iterations of this result.

2. The Laplace transform of $n^{-1} Z_{a+n}$ is given by $E[e^{-\lambda n^{-1} Z_{a+n}}] = \left(1 + \frac{\lambda}{n}\right)^{-(a+n)}$

and we easily check that it converges towards $e^{-\lambda}$ for any $\lambda \in \mathbb{R}_+$. This proves that $n^{-1} Z_{a+n}$ converges in law towards the constant 1 as n goes to $+\infty$.

A less computational proof consists of writing: $Z_{a+n} \overset{\text{(law)}}{=} Z_a + X_1 + \cdots + X_n$, where on the right hand side of this equality, the $n + 1$ r.v.s are independent, each of the X_is having exponential law with parameter 1. Then applying the law of large numbers, we get $\frac{Z_{a+n}}{n} \to 1 = E[X_1]$ in law as $n \to \infty$.

3. As we already noticed in question 6 of Exercise **4.2**, for any n, the r.v. $U^{\frac{1}{a+(n-1)}}$ is distributed as $Z_{a+(n-1),1}$, so from question 1, we have

$$Z_a \overset{\text{(law)}}{=} \left(nU_1^{\frac{1}{a}} \ldots U_n^{\frac{1}{a+(n-1)}}\right) \frac{Z_{a+n}}{n},$$

where Z_{a+n} is independent of $(U_1, \ldots, U_n)$. We deduce the result from this independence and the convergence in law established in question 2.

4. By the same arguments as in question 1, we show that

$$Z_a \overset{\text{(law)}}{=} Z_{a,r} Z_{a+r,r} \ldots Z_{a+(n-1)r,r} Z_{a+nr},$$

and as in question 2, we show that $n^{-1} Z_{a+nr}$ converges in law to the constant r as n goes to $+\infty$. Then we only need to use the same reasoning as in question 3.

Solution to Exercise 4.7

1. From the identities: $\left(\frac{A}{L}, \frac{1-A}{L}\right) \overset{\text{(law)}}{=} \left(\frac{1}{Z_b}, \frac{1}{Z_a}\right)$ and $M = \frac{L}{A(1-A)}$, we deduce

$$(A, M) \overset{\text{(law)}}{=} \left(\frac{Z_a}{Z_a + Z_b}, Z_a + Z_b\right),$$

and we know from question 9 of Exercise **4.2** that the right hand side of the above identity is distributed as $(Z_{a,b}, Z_{a+b})$, these two variables being independent.

2. (a) From $Q_\beta(\Omega) = 1$, we obtain $c_\beta = E[L^\beta]^{-1}$. Since $L = MA(1 - A)$, we have from the independence between M and A: $E[L^\beta] = E[M^\beta]E[A^\beta(1 - A)^\beta]$. Now, let us compute $E[A^\beta(1 - A)^\beta]$: first we have $A^\beta(1 - A)^\beta = \frac{Z_a^\beta Z_b^\beta}{(Z_a + Z_b)^{2\beta}}$. Since we can write $(Z_a Z_b)^\beta$ as $(Z_a Z_b)^\beta = \frac{Z_a^\beta Z_b^\beta}{(Z_a + Z_b)^{2\beta}}(Z_a + Z_b)^{2\beta} = A^\beta(1 - A)^\beta(Z_a + Z_b)^{2\beta}$ and from the independence between $\left(\frac{Z_a}{Z_a + Z_b}, \frac{Z_b}{Z_a + Z_b}\right)$ and $Z_a + Z_b$, established in question 9 of Exercise **4.2**, we have:

$$\begin{aligned}
E[A^\beta(1 - A)^\beta] &= E[(Z_a Z_b)^\beta]E[(Z_a + Z_b)^{2\beta}]^{-1} \\
&= E[Z_a^\beta]E[Z_b^\beta]E[Z_{a+b}^{2\beta}]^{-1} \\
&= \frac{\Gamma(a + \beta)}{\Gamma(a)} \frac{\Gamma(b + \beta)}{\Gamma(b)} \frac{\Gamma(a + b)}{\Gamma(a + b + 2\beta)}.
\end{aligned}$$

Finally, with $E[M^\beta] = \frac{\Gamma(a+b+\beta)}{\Gamma(a+b)}$, we have:

$$c_\beta = \frac{\Gamma(a)\Gamma(b)}{\Gamma(a+\beta)\Gamma(b+\beta)} \frac{\Gamma(a+b+2\beta)}{\Gamma(a+b+\beta)} = \frac{\Gamma(a)\Gamma(b)}{\Gamma(a+b+\beta)B(a+\beta,b+\beta)}.$$

(b) Let us compute the joint Mellin transform of the pair of r.v.s (A, M) under Q_β, that is: $E_{Q_\beta}[A^k M^n]$, for every $k, n \geq 0$.

$$\begin{aligned}
E_{Q_\beta}[A^k M^n] &= c_\beta E[A^k M^n L^\beta] \\
&= c_\beta E[A^{k+\beta}(1-A)^\beta M^{n+\beta}] \\
&= c_\beta E[A^{k+\beta}(1-A)^\beta]E[M^{n+\beta}].
\end{aligned}$$

As in 2 (a), by writing $Z_a^{k+\beta} Z_b^\beta = A^{k+\beta}(1-A)^\beta (Z_a + Z_b)^{k+2\beta}$, we have

$$\begin{aligned}
E[A^{k+\beta}(1-A)^\beta] &= E[Z_a^{k+\beta}]E[Z_b^\beta]E[Z_{a+b}^{k+2\beta}]^{-1} \\
&= \frac{\Gamma(a+k+\beta)}{\Gamma(a)} \frac{\Gamma(b+k+\beta)}{\Gamma(b)} \frac{\Gamma(a+b)}{\Gamma(a+b+k+2\beta)},
\end{aligned}$$

so that with $E[M^{n+\beta}] = \frac{\Gamma(a+b+n+\beta)}{\Gamma(a+b)}$, we obtain:

$$\begin{aligned}
E_{Q_\beta}[A^k M^n] &= c_\beta \frac{\Gamma(a+k+\beta)}{\Gamma(a)} \frac{\Gamma(b+k+\beta)}{\Gamma(b)} \frac{\Gamma(a+b+n+\beta)}{\Gamma(a+b+k+2\beta)} \\
&= \frac{c_\beta}{\Gamma(a)\Gamma(b)} B(a+k+\beta,b+\beta)\Gamma(a+b+n+\beta) \\
&= \frac{B(a+k+\beta,b+\beta)}{B(a+\beta,b+\beta)} \frac{\Gamma(a+b+n+\beta)}{\Gamma(a+b+\beta)} \\
&= E[Z_{a+b,b+\beta}^k]E[Z_{a+b}^n].
\end{aligned}$$

Thus, under Q_β, the r.v. (A, M) has the same joint Mellin transform as the pair $(Z_{a+b,b+\beta}, Z_{a+b})$, hence their laws coincide.

(c) For f be a bounded Borel function, we have

$$\begin{aligned}
E_{Q_\beta}\left[f\left(\frac{A}{L}, \frac{1-A}{L}\right)\right] &= c_\beta E\left[f\left(\frac{1}{Z_b}, \frac{1}{Z_a}\right)L^\beta\right] \\
&= c_\beta E\left[f\left(\frac{1}{Z_b}, \frac{1}{Z_a}\right)\frac{(Z_a Z_b)^\beta}{(Z_a + Z_b)^\beta}\right] \\
&= \frac{c_\beta}{\Gamma(a)\Gamma(b)} \int_{\mathbb{R}_+^2} f\left(\frac{1}{x}, \frac{1}{y}\right)\frac{(xy)^\beta}{(x+y)^\beta}x^{a-1}y^{b-1}e^{-(x+y)}\,dxdy \\
&= \frac{c_\beta}{\Gamma(a)\Gamma(b)} \int_{\mathbb{R}_+^2} f(s,t)(s+t)^{-\beta}s^{-(a+1)}t^{-(b+1)}e^{-\frac{s+t}{st}}\,dsdt,
\end{aligned}$$

which gives the density g of $\left(\frac{A}{L}, \frac{1-A}{L}\right)$ under Q_β:

$$g(s,t) = \frac{c_\beta}{\Gamma(a)\Gamma(b)}(s+t)^{-\beta}s^{-(a+1)}t^{-(b+1)}e^{-\frac{s+t}{st}}\, \mathbb{1}_{\{s\geq 0,\, t\geq 0\}}.$$

Solution to Exercise 4.8

1. From the identity in law (4.5.4), we have

$$Z \overset{\text{(law)}}{=} \sqrt{2Z}\sqrt{2Z_{\frac{1}{2}}},$$

where $Z_{\frac{1}{2}}$ is a gamma r.v. with parameter $1/2$ which is independent of Z. To conclude, it suffices to check that $\sqrt{2Z_{\frac{1}{2}}} \overset{\text{(law)}}{=} |N|$, which is easy.

2. Let $Z_1, Z_2, \ldots, Z_p$ be exponentially distributed r.v.s with parameter 1 such that the $2p$ r.v.s $Z_1, Z_2, \ldots, Z_p, N_1, N_2, \ldots, N_p$ are independent. Put $Z = Z_0$. From above, the identities in law: $Z_{i-1} \overset{\text{(law)}}{=} \sqrt{2Z_i}|N_i|$ hold for any $i = 1, \ldots, p$.

Replacing Z_1 by $\sqrt{2Z_2}|N_2|$ in the identity $Z \overset{\text{(law)}}{=} \sqrt{2Z_1}|N_1|$ and from the independence hypothesis, we obtain: $Z \overset{\text{(law)}}{=} Z_2^{\frac{1}{4}} 2^{\frac{1}{2}+\frac{1}{4}}\left(|N_1||N_2|^{\frac{1}{2}}\right)$. We get the result by successive iterations of this process.

3. It is clear that the term $Z^{\frac{1}{2p}} 2^{\frac{1}{2}+\cdots+\frac{1}{2p}}$ in equation (4.8.2) converges almost surely to 2. Moreover, since $\frac{1}{2}Z^{\frac{1}{2p}} 2^{\frac{1}{2}+\cdots+\frac{1}{2p}}\left(|N_1||N_2|^{\frac{1}{2}} \ldots |N_p|^{\frac{1}{2^{p-1}}}\right) \overset{\text{(law)}}{=} \frac{1}{2}Z$ according to equation (4.8.2), then $\left(|N_1||N_2|^{\frac{1}{2}} \ldots |N_p|^{\frac{1}{2^{p-1}}}\right)$ converges in law to $\frac{1}{2}Z$.

Solution to Exercise 4.9

1. For every $t \geq 0$, we have

$$
\begin{aligned}
P(T_1 + T_2 > t) &= \int_0^\infty \lambda_1 e^{-\lambda_1 s} P(s + T_2 > t)\, ds \\
&= \int_0^t \lambda_1 e^{-\lambda_1 s} e^{-(\lambda_1 - \lambda_2)s}\, ds + \int_t^\infty \lambda_1 e^{-\lambda_1 s}\, ds \\
&= \frac{\lambda_2}{\lambda_2 - \lambda_1} e^{-\lambda_1 t} - \frac{\lambda_1}{\lambda_2 - \lambda_1} e^{-\lambda_2 t} \\
&= \frac{\lambda_2}{\lambda_2 - \lambda_1} P(T_1 > t) - \frac{\lambda_1}{\lambda_2 - \lambda_1} P(T_2 > t).
\end{aligned}
$$

This proves that the probability measure $P(T_1 + T_2 \in dt)$ is a linear combination of the probability measures $P(T_1 \in dt)$ and $P(T_2 \in dt)$ with coefficients $\alpha = \frac{\lambda_2}{\lambda_2 - \lambda_1}$ and $\beta = -\frac{\lambda_1}{\lambda_2 - \lambda_1}$.

2. (a) This follows from the simple calculation:

$$
\begin{aligned}
P(T_1 > T_3) &= \int_{\{s \geq t\}} \lambda_1(\lambda_2 - \lambda_1) e^{-\lambda_1 s} e^{-(\lambda_2 - \lambda_1)t}\, ds\, dt \\
&= \lambda_1(\lambda_2 - \lambda_1) \int_0^\infty e^{-(\lambda_2 - \lambda_1)t} \left(\int_t^\infty e^{-\lambda_1 s}\, ds\right) dt = \frac{\lambda_2 - \lambda_1}{\lambda_2}.
\end{aligned}
$$

(b) Let us first compute the law of (T_1, T_3) conditionally on $(T_1 > T_3)$. Let f be a bounded Borel function defined on $\mathbb{R}^2$, we have from above:

$$E[f(T_1, T_3) \mid T_1 > T_3] = \frac{1}{P(T_1 > T_3)} E[f(T_1, T_3) \mathbb{1}_{\{T_1 > T_3\}}]$$

$$= \int_{\mathbb{R}^2} f(s, t) \lambda_1 \lambda_2 e^{-\lambda_1 s} e^{-(\lambda_2 - \lambda_1)t} \mathbb{1}_{\{0 \leq t \leq s\}} \, ds \, dt .$$

So, conditionally on $(T_1 > T_3)$, the pair of r.v.s (T_1, T_3) has density

$$\lambda_1 \lambda_2 e^{-\lambda_1 s} e^{-(\lambda_2 - \lambda_1)t} \mathbb{1}_{\{0 \leq t \leq s\}} ,$$

on $\mathbb{R}^2$. Now we compute the law of $(T_1 + T_2, T_2)$:

$$E[f(T_1 + T_2, T_2)] = \int_{\mathbb{R}_+^2} f(x + y, y) \lambda_1 \lambda_2 e^{-\lambda_1 x} e^{-\lambda_2 y} \, dx \, dy$$

$$= \int_{\mathbb{R}^2} f(s, t) \lambda_1 \lambda_2 e^{-\lambda_1 s} e^{-(\lambda_2 - \lambda_1)t} \mathbb{1}_{\{0 \leq t \leq s\}} \, ds \, dt .$$

We note that the pair of r.v.s $(T_1 + T_2, T_2)$ has the same law as (T_1, T_3) conditionally on $(T_1 > T_3)$.

(c) From the result we just proved, we can write:

$$E[f(T_1) \mathbb{1}_{\{T_1 > T_3\}}] = E[f(T_1 + T_2)] P[T_1 > T_3] = \frac{\lambda_2 - \lambda_1}{\lambda_2} E[f(T_1 + T_2)]. \quad (4.9.a)$$

On the other hand, inverting T_1 and T_3 in question 2 (b), we obtain that the pair of r.v.s $(T_3 + T_2, T_2)$ has the same law as the pair of r.v.s (T_3, T_1), conditionally on $(T_3 > T_1)$. This implies

$$E[f(T_1) \mathbb{1}_{\{T_1 < T_3\}}] = E[f(T_2)] P[T_3 > T_1] = \frac{\lambda_1}{\lambda_2} E[f(T_2)]. \quad (4.9.b)$$

Writing $E[f(T_1)] = E[f(T_1) \mathbb{1}_{\{T_1 > T_3\}}] + E[f(T_1) \mathbb{1}_{\{T_1 < T_3\}}]$ and combining equations (4.9.a) and (4.9.b) gives:

$$E[f(T_1 + T_2)] = \frac{\lambda_2}{\lambda_2 - \lambda_1} E[f(T_1)] - \frac{\lambda_1}{\lambda_2 - \lambda_1} E[f(T_2)] ,$$

which proves the identity of question 1.

3. It can be proved by induction that formula (4.9.1) is satisfied with the $\alpha_i^{(n)'}$s given in (4.9.2). Question 1 gives the result at rank $n = 2$. Now suppose that the result is true at some rank $n \in \mathbb{N}$ and let T_{n+1} be independent of $T_1, T_2, \ldots, T_n$ and

such that $P(T_{n+1} > t) = \exp(-\lambda_{n+1}t)$, λ_{n+1} being distinct from $\lambda_1, \ldots, \lambda_n$. For all $t \geq 0$, we have:

$$P\left(\sum_{i=1}^{n+1} T_i > t\right) = \int_0^\infty P\left(\sum_{i=1}^n T_i > t - s\right) \lambda_{n+1} e^{-\lambda_{n+1}s} \, ds$$

$$= \int_0^t P\left(\sum_{i=1}^n T_i > t - s\right) \lambda_{n+1} e^{-\lambda_{n+1}s} \, ds + \int_t^\infty \lambda_{n+1} e^{-\lambda_{n+1}s} \, ds$$

$$= \int_0^t \sum_{i=1}^n \alpha_i^{(n)} P(T_i > t - s) \lambda_{n+1} e^{-\lambda_{n+1}s} \, ds + e^{-\lambda_{n+1}t}.$$

Recall that $\sum_{i=1}^n \alpha_i^{(n)} = 1$. Using the fact that for any $i = 1, 2, \ldots, n$, the constants $a_i = \frac{\lambda_{n+1}}{\lambda_{n+1} - \lambda_i}$ and $b_i = \frac{\lambda_i}{\lambda_i - \lambda_{n+1}}$ satisfy $P(T_i + T_{n+1} > t) = a_i P(T_i > t) + b_i P(T_{n+1} > t)$, we can write the above identity in the following manner:

$$P\left(\sum_{i=1}^{n+1} T_i > t\right) = \sum_{i=1}^n \alpha_i^{(n)} \left(\int_0^t P(T_i > t - s) \lambda_{n+1} e^{-\lambda_{n+1}s} \, ds + e^{-\lambda_{n+1}t}\right)$$

$$= \sum_{i=1}^n \alpha_i^{(n)} P(T_i + T_{n+1} > t)$$

$$= \sum_{i=1}^n \alpha_i^{(n)} (a_i P(T_i > t) + b_i P(T_{n+1} > t)) = \sum_{i=1}^{n+1} \alpha_i^{(n+1)} P(T_i > t),$$

where the $\alpha_i^{(n+1)}$s are determined via the last equality by $\alpha_i^{(n+1)} = \alpha_i^{(n)} a_i$, $i = 1, \ldots, n$, and $\alpha_{n+1}^{(n+1)} = \sum_{i=1}^n \alpha_i^{(n)} b_i$.

This equality being valid for all $t \geq 0$, we proved (4.9.1) and (4.9.2) at rank $n+1$.

Solution to Exercise 4.10

1. According to question 1 (or 9) of Exercise **4.2**, the pair $\left(\frac{T}{T+T'}, T + T'\right)$ is distributed as $(\tilde{U}, Z_2)$, where $\tilde{U}$ and Z_2 are independent, $\tilde{U}$ is uniformly distributed on $[0, 1]$ and Z_2 has density $t e^{-t} \mathbb{1}_{\{t \geq 0\}}$ (the reader may also want to give a simple direct proof).

2. Put $\tilde{U} = \frac{1}{2}(1 + U)$, then $\tilde{U}$ is uniformly distributed on $[0, 1]$ and independent of (T, T'). Moreover, we have the identity:

$$\left(\log \frac{1 + U}{1 - U}, U(T + T')\right) = \left(\log \frac{\tilde{U}}{1 - \tilde{U}}, (2\tilde{U} - 1)(T + T')\right).$$

To conclude, note that the identity in law

$$\left(\log \frac{\tilde{U}}{1 - \tilde{U}}, (2\tilde{U} - 1)(T + T')\right) \stackrel{\text{(law)}}{=} \left(\log \frac{T}{T'}, T - T'\right)$$

may be obtained from the identity $\left(\frac{T}{T+T'}, T+T'\right) \overset{(\text{law})}{=} (\tilde{U}, T+T')$ established in the previous question, from a one-to-one transformation.

3. The characteristic function of (X, Y) is

$E[\exp i(\nu X + \mu Y)]$

$= \int_0^\infty \exp(i\nu \log t) \exp(-t(1 - i\mu)) \, dt \int_0^\infty \exp(-i\nu \log t') \exp(-t'(1 + i\mu)) \, dt'$

$= (1 - i\mu)^{-(1+i\nu)} (1 + i\mu)^{-(1-i\mu)} \int_0^\infty \exp(i\nu \log s - s) \, ds \int_0^\infty \exp(-i\nu \log s' - s') \, ds'$

$= (1 - i\mu)^{-(1+i\nu)} (1 + i\mu)^{-(1-i\mu)} \Gamma(1 + i\nu)\Gamma(1 - i\nu)$

$= (1 - i\mu)^{-(1+i\nu)} (1 + i\mu)^{-(1-i\mu)} i\nu\Gamma(i\nu)\Gamma(1 - i\nu)$

$= (1 - i\mu)^{-(1+i\nu)} (1 + i\mu)^{-(1-i\mu)} \frac{\pi\nu}{\sinh(\pi\nu)},$

where Γ is the extension in the complex plane of the classical gamma function.

4. Deriving with respect to μ the expression obtained in question 3, we have:

$E[Y \exp i(\nu X + \mu Y)]$

$= \frac{\pi\nu}{\sinh(\pi\nu)} \frac{(1 - i\nu)(1 + i\mu)^{-i\nu}(1 - i\mu)^{i\nu+1} - (i\nu + 1)(1 - i\mu)^{i\nu}(1 + i\mu)^{1-i\nu}}{(1 - i\mu)^{2(i\nu+1)}(1 + i\nu)^{2(1-i\nu)}}.$

Setting, $\nu = -\mu$ in the right-hand side of the above equality, we obtain the result.

The relation

$$E[Y \exp i\mu(Y - X)] = E[E[Y \mid Y - X] \exp i\mu(Y - X)] = 0, \quad \mu \in \mathbb{R}$$

characterizes $E[Y \mid Y - X]$ and shows that this r.v. equals 0.

5. The first part of the question is a particular case of question 2 of Exercise **4.9**. To show the second part, first observe that

$$(X, Y) \overset{(\text{law})}{=} \left(\log \frac{T}{T'}, T - T'\right) \overset{(\text{law})}{=} \left(\log \frac{T'}{T}, T' - T\right) \overset{(\text{law})}{=} -(X, Y) \quad \text{and}$$

$$(|X|, |Y|) = (X, Y)\mathbb{1}_{\{X>0, Y>0\}} - (X, Y)\mathbb{1}_{\{X\leq 0, Y\leq 0\}}.$$

The two above equalities imply that

$$(|X|, |Y|) \overset{(\text{law})}{=} ((X, Y) \mid X > 0, Y > 0),$$

which, from the definition of (X, Y), can be written as:

$$(|X|, |Y|) \overset{(\text{law})}{=} \left(\left(\log \frac{T}{T'}, T - T'\right) \mid T > T'\right).$$

From the first part of the question, we finally have:

$$(|X|, |Y|) \overset{(\text{law})}{=} \left(\log\left(1 + \frac{2T'}{T}\right), T'\right).$$

Solution to Exercise 4.11

1. From the identity in law (4.2.3), we have $R \overset{(\text{law})}{=} \sqrt{2Z_{\frac{3}{2}}}$, where $Z_{\frac{3}{2}}$ is gamma distributed with parameter $\frac{3}{2}$, so we easily compute the law of R: $P(R \in dr) = \sqrt{\frac{2}{\pi}} r^2 e^{-\frac{r^2}{2}} \, dr$.

2. If f is a bounded Borel function defined on $\mathbb{R}^2$, then from the change of variables $(r, u) = (r, \frac{x}{r})$, we obtain:

$$E[f(R, U)] = \int_{\mathbb{R}^2_+} f\left(r, \frac{x}{r}\right) \frac{1}{r} \mathbb{1}_{[0, r]}(x) \sqrt{\frac{2}{\pi}} r^2 e^{-\frac{r^2}{2}} \, dx \, dr$$

$$= \int_0^1 \int_0^\infty f(r, u) \sqrt{\frac{2}{\pi}} r^2 e^{-\frac{r^2}{2}} \, dr \, du \,.$$

This shows that R and U are independent and U is uniformly distributed over $[0, 1]$.

3. If f is as above, then $E[f(X, Y) \mid R = r] = \frac{1}{r} \int_0^r f(x, r-x) \, dx = \frac{1}{r} \int_0^r f(r-y, y) \, dy$. This shows that conditionally on $R = r$, Y is uniformly distributed over $[0, r]$. Integrating this identity with respect to the law of R, we obtain:

$$E[f(X, Y)] = \int_0^\infty \frac{1}{r} \int_0^r f(r-y, y) \, dy \sqrt{\frac{2}{\pi}} r^2 e^{-\frac{r^2}{2}} \, dr$$

$$= \int_0^\infty \int_0^\infty f(x, y) \sqrt{\frac{2}{\pi}} (x+y) e^{-\frac{(x+y)^2}{2}} \, dx \, dy \,,$$

which gives the density of the law of (X, Y): $\sqrt{\frac{2}{\pi}} (x+y) e^{-\frac{(x+y)^2}{2}} \mathbb{1}_{[0, +\infty)^2}(x, y)$.

4. From the law of the pair (X, Y), we deduce:

$$E[f(G, T)] = \int_0^\infty \int_0^\infty f(x-y, xy) \sqrt{\frac{2}{\pi}} (x+y) e^{-\frac{(x+y)^2}{2}} \, dx \, dy$$

$$= \int_{\mathbb{R}^2} f(g, t) \frac{1}{\sqrt{2\pi}} e^{-\frac{g^2}{2}} \, dg \, 2e^{-2t} \mathbb{1}_{[0, \infty)}(t) \, dt \,.$$

We conclude that G and T are independent. Moreover, G is a centered normal random variable with variance 1 and T is exponentially distributed with parameter 2.

5. From (4.11.1) and the independence between G and T, conditionally on $T = t$, $\frac{V^2}{1-V^2}$ has the same law as $\frac{G^2}{2t}$, that is: $P\left(\frac{V^2}{1-V^2} \in dz \mid T = t\right) = \left(\frac{t}{\pi}\right)^{\frac{1}{2}} z^{-\frac{1}{2}} e^{-tz} \, dz$.

6. From question 5, we know that G^2 is gamma distributed with parameter $\frac{1}{2}$. From question 4, $2T$ is exponentially distributed with parameter 1, and A is beta

distributed with parameters $\frac{1}{2}$ and $\frac{1}{2}$, so equation (4.11.2) is given by equation (4.2.1) in the particular case $a = b = \frac{1}{2}$. What makes the difference between (4.11.1) and (4.11.2) is the fact that in (4.11.1) the variable $\frac{V^2}{1-V^2}$ is not independent of T whereas it is the case for A in (4.11.2).

Solution to Exercise 4.12

1. Since Θ is uniformly distributed on $[0, 2\pi)$, we have $2\Theta \pmod{2\pi} \overset{(\text{law})}{=} \Theta$, hence $\tan(2\Theta) \overset{(\text{law})}{=} \tan\Theta$.

2.(i) Let Θ be the argument of the complex normal variable $Z = N + iN'$. As we already noticed in Exercise **3.5**, Θ is uniformly distributed on $[0, 2\pi)$. So, the identity: $\tan\Theta = \frac{N}{N'}$ gives the result.

(ii) Let f be a bounded Borel function. In the equality

$$E[f(\tan\Theta)] = \frac{1}{2\pi} \int_0^{2\pi} f(\tan x)\,dx = \frac{1}{\pi} \int_0^\pi f(\tan x)\,dx,$$

taking $y = \tan x$, we obtain $E[f(\tan\Theta)] = \int_{-\infty}^{+\infty} f(y)\,\frac{dy}{\pi(1+y^2)}$.

3. Thanks to the well-known formula $\tan 2a = 2\dfrac{\tan a}{1 - \tan^2 a}$, and applying question 1, we have

$$\tan(2\Theta) = 2\frac{\tan\Theta}{1 - \tan^2\Theta} \overset{(\text{law})}{=} \tan\Theta,$$

so that if C is a Cauchy r.v., then $C \overset{(\text{law})}{=} \frac{2C}{1-C^2}$. Since $C \overset{(\text{law})}{=} \frac{1}{C}$, which may be deduced from question 2 (i), we obtain:

$$C \overset{(\text{law})}{=} \frac{1 - C^2}{2C} = \frac{1}{2}\left(\frac{1}{C} - C\right).$$

To obtain the second equality, first recall from Exercises **3.7** and **3.8** that, N and N' being as in question 2, $N + N'$ and $N - N'$ are independent, hence $N + N' \overset{(\text{law})}{=} N - N' \overset{(\text{law})}{=} \sqrt{2}N$. So, from question 2 (i), we deduce:

$$C \overset{(\text{law})}{=} \frac{N' + N}{N' - N} = \frac{1 + \frac{N'}{N}}{1 - \frac{N'}{N}} \overset{(\text{law})}{=} \frac{1 + C}{1 - C}.$$

Solution to Exercise 4.13

a) Recall that the law of a variable Z taking values in $[0,1)$ is characterized by its discrete Fourier transform:

$$\varphi_2(p) = E[\exp(2i\pi pZ)], \quad p \in \mathbb{Z}$$

and that Z is uniformly distributed on $[0,1)$ if and only if $\varphi_2(p) = \mathbb{1}_{\{p=0\}}$. As a consequence of this, it follows immediately that $\{mU+nV\}$ is uniformly distributed.

b) A necessary and sufficient condition for the variables $\{mU+nV\}$ and $\{m'U+n'V\}$ to be independent is that

$$E[\exp[2i\pi(p\{mU+nV\}+q\{m'U+n'V\})]]$$
$$= E[\exp(2i\pi p\{mU+nV\})]E[\exp(2i\pi q\{m'U+n'V\})],$$

for any $p,q \in \mathbb{Z}$.

Observe that since

$$\exp[2i\pi(p\{mU+nV\}+q\{m'U+n'V\})] = \exp[2i\pi(p(mU+nV)+q(m'U+n'V))],$$

the characteristic function of interest to us is:

$$E[\exp[2i\pi(p\{mU+nV\}+q\{m'U+n'V\})]]$$
$$= \int_{[0,1]^2} \exp[2i\pi(p(mx+ny)+q(m'x+n'y))]\,dx\,dy.$$

The above expression equals 0 if $pm + qm' \neq 0$ or $pn + qn' \neq 0$ and it equals 1 if $pm + qm' = 0$ and $pn + qn' = 0$. But there exist some integers $p,q \in \mathbb{Z} \setminus \{(0,0)\}$ such that $pm + qm' = 0$ and $pn + qn' = 0$ if and only if $nm' = n'm$ (in which case take e.g. $q = nm, p = -nm' = -n'm$).

We conclude that a necessary and sufficient condition for the variables $\{mX+nY\}$ and $\{m'X + n'Y\}$ to be independent is that $nm' \neq n'm$.

Solution to Exercise 4.14

1. Since the uniform law on $[0,1]$ is characterized by $E[\exp(2i\pi pU)] = 0$, for every integer $p \in \mathbb{Z} \setminus \{0\}$, it is enough to check that:

$$E[\exp(2i\pi p\{U + Y\})] = E[\exp(2i\pi pU)]E[\exp(2i\pi pY)]$$
$$= 0.$$

2. Assume that $\{V + Y\} \overset{(\text{law})}{=} V$, then for all $p \in \mathbb{Z} \backslash \{0\}$, the equality

$$E\left[\exp(2i\pi p(V + Y))\right] = E\left[\exp(2i\pi pV)\right]$$

implies

$$E\left[\exp(2i\pi pV)\right]\left(E\left[\exp(2i\pi pY)\right] - 1\right) = 0. \tag{4.14.a}$$

Now take $Y = \frac{1}{2p}$, then $E[\exp(2i\pi pY) - 1] \equiv -2$ and, from (4.14.a), $E[\exp(2i\pi pV)] = 0$, which implies that V is uniform.

Solution to Exercise 4.15

1. Let X be an r.v. such that $|X| \leq 1$, a.s. and suppose that X is infinitely divisible. For each $n \geq 1$, there are i.i.d. r.v.s X_n^i, $i = 1, \ldots n$ such that $X = X_n^1 + \cdots + X_n^n$. Then these r.v.s must themselves be bounded by $1/n$, i.e. $|X_n^i| \leq 1/n$, a.s. for all $i = 1, \ldots, n$. Indeed,

$$P(X_n^1 > 1/n, \ldots, X_n^n > 1/n) = P(X_n^1 > 1/n)^n \leq P(X > 1) = 0.$$

We show that $P(X_n^1 < -1/n) = 0$ via the same argument. Denote by φ the Laplace transform of X. From the infinite divisibility of X, $\varphi^{1/n}$ is the Laplace transform of the X_n^is. Let μ_n be their common law and write

$$n(\varphi^{1/n}(\lambda) - 1) = n \int_{-1/n}^{1/n} (e^{-\lambda x} - 1) \, d\mu_n(x)$$

$$= n \int_{-1/n}^{1/n} (e^{-\lambda x} - 1 + \lambda x) \, d\mu_n(x) - n\lambda \int_{-1/n}^{1/n} x \, d\mu_n(x)$$

$$= n \int_{-1/n}^{1/n} (e^{-\lambda x} - 1 + \lambda x) \, d\mu_n(x) - \lambda E(X).$$

Then since for any positive real number a, $\lim_{n \to +\infty} n(a^{1/n} - 1) = \log a$, and also, $\varphi(\lambda) > 0$, for all $\lambda \geq 0$, we have

$$\log \varphi(\lambda) = \lim_{n \to +\infty} n \int_{-1/n}^{1/n} (e^{-\lambda x} - 1 + \lambda x) \, d\mu_n(x) - \lambda E(X).$$

But for all n and $x \in [-1/n, 1/n]$, $|e^{-\lambda x} - 1 + \lambda x| \leq \frac{\lambda^2}{2n^2}$, so that

$$\lim_{n \to +\infty} n \int_{-1/n}^{1/n} (e^{-\lambda x} - 1 + \lambda x) \, d\mu_n(x) = 0$$

and $\log \varphi(\lambda) = -\lambda E(X)$, for all $\lambda \geq 0$ which means that $X = E(X)$, a.s.

2. The formula for $\mu_a(dx)$ is:

$$\mu_a(dx) = \frac{e^{-ax}}{x}\left(dx - \sum_{j=1}^{\infty}\delta_j(dx)\right). \tag{4.15.a}$$

Note that $\int|\mu_a(dx)|x < \infty$. The formula (4.15.a) follows from the fact that $e_a = U_a + N_a$, with U_a and N_a independent. The r.v.s e_a and N_a are infinitely divisible; in particular, N_a is a geometric random variable: $P(N_a = n) = (1 - e^{-a})e^{-an}, n = 1, 2, \ldots$ Consequently, $\mu_a(dx) = \nu_a(dx) - n_a(dx)$, with $\nu_a(dx)$ and $n_a(dx)$ denoting the standard Lévy measures of e_a and N_a.

The expressions of ν_a and n_a are easily found:

$$\nu_a(dx) = \frac{e^{-ax}}{x}dx \quad \text{and} \quad n_a(dx) = \sum_{j=1}^{\infty}\frac{e^{-ja}}{j}\delta_j(dx),$$

which proves (4.15.a).

3. Suppose that V_1 and V_2 are i.i.d. and satisfy $U \overset{(\text{law})}{=} V_1 + V_2$. Let j be 1 or 2. From the inequalities

$$P(V_j < 0)^2 \le P(U < 0) = 0 \quad \text{and} \quad P(V_j > 1/2)^2 \le P(U > 1) = 0,$$

we deduce that $0 \le V_j \le 1/2$, a.s. But for all $p \in \mathbb{Z} \setminus \{0\}$,

$$E(e^{2i\pi pU}) = \left[E(e^{2i\pi pV_k})\right]^2 = 0, \quad k = 1, 2,$$

hence from the characterization of the uniform law on $[0, 1]$ (see Solution **4.13**), we deduce that V_k is necessarily uniformly distributed over $[0, 1]$. It is a contradiction with $P(0 \le V_j \le 1/2) = 1$.

4. Let $X = X_1 + \ldots + X_k$, where X_j are independent and uniformly distributed over $[0, 1]$. Suppose that there exist $Y_1, \ldots, Y_{k+1}$, i.i.d. such that $X \overset{(\text{law})}{=} Y_1 + \cdots + Y_{k+1}$. Then from the inequalities

$$P(Y_j < 0)^{k+1} \le P(X < 0) = 0 \quad \text{and} \quad P(Y_j > k/(k+1))^{k+1} \le P(X > k) = 0,$$

we deduce that $0 \le Y_j \le k/(k+1)$, a.s. But for all $p \in \mathbb{Z} \setminus \{0\}$,

$$E(e^{2i\pi pX}) = \left[E(e^{2i\pi pX_j})\right]^k = \left[E(e^{2i\pi pY_j})\right]^{k+1} = 0,$$

so Y_j is uniformly distributed, which contradicts $P(0 \le Y_j \le k/(k+1)) = 1$.

5. Set $V_k = (U_1 + \ldots + U_k)/k$, so that $V_{k+1} = \frac{k}{k+1} V_k + \frac{1}{k+1} U_{k+1}$. Since V_k and U_{k+1} are independent, we have for all bounded measurable function g,

$$
\begin{aligned}
E(g(V_{k+1})) &= E\left(g\left(\frac{k}{k+1} V_k + \frac{1}{k+1} U_{k+1}\right)\right) \\
&= \int_0^1 \int_0^1 g\left(\frac{k}{k+1} v + \frac{1}{k+1} u\right) f_k(v)\, dv\, du \\
&= \int_0^1 \int_{u/(k+1)}^1 g(w) f_k\left(\frac{k+1}{k} w - \frac{1}{k} u\right) \frac{k+1}{k}\, dw\, du \\
&= \int_0^1 g(w) \int_0^{((k+1)w)\wedge 1} f_k\left(\frac{k+1}{k} w - \frac{1}{k} u\right) \frac{k+1}{k}\, du\, dw
\end{aligned}
$$

From the last equality, we derive the expression of the density function of V_{k+1} in terms of that of V_k: $f_{k+1}(x) = \frac{k+1}{k} \int_0^{((k+1)x)\wedge 1} f_k\left(\frac{(k+1)x-u}{k}\right) du\, \mathbb{1}_{\{x\in[0,1]\}}$.

Solution to Exercise 4.16

1. If f is any bounded Borel function, then $E[f(\cos(\Theta))] = \frac{1}{\pi} \int_0^\pi f(\cos\theta)\, d\theta = \frac{1}{\pi} \int_{-1}^1 f(x) \frac{dx}{\sqrt{1-x^2}}$, where we put $x = \cos(\theta)$ in the last equality. So, X has density $\frac{1}{\pi\sqrt{1-x^2}}$, on the interval $[-1,1]$.

2. Applying Exercise **4.13** for $m = 1$, $n = 1$, $m' = 1$, $n' = -1$ and $\Theta = \pi U$, $\Theta' = \pi V$ shows that $(\cos(\Theta + \Theta'), \cos(\Theta - \Theta')) \overset{(\text{law})}{=} (\cos\Theta, \cos\Theta')$.

3. This identity in law follows immediately from question 2 and the formula $\cos(\Theta + \Theta') + \cos(\Theta - \Theta') = 2\cos(\Theta)\cos(\Theta')$.

Solution to Exercise 4.17

1. See question 7 of Exercise **4.2**.

2. It suffices to check that the inverse Fourier transform of the function $\exp(-|\lambda|)$, $\lambda \in \mathbb{R}$, corresponds to the density of the standard Cauchy variable. Indeed, we have:

$$
\frac{1}{2\pi} \int_{\mathbb{R}} e^{-i\lambda x} e^{-|\lambda|}\, d\lambda = \frac{1}{2\pi} \left(\frac{1}{1 - ix} + \frac{1}{1 + ix}\right) = \frac{1}{\pi(1 + x^2)},
$$

for all $x \in \mathbb{R}$.

3. We proved in question 2 of Exercise **4.12** that if N and N' are independent, centered Gaussian r.v.s, with variance 1, then $C \overset{(\text{law})}{=} \frac{N}{N'}$. So, conditioning on N' and applying the result of question 1, we can write:

$$E[\exp(i\lambda C)] = E\left[\exp\left(i\lambda\frac{N}{N'}\right)\right] = E\left[\exp\left(-\frac{\lambda^2}{2}\frac{1}{N'^2}\right)\right] = E\left[\exp\left(-\frac{\lambda^2}{2}T\right)\right].$$

4. A direct consequence of the result of question 7 (b), Exercise **4.2**, is that $\frac{1}{n^2}\sum_{i=1}^{n} T_i$ has the same law as T, which disagrees with the law of large numbers. Indeed, if the T_is had finite expectation, then the law of large numbers would imply that $\frac{1}{n}\sum_{i=1}^{n} T_i$ converges almost surely to this expectation and $\frac{1}{n^2}\sum_{i=1}^{n} T_i$ would converge almost surely to 0.

5. For every bounded Borel function f, defined on $\mathbb{R}^n$, we have:

$$E\left[f\left(\frac{N_1}{N_0}, \ldots, \frac{N_n}{N_0}\right)\right]$$

$$= \int_{\mathbb{R}^{n+1}} f\left(\frac{x_1}{x_0}, \ldots, \frac{x_n}{x_0}\right) \frac{1}{(2\pi)^{\frac{n+1}{2}}} \exp -\frac{1}{2}\left(\sum_{j=0}^{n} x_j^2\right) dx_0 dx_1 \ldots dx_n$$

$$= \int_{\mathbb{R}^n} f(y_1, \ldots, y_n) \left[\int_{\mathbb{R}} dx_0 \frac{|x_0|^n}{(2\pi)^{\frac{n+1}{2}}} \exp -\frac{x_0^2}{2}\left(1 + \sum_{j=1}^{n} y_j^2\right)\right] dy_1 \ldots dy_n$$

$$= \int_{\mathbb{R}^n} f(y_1, \ldots, y_n) \left(\int_{\mathbb{R}} dx_0 \frac{|x_0|^n}{\sqrt{2\pi}} \exp -\frac{x_0^2}{2}\right) \frac{1}{(2\pi)^{\frac{n}{2}}}\left(1 + \sum_{j=1}^{n} y_j^2\right)^{-\frac{n+1}{2}} dy_1 \ldots dy_n$$

$$= \int_{\mathbb{R}^n} f(y_1, \ldots, y_n) \frac{\Gamma(\frac{n+1}{2})}{\pi^{\frac{n+1}{2}}}\left(1 + \sum_{j=1}^{n} y_j^2\right)^{-\frac{n+1}{2}} dy_1 \ldots dy_n.$$

Hence, the density of the law of $\left(\frac{N_1}{N_0}, \ldots, \frac{N_n}{N_0}\right)$ is given by:

$$\Gamma\left(\frac{n+1}{2}\right)\left[\pi\left(1 + \sum_{j=1}^{n} y_j^2\right)\right]^{-\frac{n+1}{2}}, \quad (y_1, \ldots, y_n) \in \mathbb{R}^n,$$

where $\Gamma(\frac{n+1}{2}) = \sqrt{\pi}2^{-\frac{n}{2}} \cdot 2 \cdot 3 \cdot \ldots \cdot (n-1)$, if n is even and $\Gamma(\frac{n+1}{2}) = \left(\frac{n-1}{2}\right)!$ if n is odd. To obtain the characteristic function of this random vector, set $T \overset{(\text{def})}{=} \frac{1}{N_0^2}$ and observe that

$$E\left[\exp\left(i\sum_{j=1}^{n} \lambda_j \frac{N_j}{N_0}\right) \Big| N_0\right] = e^{-\frac{1}{2N_0^2}\sum_{j=1}^{n} \lambda_j^2} = e^{-\frac{1}{2}\left(\sum_{j=1}^{n} \lambda_j^2\right)T},$$

so that from questions 1 and 3

$$E\left[\exp\left(i\sum_{j=1}^{n}\lambda_j\frac{N_j}{N_0}\right)\right] = E\left[e^{-\frac{1}{2}\left(\sum_{j=1}^{n}\lambda_j^2\right)T}\right] = \exp\left(-\left(\sum_{j=1}^{n}\lambda_j^2\right)^{\frac{1}{2}}\right), \qquad \text{(4.17.a)}$$

for every $(\lambda_1,\ldots,\lambda_n) \in \mathbb{R}^n$.

6. Put $X = \left(\frac{N_1}{N_0},\ldots,\frac{N_n}{N_0}\right)$. We obtain

$$E[\exp(it\langle\theta, X\rangle)] = \exp(-|t|),$$

for all $t \in \mathbb{R}$, by taking $\theta = \frac{\lambda}{|\lambda|}t$, where $\lambda = (\lambda_1,\ldots,\lambda_n) \in \mathbb{R}^n$, in the equation (4.17.a). Since the law ν_X of X is characterized by (4.17.a), this proves that ν_X is the unique distribution such that for any $\theta \in \mathbb{R}^n$, with $|\theta| = 1$, $\langle\theta, X\rangle$ is a standard Cauchy r.v..

Solution to Exercise 4.18

0. *Preliminaries.* (i) and (ii) follow directly from the definition of X and the independence between A and N.

1. Since each of the variables A, N and X is almost surely different from 0, we can write: $\frac{1}{X} \overset{\text{(law)}}{=} \frac{1}{\sqrt{A}}\frac{1}{N}$, which implies: $\frac{1}{|X|^\alpha} \overset{\text{(law)}}{=} \frac{1}{A^{\frac{\alpha}{2}}}\frac{1}{|N|^\alpha}$, for every $\alpha \geq 0$. Moreover, each of the variables $\frac{1}{|X|^\alpha}$, $\frac{1}{A^{\frac{\alpha}{2}}}$ and $\frac{1}{|N|^\alpha}$ being positive, their expectation is defined, so formula (4.18.1) follows from the independence between A and N.

From equation (4.18.1), we deduce that $E\left[\frac{1}{|X|^\alpha}\right]$ is finite whenever both $E\left[\frac{1}{A^{\frac{\alpha}{2}}}\right]$ and $E\left[\frac{1}{|N|^\alpha}\right]$ are finite, that is when $\alpha < 1$ and $A^{-\frac{\alpha}{2}}$ is integrable.

2. Applying the fact that for every $r > 0$, $\alpha > 0$, $r^{-\frac{\alpha}{2}} = \frac{1}{\Gamma(\frac{\alpha}{2})}\int_0^\infty dt\, t^{\frac{\alpha}{2}-1}e^{-tr}$ and Fubini's theorem, we obtain:

$$E\left[\frac{1}{A^{\frac{\alpha}{2}}}\right] = \frac{1}{\Gamma\left(\frac{\alpha}{2}\right)}\int_0^\infty dt\, t^{\frac{\alpha}{2}-1}E[e^{-tA}].$$

Now, we deduce from the expression of the characteristic function of N, and the independence between this variable and A that $E[e^{-tA}] = E[e^{i\sqrt{2tA}N}] = E[e^{i\sqrt{2t}X}]$. Finally, one has

$$E\left[\frac{1}{A^{\frac{\alpha}{2}}}\right] = \frac{1}{\Gamma\left(\frac{\alpha}{2}\right)}\int_0^\infty dt\, t^{\frac{\alpha}{2}-1}E[e^{i\sqrt{2t}X}] = \frac{1}{\Gamma\left(\frac{\alpha}{2}\right)2^{\frac{\alpha}{2}-1}}\int_0^\infty dx\, x^{\alpha-1}E[e^{ixX}]$$

$$= \frac{1}{\Gamma\left(\frac{\alpha}{2}\right)2^{\frac{\alpha}{2}-1}}\int_0^\infty dx\, x^{\alpha-1}E[\cos(xX)],$$

where the last equality is due to the fact that the imaginary part of the previous integral vanishes.

3.1. (iii) An obvious calculation gives:

$$E\left[\frac{1}{|N|^\alpha}\right] = \sqrt{\frac{2}{\pi}} \int_0^\infty dx\, \frac{1}{x^\alpha} e^{-\frac{x^2}{2}} = \frac{1}{\sqrt{2^\alpha \pi}} \int_0^\infty dy\, \frac{1}{y^{\frac{\alpha}{2}+\frac{1}{2}}} e^{-y} = \frac{1}{\sqrt{2^\alpha \pi}} \Gamma\left(\frac{1}{2} - \frac{\alpha}{2}\right).$$

$$(4.18.a)$$

3.2. First note that from the hypothesis, $|X|$ is almost surely strictly positive and $\int_0^t dx\, x^{\alpha-1} \cos(xX) = \frac{1}{|X|^\alpha} \int_0^{t|X|} dy\, y^{\alpha-1} \cos(y)$. Taking the expectation of each member, and applying Fubini's Theorem, we obtain:

$$\int_0^t dx\, x^{\alpha-1} E[\cos(xX)] = E\left[\frac{1}{|X|^\alpha} \int_0^{t|X|} dy\, y^{\alpha-1} \cos(y)\right].$$

Moreover, from 3.1 (ii), $\lim_{t\to\infty} \frac{1}{|X|^\alpha} \int_0^{t|X|} dy\, y^{\alpha-1} \cos(y) = \frac{a_\alpha}{|X|^\alpha}$, a.s. Since $E\left[\frac{1}{|X|^\alpha}\right] < \infty$ and $\int_0^s dy\, y^{\alpha-1} \cos(y)$ is bounded in s, we can apply Lebesgue's Theorem of dominated convergence to the variable $\frac{1}{|X|^\alpha} \int_0^{t|X|} dy\, y^{\alpha-1} \cos(y)$. Therefore, from (4.18.2) we have:

$$E\left[\frac{1}{A^{\frac{\alpha}{2}}}\right] = \lim_{t\to\infty} \frac{1}{\Gamma\left(\frac{\alpha}{2}\right) 2^{\frac{\alpha}{2}-1}} \int_0^t dx\, x^{\alpha-1} E[\cos(xX)]$$

$$= \lim_{t\to\infty} \frac{1}{\Gamma\left(\frac{\alpha}{2}\right) 2^{\frac{\alpha}{2}-1}} E\left[\frac{1}{|X|^\alpha} \int_0^{t|X|} dy\, y^{\alpha-1} \cos(y)\right] = \frac{a_\alpha}{\Gamma\left(\frac{\alpha}{2}\right) 2^{\frac{\alpha}{2}-1}} E\left[\frac{1}{|X|^\alpha}\right].$$

3.3. We easily deduce from the previous questions that

$$a_\alpha = E\left[\frac{1}{|N|^\alpha}\right]^{-1} \Gamma\left(\frac{\alpha}{2}\right) 2^{\frac{\alpha}{2}-1} = 2^{\alpha-1} \sqrt{\pi} \frac{\Gamma\left(\frac{\alpha}{2}\right)}{\Gamma\left(\frac{1}{2} - \frac{\alpha}{2}\right)}.$$

Preliminary 3.1 (i) implies

$$\Gamma\left(\frac{\alpha}{2}\right) = \frac{\sqrt{2\pi}\, 2^{\frac{1}{2}-\alpha} \Gamma(\alpha)}{\Gamma\left(\frac{\alpha}{2} + \frac{1}{2}\right)} \quad \text{and} \quad \Gamma\left(\frac{\alpha}{2} + \frac{1}{2}\right) \Gamma\left(\frac{1}{2} - \frac{\alpha}{2}\right) = \frac{\pi}{\sin\pi\left(\frac{\alpha}{2} + \frac{1}{2}\right)} = \frac{\pi}{\cos\left(\frac{\pi\alpha}{2}\right)},$$

which yields $a_\alpha = \Gamma(\alpha) \cos\left(\frac{\pi\alpha}{2}\right)$.

3.4. Recall the hint for question 2: $\frac{1}{r^\alpha} = \frac{1}{\Gamma(\alpha)} \int_0^\infty x^{\alpha-1} e^{-rx}\, dx$ for all $r > 0$ and $\alpha > 0$. This formula may be extended to any complex number in the domain $H = \{z : \text{Re}(z) > 0\}$, by analytical continuation, that is

$$\frac{1}{z^\alpha} = \frac{1}{\Gamma(\alpha)} \int_0^\infty x^{\alpha-1} e^{-zx}\, dx, \quad z \in H.$$

Set $z = \lambda + i\mu$, then identifying the real parts of each side of the above equality, we have:

$$\frac{\cos\left(\frac{\pi\alpha}{2}\right)}{\mu^\alpha} = \lim_{\lambda \to 0} \frac{1}{\Gamma(\alpha)} \int_0^\infty x^{\alpha-1} e^{-\lambda x} \cos(\mu x)\, dx,$$

hence by putting $y = \mu x$:

$$\Gamma(\alpha) \cos\left(\frac{\pi\alpha}{2}\right) = \lim_{\lambda \to 0} \frac{1}{\Gamma(\alpha)} \int_0^\infty y^{\alpha-1} e^{-\lambda y} \cos(y)\, dy.$$

It remains for us to check that $\lim_{\lambda \to 0} \frac{1}{\Gamma(\alpha)} \int_0^\infty y^{\alpha-1} e^{-\lambda y} \cos(y)\, dy = \frac{1}{\Gamma(\alpha)} \int_0^\infty y^{\alpha-1} \cos(y)\, dy$. To this end, write

$$\int_{\frac{\pi}{2}}^\infty y^{\alpha-1} e^{-\lambda y} \cos(y)\, dy = \sum_{k \geq 0} \int_{(2k+1)\frac{\pi}{2}}^{(2k+3)\frac{\pi}{2}} y^{\alpha-1} e^{-\lambda y} \cos(y)\, dy$$

$$= \sum_{k \geq 0} \int_{(4k+1)\frac{\pi}{2}}^{(4k+3)\frac{\pi}{2}} y^{\alpha-1} e^{-\lambda y} \cos(y)\, dy + \int_{(4k+3)\frac{\pi}{2}}^{(4k+5)\frac{\pi}{2}} y^{\alpha-1} e^{-\lambda y} \cos(y)\, dy$$

$$= \sum_{k \geq 0} \int_{(4k+1)\frac{\pi}{2}}^{(4k+3)\frac{\pi}{2}} (y^{\alpha-1} - (y+\pi)^{\alpha-1} e^{-\lambda \pi}) e^{-\lambda y} \cos(y)\, dy.$$

Observe that when $0 < \alpha < 1$, each term of the above series is negative and we can state:

$$\sum_{k \geq 0} \int_{(4k+1)\frac{\pi}{2}}^{(4k+3)\frac{\pi}{2}} (y^{\alpha-1} - (y+\pi)^{\alpha-1} e^{-\lambda \pi}) \cos(y)\, dy \leq \int_{\frac{\pi}{2}}^\infty y^{\alpha-1} e^{-\lambda y} \cos(y)\, dy$$

$$\leq \sum_{k \geq 0} \int_{(4k+1)\frac{\pi}{2}}^{(4k+3)\frac{\pi}{2}} (y^{\alpha-1} - (y+\pi)^{\alpha-1}) e^{-\lambda y} \cos(y)\, dy.$$

By monotone convergence, both series of the above inequalities converge towards: $\sum_{k \geq 0} \int_{(4k+1)\frac{\pi}{2}}^{(4k+3)\frac{\pi}{2}} (y^{\alpha-1} - (y+\pi)^{\alpha-1}) \cos(y)\, dy = \int_{\frac{\pi}{2}}^\infty y^{\alpha-1} \cos(y)\, dy$, as $\lambda \to 0$ and the result follows.

4. The equivalent form of equation (4.18.1) is

$$E\left[\frac{1}{|Y|^\alpha}\right] = E\left[\frac{1}{B^\alpha}\right] E\left[\frac{1}{|C|^\alpha}\right], \tag{4.18.b}$$

for every $\alpha \geq 0$. Now, recall the expression for the characteristic function of C: $E[e^{itC}] = \exp(-|t|)$, $t \in \mathbb{R}$. So, by the same arguments as in question 2, we obtain that for every $\alpha > 0$, $E\left[\frac{1}{B^\alpha}\right] = \frac{1}{\Gamma(\alpha)} \int_0^\infty dt\, t^{\alpha-1} E[e^{-tB}] = \frac{1}{\Gamma(\alpha)} \int_0^\infty dt\, t^{\alpha-1} E[e^{itBC}] = \frac{1}{\Gamma(\alpha)} \int_0^\infty dt\, t^{\alpha-1} E[e^{itY}] = \frac{1}{\Gamma(\alpha)} \int_0^\infty dt\, t^{\alpha-1} E[\cos(tY)]$. Now, suppose that $E\left[\frac{1}{|Y|^\alpha}\right] < +\infty$, then as in question 3.2, we have:

$$E\left[\frac{1}{B^\alpha}\right] = \frac{a_\alpha}{\Gamma(\alpha)} E\left[\frac{1}{|Y|^\alpha}\right]. \tag{4.18.c}$$

On the other hand, thanks to the identity in law $C \overset{(\text{law})}{=} \frac{N}{N'}$ established in question 2 of Exercise **4.12** (here, N and N' are independent centered Gaussian variables, with variance 1) and from formula (4.18.a), we obtain

$$E\left[\frac{1}{|C|^\alpha}\right] = E\left[\frac{|N|^\alpha}{|N'|^\alpha}\right] = E[|C|^\alpha] = \frac{1}{\pi}\Gamma\left(\frac{1}{2} - \frac{\alpha}{2}\right)\Gamma\left(\frac{1}{2} + \frac{\alpha}{2}\right) = \frac{1}{\cos\left(\frac{\pi\alpha}{2}\right)}.$$
$$(4.18.\text{d})$$

So that (4.18.b), (4.18.c) and (4.18.d) yield another proof of (4.18.4).

Solution to Exercise 4.19

1. For any Borel function $f : \mathbb{R}_+ \to \mathbb{R}_+$, we have:

$$E\left[f\left((Z_{\frac{2}{\mu}})^{\frac{1}{\mu}}\right)\right] = \frac{1}{\Gamma(\frac{2}{\mu})}\int_0^\infty f(t^{\frac{1}{\mu}})t^{\frac{2}{\mu}-1}e^{-t}\,dt$$

$$= \frac{\mu}{\Gamma(\frac{2}{\mu})}\int_0^\infty f(s)s^{\gamma-1}e^{-s^\mu}\,ds.$$

Using the expression of the Laplace transform of T_μ, we can write the above expression as $\frac{\mu}{\Gamma(\frac{2}{\mu})}\int_0^\infty f(s)s^{\gamma-1}E[e^{-sT_\mu}]\,ds$, which gives with obvious changes of variable:

$$\frac{\mu}{\Gamma(\frac{2}{\mu})}E\left[\int_0^\infty f(s)s^{\gamma-1}e^{-sT_\mu}\,ds\right] = \frac{\mu}{\Gamma(\frac{2}{\mu})}E\left[\int_0^\infty f\left(\frac{t}{T_\mu}\right)\frac{t^{\gamma-1}}{T_\mu^\gamma}e^{-t}\,dt\right]$$

$$= \frac{\mu\Gamma(\gamma)}{\Gamma(\frac{2}{\mu})}E\left[f\left(\frac{Z_\gamma}{T_\mu}\right)\frac{1}{T_\mu^\gamma}\right],$$

and the result follows.

2. From (4.19.3), we deduce for $\lambda > 0$:

$$E[Z^{\frac{\lambda}{\mu}}] = E[Z_\mu^\lambda]\Gamma(\mu+1)E\left[\frac{1}{(T_\mu)^{\lambda+\mu}}\right],$$

so that:

$$E\left[\frac{1}{(T_\mu)^{\mu(\theta+1)}}\right] = \frac{E[Z^\theta]}{E[Z_\mu^{\theta\mu}]\Gamma(\mu+1)}$$

$$= \frac{\Gamma(1+\theta)}{\mu\Gamma(\mu(1+\theta))}.$$

Hence $E\left[\frac{1}{(T_\mu)^{\mu\sigma}}\right] = \frac{\Gamma(1+\sigma)}{\Gamma(1+\mu\sigma)}$. Then, we obtain the result by analytical continuation.

Solution to Exercise 4.20

1. We first note the following general fact. For every Borel function $f : (0, \infty) \to \mathbb{R}_+$, one has:

$$
\begin{aligned}
E[f(Z_\mu X)] &= \int_0^\infty \frac{1}{\Gamma(\mu)} s^{\mu-1} E[f(sX)] e^{-s} \, ds \\
&= E\left[\int_0^\infty \frac{1}{\Gamma(\mu)} \frac{t^{\mu-1}}{X^\mu} f(t) e^{-\frac{t}{X}} \, dt \right] \\
&= \int_0^\infty \frac{1}{\Gamma(\mu)} t^{\mu-1} f(t) E\left[\frac{e^{-\frac{t}{X}}}{X^\mu} \right] dt \,.
\end{aligned}
\tag{4.20.a}
$$

Now, assume (i) holds and observe that an equivalent form of equation (4.20.1) is $E[cg(Y)] = E\left[g\left(\frac{1}{X} \right) \frac{1}{X^\mu} \right]$, for every Borel function $g : (0, \infty) \to \mathbb{R}_+$, so that, the above expression is

$$
\int_0^\infty \frac{1}{\Gamma(\mu)} t^{\mu-1} f(t) E[ce^{-tY}] \, dt \,,
$$

and (ii) follows.

Conversely if (ii) holds then on the one hand, we have

$$
E[f(Z_\mu X)] = \int_0^\infty \frac{1}{\Gamma(\mu)} t^{\mu-1} f(t) E[ce^{-tY}] \, dt
$$

and, on the other hand, (4.20.a) allows us to identify $E[ce^{-tY}]$ as $E\left[\frac{1}{X^\mu} e^{-\frac{t}{X}} \right]$ for every $t \geq 0$ and (i) follows.

(i)$\Leftrightarrow$(iii): suppose (i) holds, then for every Borel function $f : (0, \infty) \to \mathbb{R}_+$:

$$
\begin{aligned}
E[f(Z^{\frac{1}{\mu}} X)] &= \int_0^\infty E[f(t^{\frac{1}{\mu}} X)] e^{-t} \, dt \\
&= \int_0^\infty f(u) \mu u^{\mu-1} E\left[\frac{e^{-\left(\frac{u}{X}\right)^\mu}}{X^\mu} \right] du \\
&= \int_0^\infty f(u) c\mu u^{\mu-1} E\left[e^{-(uY)^\mu} \right] du \,,
\end{aligned}
$$

and (iii) follows.

If (iii) holds, then an obvious change of variables yields for every $\lambda > 0$:

$$
E[f(\lambda^{-\frac{1}{\mu}} Z^{\frac{1}{\mu}} X)] = \int_0^\infty c\mu u^{\mu-1} f(\lambda^{-\frac{1}{\mu}} u) E[e^{-(uY)^\mu}] \, du = \int_0^\infty E\left[\frac{c}{Y^\mu} f\left(\frac{t}{Y} \right) \right] \lambda e^{-\lambda t} \, dt \,,
$$

on one hand, and $E[f(\lambda^{-\frac{1}{\mu}} Z^{\frac{1}{\mu}} X)] = \int_0^\infty E\left[f\left(t^{\frac{1}{\mu}} X \right) \right] \lambda e^{-\lambda t} \, dt$ on the other hand. Equating these expressions, and using the injectivity of the Laplace transform, we obtain (i).

2. Suppose X, Z, Z_μ and T_μ are independent. Equation (4.19.3) allows us to write:

$$E[f(Z^{\frac{1}{\mu}}X)] = E\left[f\left(\frac{Z_\mu X}{T_\mu}\right)\frac{\Gamma(\mu+1)}{T_\mu^\mu}\right].$$

So the equivalence between (ii) and (iii) is a consequence of the following equalities,

$$E\left[f\left(\frac{Z_\mu X}{T_\mu}\right)\frac{\Gamma(\mu+1)}{T_\mu^\mu}\right] = E\left[\int_0^\infty dt\,\frac{c}{\Gamma(\mu)}t^{\mu-1}e^{-tY}f\left(\frac{t}{T_\mu}\right)\frac{\Gamma(\mu+1)}{T_\mu^\mu}\right]$$

$$= \int_0^\infty du\,c\mu u^{\mu-1}f(u)E[\exp(-T_\mu uY)]$$

$$= \int_0^\infty du\,c\mu u^{\mu-1}f(u)E[\exp(-uY)^\mu] = E[f(Z^{\frac{1}{\mu}}X)],$$

where we set $u = \frac{t}{T_\mu}$ in the second written expression.

Solution to Exercise 4.21

1. Let Z, T_ν, T_μ be as in the statement of question 1. For all $t \geq 0$, we have:

$$P(Z^{\frac{1}{\mu}} > tT_\nu) = E\left(\int_{t^\mu T_\nu^\mu}^\infty e^{-s}\,ds\right) = E[\exp(-t^\mu T_\nu^\mu)]$$

$$= E[\exp(-tT_\nu T_\mu)] = P(Z > tT_\nu T_\mu).$$

The third equality following from the expression of the Laplace transform of T_μ, given in Exercise **4.19**. This proves the identity in law

$$\frac{Z^{\frac{1}{\mu}}}{T_\nu} \overset{\text{(law)}}{=} \frac{Z}{T_\mu T_\nu},$$

and the other identity follows by exchanging the role of μ and ν. We deduce (4.21.2), from the above identity in law and the fact that $\frac{1}{T_\nu}$ is simplifiable (see Exercise **1.13**).

2. Choosing $a = \frac{1}{n}$ in (4.5.4) and $\mu = \frac{1}{n}$ in (4.21.2), we obtain:

$$Z^n \overset{\text{(law)}}{=} n^n Z_{\frac{1}{n}}\ldots Z_{\frac{n-1}{n}}Z \overset{\text{(law)}}{=} \frac{Z}{T_{\frac{1}{n}}}.$$

Since Z is simplifiable, we can write:

$$n^n Z_{\frac{1}{n}}\ldots Z_{\frac{n-1}{n}} \overset{\text{(law)}}{=} \frac{1}{T_{\frac{1}{n}}},$$

which leads directly to the required formula.

3. When we plug the representations:

$$Z^n \overset{\text{(law)}}{=} n^n Z_{\frac{1}{n}}\ldots Z_{\frac{n-1}{n}}Z$$

$$Z^m \overset{\text{(law)}}{=} m^m Z_{\frac{1}{m}}\ldots Z_{\frac{m-1}{m}}Z,$$

into the formula

$$Z^n \overset{\text{(law)}}{=} \frac{Z^m}{\left(T_{\frac{m}{n}}\right)^m},$$

we obtain

$$n^n Z_{\frac{1}{n}} \ldots Z_{\frac{n-1}{n}} \overset{\text{(law)}}{=} m^m Z_{\frac{1}{m}} \ldots Z_{\frac{m-1}{m}} \frac{1}{\left(T_{\frac{m}{n}}\right)^m}. \tag{4.21.a}$$

(We simplified by Z on both sides.) Then we may use the beta–gamma algebra (formula (4.2.1)) to write:

$$Z_{\frac{k}{n}} \overset{\text{(law)}}{=} Z_{\frac{k}{n}, \frac{k}{m} - \frac{k}{n}} Z_{\frac{k}{m}}, \quad k \leq n,$$

where, on the right hand side, $Z_{\frac{k}{n}, \frac{k}{m} - \frac{k}{n}}$ is independent of $Z_{\frac{k}{m}}$. Then (4.21.a) simplifies into:

$$\left(\frac{m}{T_{\frac{m}{n}}}\right)^m \overset{\text{(law)}}{=} n^n \left(\prod_{k=1}^{m-1} Z_{\frac{k}{n}, k\left(\frac{1}{m} - \frac{1}{n}\right)}\right) \left(\prod_{k=m}^{n-1} Z_{\frac{k}{n}}\right),$$

which is the desired formula.

4. Raising formula (4.8.2) to the power 2^p gives:

$$Z^{2^p} \overset{\text{(law)}}{=} 2^{2^p - 1} \left(N_1^{2^p} N_2^{2^{p-1}} \ldots N_p^2\right) Z, \tag{4.21.b}$$

where $N_1, \ldots, N_p$ are independent centered Gaussian variables with variance 1. Combining this identity with equation (4.21.2) for $\mu = \frac{1}{2^p}$ and simplifying by Z gives formula (4.21.5).

5. (a) Equation (4.21.6) follows from either formula (4.21.3) for $n = 2$ or formula (4.21.5) for $p = 1$ and has already been noticed in question 1 of Exercise **4.17**: precisely, one has $T_{\frac{1}{2}} \overset{\text{(law)}}{=} \frac{1}{2N^2}$, which yields (4.21.6).

Equation (4.21.7) follows from formula (4.21.3) for $n = 3$:

$$P\left(\frac{1}{T_{\frac{1}{3}}} > x\right) = P\left(Z_{\frac{2}{3}} > \frac{x}{27 Z_{\frac{1}{3}}}\right)$$

$$= \frac{1}{\Gamma\left(\frac{2}{3}\right)} E\left[\int_{x(27 Z_{\frac{1}{3}})^{-1}}^{\infty} dy \frac{e^{-y}}{y^{\frac{1}{3}}}\right].$$

Hence, we obtain:

$$P\left(\left(T_{\frac{1}{3}}\right)^{-1} \in dx\right) = \frac{dx}{\Gamma\left(\frac{2}{3}\right)} E\left[\frac{1}{27 Z_{\frac{1}{3}}} \left(\frac{x}{27 Z_{\frac{1}{3}}}\right)^{-\frac{1}{3}} \exp\left(-\frac{x}{27 Z_{\frac{1}{3}}}\right)\right]$$

$$= \frac{dx}{9\,\Gamma\left(\frac{1}{3}\right)\Gamma\left(\frac{2}{3}\right)} \frac{1}{x^{\frac{1}{3}}} \int_0^{\infty} \frac{dy}{y^{\frac{4}{3}}} \exp\left(-\left(\frac{x}{27y} + y\right)\right),$$

which yields (4.21.7), after some elementary transformations.

(b) We use the representation (4.21.5) for $p = 2$:

$$T_{\frac{1}{4}} \overset{(\mathrm{law})}{=} \frac{1}{8N_1^2 N_2^4}.$$

First we write

$$N_1 + iN_2 = R\exp(i\Theta),$$

where Θ and R are independent, Θ is uniform on $[0, 2\pi)$, and from question 3 of Exercise **4.2**, $R = (N_1^2 + N_2^2)^{\frac{1}{2}} \overset{(\mathrm{law})}{=} \sqrt{2Z}$. This gives

$$T_{\frac{1}{4}} \overset{(\mathrm{law})}{=} \frac{1}{8R^6(\cos\Theta)^2(\sin\Theta)^4} \overset{(\mathrm{law})}{=} \frac{1}{4^3 Z^3 A(1-A)^2},$$

where $A = \cos^2\Theta$ is independent of Z and has law $P(A \in da) = \frac{1}{\pi}\frac{da}{\sqrt{a(1-a)}}$, $a \in [0,1]$. To obtain (4.21.8), it remains to write:

$$P\left(\frac{1}{4(T_{\frac{1}{4}})^{\frac{1}{3}}} > \frac{1}{4x^{\frac{1}{3}}}\right) = P(T_{\frac{1}{4}} < x) = P\left(Z > \frac{1}{4x^{\frac{1}{3}}A^{\frac{1}{3}}(1-A)^{\frac{2}{3}}}\right)$$

$$= E\left[\exp\left(-\frac{1}{4x^{\frac{1}{3}}A^{\frac{1}{3}}(1-A)^{\frac{2}{3}}}\right)\right].$$

(c) From (4.21.4), we have

$$T \overset{(\mathrm{law})}{=} Z_{\frac{1}{3},\frac{1}{6}} Z_{\frac{2}{3}}.$$

(We refer to Exercise **4.2** for the law of $Z_{\frac{1}{3},\frac{1}{6}}$.) Then, we obtain:

$$P(T > x) = P(Z_{\frac{2}{3}} > xZ_{\frac{1}{3},\frac{1}{6}}^{-1}) = \frac{1}{\Gamma\left(\frac{2}{3}\right)}E\left[\int_{xZ_{\frac{1}{3},\frac{1}{6}}^{-1}} dt\, t^{\frac{2}{3}-1}e^{-t}\right],$$

hence, denoting here Z for $Z_{\frac{1}{3},\frac{1}{6}}$, we obtain:

$$P(T \in dx) = \frac{1}{\Gamma\left(\frac{2}{3}\right)}E\left[\frac{1}{Z}\left(\frac{Z}{x}\right)^{\frac{1}{3}}\exp\left(-\frac{x}{Z}\right)\right] = \frac{C}{x^{\frac{1}{3}}}\int_0^1 dz\,\frac{e^{-\frac{x}{z}}}{z^{\frac{4}{3}}(1-z)^{\frac{5}{6}}},$$

with $C = \left(\Gamma\left(\frac{2}{3}\right)B\left(\frac{1}{3},\frac{1}{6}\right)\right)^{-1}$.

Solution to Exercise 4.22

To prove identity (4.22.2) (i), it is enough to show that for every $\gamma > 0$,

$$E[|\Delta_n(X_1,\ldots,X_n)|^{2\gamma}] = \prod_{j=2}^n E\left[\left(\frac{1}{T_{\frac{1}{j}}}\right)^{\gamma}\right],$$

but from identity (4.21.2), (or (4.19.2)), one has

$$E\left[\left(\frac{1}{T_{\frac{1}{j}}}\right)^{\gamma}\right] = \frac{E[Z^{j\gamma}]}{E[Z^{\gamma}]} = \frac{\Gamma(1+j\gamma)}{\Gamma(1+\gamma)},$$

for every j, and the result follows thanks to (4.22.1).

A direct application of (4.21.3) shows that for every $j = 2, 3, \ldots, n$:

$$\frac{1}{T_{\frac{1}{j}}} \overset{\text{(law)}}{=} (j^j) \prod_{1 \le i < j} Z_{\frac{i}{j}}.$$

Since the r.v.s $(j^j) \prod_{1 \le i < j} Z_{\frac{i}{j}}$, $j = 2, 3, \ldots, n$, on the right hand side, are indepen-dent, their product is equal in law to the product of $\frac{1}{T_{\frac{1}{j}}}$, $j = 2, 3, \ldots, n$ and we have obtained (4.22.2) (ii).

Solution to Exercise 4.23

1. Let Z be an exponential variable with mean 1, such that T_{μ}, T_{μ}' and Z are independent. First, for every $s \ge 0$, we have $E[\frac{1}{1+sX}] = E[e^{-sXZ}]$. Moreover, the identity in law (4.21.2) implies that for every $s \ge 0$:

$$E\left[e^{-sXZ}\right] = E\left[e^{-sT_{\mu}Z^{\frac{1}{\mu}}}\right] = E\left[e^{-s^{\mu}Z}\right],$$

and the last term clearly equals $\frac{1}{1+s^{\mu}}$.

2. From the definition of X we have: $E[X^s] = E[T_{\mu}^s]E[T_{\mu}^{-s}]$. The identity (4.19.4) implies $E[T_{\mu}^{-s}] = \frac{\Gamma\left(\frac{s}{\mu}\right)}{\mu\Gamma(s)}$, for $s > 0$, as well as $E[T_{\mu}^s] = \frac{\Gamma\left(1-\frac{s}{\mu}\right)}{\Gamma(1-s)}$, for $s < \mu$. So that with the formula of complements, as given in Exercise **4.18**, we obtain

$$E[X^s] = \frac{1}{\mu}\frac{\Gamma\left(\frac{s}{\mu}\right)\Gamma\left(1-\frac{s}{\mu}\right)}{\Gamma(s)\Gamma(1-s)} = \frac{\sin \pi s}{\mu \sin \frac{\pi s}{\mu}}, \quad 0 < s < \mu.$$

3. Now observe that the expression (4.23.4) admits an analytical continuation in the domain $\{z \in \mathbf{C} : 0 < \mathrm{Re}(z) < \mu\}$, hence the characteristic function of $\log X^{\mu}$ is given by

$$E[X^{i\mu t}] = E[\exp it \log X^{\mu}] = \frac{\sinh \pi \mu t}{\mu \sinh \pi t}, \quad t \in \mathbb{R}.$$

Thanks to the integral computed above, we deduce that

$$P(\log X^{\mu} \in dx) = \frac{1}{\pi\mu}\frac{\sin \pi\mu}{\cosh x + \cos \pi\mu} dx,$$

from which we deduce (4.23.3) by an obvious change of variables. Finally, again some elementary change of variables allows us to deduce the expression of $\frac{T_\mu}{T'_\mu}$ in terms of a Cauchy variable from (4.23.3).

Solution to Exercise 4.24

Part A.

Let X be an r.v. such that: $E[X^n] = \frac{\mu(n)}{\nu(n)}$. We deduce from our definition of μ and ν that $E[V^n] = E[(WX)^n]$, for every n; but since V is moments determinate, we have $V \stackrel{(\text{law})}{=} WX$. On the other hand, we know that $V \stackrel{(\text{law})}{=} WR$, hence, $WX \stackrel{(\text{law})}{=} WR$. Since W is simplifiable, it follows that $X \stackrel{(\text{law})}{=} R$.

Part B. Examples.

(i) Write $V \stackrel{(\text{law})}{=} ZZ'$, where Z and Z' are two independent exponential variables and $W \stackrel{(\text{law})}{=} Z_\alpha Z_\beta$, where Z_α and Z_β are two independent gamma variables, with respective parameters α and β. The sequences of moments given in (i) are those of V and W respectively. Moreover, from the beta–gamma algebra (see Exercise **4.2**), we have

$$Z \stackrel{(\text{law})}{=} Z_{1,\alpha-1} Z_\alpha \quad \text{and} \quad Z' \stackrel{(\text{law})}{=} Z_{1,\beta-1} Z_\beta,$$

consequently, $R \stackrel{(\text{law})}{=} Z_{1,\alpha-1} Z_{1,\beta-1}$, whose density is

$$(\alpha-1)(\beta-1) \int_r^1 (1-t)^{\alpha-2} \frac{(t-r)^{\beta-2}}{t^{\beta-1}} \, dt, \quad (0 < r < 1).$$

(ii) Here, the moments $\mu(n)$ are those of $V \stackrel{(\text{law})}{=} Z^2$; then the duplication formula for the gamma function translates probabilistically as: $Z^2 \stackrel{(\text{law})}{=} 2ZN^2$ (see Exercise **4.5**). Hence, since $W \stackrel{(\text{law})}{=} Z$, we get $R \stackrel{(\text{law})}{=} 2N^2 \stackrel{(\text{law})}{=} Z_{1/2}$.

(iii) Now, the moments $\mu(n)$ are those of $V \stackrel{(\text{law})}{=} Z^3$, and the moments $\nu(n)$ are those of $W \stackrel{(\text{law})}{=} Z^2$. From the discussion in Exercise **4.21**, and more particularly (4.21.2) for $\mu = \frac{2}{3}$, we find that $R = \left(\frac{1}{T_{2/3}}\right)^2$ is a desired solution. Note that, here, following the comment (b) in Exercise **1.11**, the law of Z^3 is not moments determinate!

Chapter 5

Convergence of random variables

$\boxed{\text{``Overhead''}}$

• A sequence of random variables $\{X_n, n \in \mathbb{N}\}$ may converge in a number of different ways, i.e.

$$\{X_n \underset{n \to \infty}{\overset{a.s.}{\longrightarrow}} \cdot\} \;\Rightarrow\; \{X_n \underset{n \to \infty}{\overset{(P)}{\longrightarrow}} \cdot\} \;\Rightarrow\; \{\mathcal{L}(X_n) \underset{n \to \infty}{\overset{(w)}{\longrightarrow}} \cdot\}$$
$$\Uparrow$$
$$\{X_n \underset{n \to \infty}{\overset{L^p}{\longrightarrow}} \cdot\}$$

• Important additional assumptions are necessary to obtain implications in reverse order. Adequate references may be found in most of the books cited in the References.

• The most well-known examples of almost sure convergence occur with *the Law of Large Numbers* (LLN), whereas *the Central Limit Theorem* (CLT) involves convergence in distribution. See Jacod and Protter [40] for a "minimalist" discussion.

• To a large extent, the "universal character" of the LLN and the CLT make them the "two pillars" of Probability Theory.

• None of our exercises (with the exception of Exercise **5.6**) discuss an important complement to the LLN and the CLT, namely the LDT: *the Large Deviations Theorem of Cramer*, and its many extensions due to Donsker–Varadhan and Freidlin–Wentzell, among others. We recommend the nice succession of exercises in Letac ([52], Exercises 408, 409 and 410), which constitutes a self-contained proof of Cramer's LDT Theorem.

---

* 5.1 Around Scheffé's lemma

Throughout this exercise, we consider a sequence (X_n) of positive, integrable, random variables such that

$$X_n \overset{(P)}{\longrightarrow} X,$$

with $E(X) < \infty$.

Then Scheffé's lemma asserts that

$$\lim_{n\to\infty} E(|X_n - X|) = 0 \quad \text{if and only if} \quad \lim_{n\to\infty} E(X_n) = E(X).$$

In this exercise, we provide a statement from which Scheffé's lemma is easily derived. Let us also consider $Y \in L^1_+$.

1. Prove that if $\lim_{n\to\infty} E(X_n) = E(X)$, then

$$E[|X_n - Y|]$$

   converges to $E[|X - Y|]$, as $n \to \infty$.

2. Derive Scheffé's lemma.

* 5.2 Convergence of sums of squares of independent Gaussian variables

Let $(X_n, n \in \mathbb{N})$ be a sequence of independent Gaussian variables, with respective mean μ_n, and variance σ_n^2.

1. Prove that $\Sigma_n X_n^2$ converges in L^1 if, and only if:

$$\Sigma_n(\mu_n^2 + \sigma_n^2) < \infty. \tag{5.2.1}$$

2. Prove that, if the condition (5.2.1) is satisfied, then: $\Sigma_n X_n^2$ converges in L^p, for every $p \in [1, \infty)$.

3. Assume that $\mu_n = 0$, for every n. Prove that:

   if $\Sigma_n \sigma_n^2 = \infty$, then $P(\Sigma_n X_n^2 = \infty) = 1$. (*See complements after Exercise 5.17*).

* 5.3 Convergence of moments and convergence in law

Let $(X_n, n \in \mathbb{N})$ be a sequence of r.v.s which take their values in $[0, 1]$.

1. Prove that, if for every $k \in \mathbb{N}$, $E(X_n^k) \xrightarrow[n\to\infty]{} \frac{1}{k+1}$, then the sequence (X_n) converges in law; identify the limit law.

2. Let $a > 0$. Solve the same question as question 1 when $\frac{1}{k+1}$ is replaced by $\frac{a}{k+a}$.

Comments and references. The result of this exercise is a very particular case (the X_ns are uniformly bounded) of application of the method of moments to prove convergence in law; precisely, if μ_n is the law of X_n, and if there exists a probability measure μ such that:

(i) the law μ is uniquely determined by its moments $\int x^k \, d\mu(x)$, $k \in \mathbb{N}$; (this is the case if $\int \exp(\alpha|x|) \, d\mu(x) < \infty$ for some $\alpha > 0$),

(ii) for every $k \in \mathbb{N}$, $\int x^k \, d\mu_n(x)$ $(= E[X_n^k])$ converges to $\int x^k \, d\mu(x)$,

then the sequence (X_n) converges in distribution to μ (see Feller [28], p. 269).

This is also discussed in Billingsley [7], Theorem 30.2, in the 1979 edition.

* 5.4 Borel test functions and convergence in law

Let (X_n) and (Y_n) be two sequences of r.v.s such that:

(i) the law of X_n does not depend on n;

(ii) $(X_n, Y_n) \xrightarrow[n \to \infty]{\text{(law)}} (X, Y)$.

1. Show that for every Borel function $\varphi : \mathbb{R} \to \mathbb{R}$, the pair $(\varphi(X_n), Y_n)$ converges in law towards: $(\varphi(X), Y)$.

2. Give an example for which the condition (ii) is satisfied, but (i) is not, and such that there exists a Borel function $\varphi : \mathbb{R} \to \mathbb{R}$ for which $(\varphi(X_n), Y_n)$ does not converge in law towards $(\varphi(X), Y)$.

Comments and references. For some applications to asymptotic results for functionals of Brownian motion, see Revuz and Yor [75], Chapter XIII.

* 5.5 Convergence in law of the normalized maximum of Cauchy variables

Let $(X_1, X_2, \ldots, X_n, \ldots)$ be a sequence of independent Cauchy variables with parameter $a > 0$, i.e.

$$P(X_i \in dx) = \frac{a \, dx}{\pi(a^2 + x^2)} \ .$$

Show that $\frac{1}{n} \left(\sup_{i \leq n} X_i \right)$ converges in law towards $\frac{1}{T}$, where T is an exponential variable, the parameter of which shall be computed in terms of a.

Comments and references: During our final references searches, we found that this exercise is in Grimmett and Stirzaker [38], p. 356!

* 5.6 Large deviations for the maximum of Gaussian vectors

Let $X = (X_1, \ldots, X_n)$ be any centered Gaussian vector.

1. Prove that for every $r \geq 0$:

$$P(\max_{1 \leq i \leq n} X_i \geq E[\max_{1 \leq i \leq n} X_i] + \sigma r) \leq e^{-\frac{r^2}{2}}, \qquad (5.6.1)$$

where $\sigma = \max_{1 \leq i \leq n} E[X_i^2]^{\frac{1}{2}}$.

Hint. Use Exercise **3.11** for a suitable choice of a Lipschitz function f.

2. Deduce from above that:

$$\lim_{r \to +\infty} \frac{1}{r^2} \log P(\max_{1 \leq i \leq n} X_i \geq r) = -\frac{1}{2\sigma^2}. \qquad (5.6.2)$$

Comments and references.

As in Exercise **3.11**, the above results may be extended to continuous time processes.

The identity (5.6.2) is a large deviation formulation of the inequality (5.6.1) and aims at investigating the supremum of Gaussian processes on a finite time interval, such as the supremum of the Brownian bridge, see Chapter 6.

R. AZENCOTT: Grandes déviations et applications. *Eighth Saint Flour Probability Summer School–1978* (Saint Flour, 1978), 1–176, *Lecture Notes in Mathematics,* **774**, Springer, Berlin, 1980.

M. LEDOUX: Isoperimetry and Gaussian analysis. *Lectures on Probability Theory and Statistics* (Saint Flour, 1994), 165–294, *Lecture Notes in Mathematics,* **1648**, Springer, Berlin, 1996.

* 5.7 A logarithmic normalization

Let $r > 0$, and consider $\pi_r(du) = r \sinh(u) e^{r(1-\cosh u)} du$, a probability on $\mathbb{R}_+$.

Let X_n be an r.v., with distribution $\pi_{1/n}$.

1. What is the law of $\frac{1}{n} \cosh X_n$?

2. Prove that: $\log(\cosh X_n) - \log n \xrightarrow[n \to \infty]{(\text{law})} Y$: Compute the law of Y.

3. Deduce therefrom that: $\dfrac{\log(\cosh X_n)}{\log n} \xrightarrow[n\to\infty]{(P)} 1$, and then that: $\dfrac{X_n}{\log n} \xrightarrow[n\to\infty]{(P)} 1$.

4. Deduce finally that $X_n - \log n \xrightarrow[n\to\infty]{(\text{law})} Y + \log \mathcal{Z}$. Give a direct proof of this result, using simply the fact that X_n is distributed with $\pi_{1/n}$.

Comments and references.

(a) The purpose of this exercise is to show that simple manipulations on a seemingly complicated sequence of laws may lead to limit results, without using the law of large numbers, or the central limit theorem.

(b) The distribution π_r occurs in connection with the distribution of the winding number of planar Brownian motion. See: M. YOR: Loi de l'indice du lacet Brownien, et distribution de Hartman-Watson. *Zeit. für Wahr.* **53** (1980), pp. 71–95.

** 5.8 A $\sqrt{n \log n}$ normalization

1. Let U be a uniform r.v. on $[0, 1]$, and ε an independent Bernoulli r.v., i.e. $P(\varepsilon = +1) = P(\varepsilon = -1) = 1/2$.

   Compute the law of $X = \varepsilon/\sqrt{U}$.

2. Let $X_1, X_2, \ldots, X_n, \ldots$ be a sequence of independent r.v.s which are distributed as X. Prove that:
$$\frac{X_1 + X_2 + \cdots + X_n}{(n \log n)^{1/2}} \xrightarrow[n\to\infty]{(\text{law})} N \,,$$

where N is Gaussian, centered, with variance 1.

3. Let $\alpha > 0$, and let $X_1, X_2, \ldots, X_n, \ldots$ be a sequence of independent r.v.s such that:
$$P(X_i \in dy) = c \frac{dy}{|y|^{3+\alpha}} \mathbb{1}_{\{|y| \geq 1\}}$$

for a certain constant c.

Give a necessary and sufficient condition on a deterministic sequence $(\varphi(n), n \in \mathbb{N})$ such that: $\varphi(n) \to \infty$, as $n \to \infty$ which implies:

$$\frac{X_1 + \cdots + X_n}{\varphi(n)} \xrightarrow[n\to\infty]{(\text{law})} N \,,$$

where N is Gaussian, centered. (*See complements after Exercise 5.17.*)

Comments and references. As a recent reference for limit theorems, we recommend the book by V.V. Petrov [68]. Of course, Feller [28] remains a classic.

* 5.9 The Central Limit Theorem involves convergence in law, not in probability

1. Consider, on a probability space $(\Omega, \mathcal{A}, P)$, a sequence $(Y_n, n \geq 1)$ of independent, equidistributed r.v.s, and X a Gaussian variable which is independent of the sequence $(Y_n, n \geq 1)$. We assume moreover that Y_1 has a second moment, and that:
$$E[Y_1] = E[X] = 0 \; ; \;\; E[Y_1^2] = E[X^2] = 1.$$
Define: $U_n = \frac{1}{\sqrt{n}}(Y_1 + \cdots + Y_n)$.

Show that: $\lim_{n \to \infty} E[\|U_n - X\|]$ exists, and compute this limit.

Hint: Use Exercise **1.4**.

2. We keep the assumptions concerning the sequence $(Y_n, n \geq 1)$.

Show that, for any fixed $p \in \mathbb{N}$, the vector
$$(Y_1, Y_2, \ldots, Y_p, U_n)$$
converges in law as $n \to \infty$. Describe the limit law.

3. Does $(U_n, n \to \infty)$ converge in probability?

Hint.   Assume that the answer is positive, and use the results of questions 1 and 2.

4. Prove that: $\bigcap_p \sigma\{Y_p, Y_{p+1}, \ldots\}$ is trivial, and give another argument to answer question 3.

Comments. See also Exercise **5.12** for some related discussion involving "non-convergence in probability".

** 5.10 Changes of probabilities and the Central Limit Theorem

Consider two probabilities P and Q defined on a measurable space $(\Omega, \mathcal{A})$ such that $Q = D \cdot P$, for $D \in L^1_+(\Omega, \mathcal{A}, P)$, i.e. Q is absolutely continuous w.r.t. P on $\mathcal{A}$.

Let $X_1, X_2, \cdots, X_n, \cdots$ be a sequence of i.i.d. r.v.s under P, with second moment; we write $m = E_P(X_1)$, and $\sigma^2 = E_P((X_1 - m)^2)$. Prove that, under Q, one has:
$$\frac{1}{\sigma\sqrt{n}} \sum_{i=1}^{n} (X_i - m) \xrightarrow{\text{(law)}} N$$
where N denotes a centered Gaussian variable, with variance 1.

Comments and references:

(a) This exercise shows that the fundamental limit theorems of Probability Theory: *the Law of Large Numbers* and *the Central Limit Theorem* are valid not only in the classical framework of i.i.d. r.v.s (with adequate integrability conditions) but also in larger frameworks. A sketch of the proof of this result can be found in P. Billingsley [7], although our proof is somewhat more direct.

    Of course, there are many other extensions of these fundamental limit theorems, the simplest of which is (arguably) the case proposed in this exercise.

(b) This exercise is also found, in fact in greater generality, in Revuz ([73], p. 170, Exercise (5.14)).

(c) (Comment by J. Pitman.) The property studied in this exercise gave rise to the notion of stable convergence, as developed by Rényi (1963) and Aldous–Eagleson (1978).

* 5.11 Convergence in law of stable(μ) variables, as $\mu \to 0$

Let $0 < \mu < 1$, and T_μ a stable(μ) variable whose Laplace transform is given by $\exp(-\lambda^\mu)$, $\lambda \geq 0$.

1. Prove that, as $\mu \to 0$, $(T_\mu)^\mu$ converges in law towards $\frac{1}{Z}$, where Z is a standard exponential variable.

2. Let T'_μ be an independent copy of T_μ. Prove that $\left(\frac{T_\mu}{T'_\mu}\right)^\mu$ converges in law towards $\frac{Z}{Z'}$, where Z and Z' are two independent copies.

3. Prove that: $\frac{Z}{Z'} \overset{\text{(law)}}{=} \frac{1}{U} - 1$, where U is uniform on $(0,1)$, and prove the result of question 2, using the explicit form of the density of $\left(\frac{T_\mu}{T'_\mu}\right)^\mu$ as given in formula (4.23.3).

Comments and references. A full discussion of this kind of asymptotic behaviour for the four-parameter family of stable variables is provided by:

N. Cressie: A note on the behaviour of the stable distributions for small index α. *Z. Wahrscheinlichkeitstheorie und Verw. Gebiete*, **33**, no. 1, 61–64 (1975/76).

** 5.12 Finite-dimensional convergence in law towards Brownian motion

Let $T_1, T_2, \ldots, T_n, \ldots$ be a sequence of positive r.v.s which have the common distribution:

$$\mu(dt) = \frac{dt}{\sqrt{2\pi t^3}} \exp\left(-\frac{1}{2t}\right).$$

[Before starting to solve this exercise, it may be helpful to have a look at the first three questions of Exercise **4.17**.]

1. Let $(\varepsilon_i; i \geq 1)$ be a sequence of positive reals.

   Give a necessary and sufficient condition which ensures that the sequence:

   $$\frac{1}{n} \sum_{i=1}^{n} \varepsilon_i^2 T_i, \quad n \in \mathbb{N},$$

   converges in law. When this condition is satisfied, what is the limit law?

   Hint. The limit law may be represented in terms of μ.

2. We now assume, *for the remainder of this exercise*, that $\varepsilon_i = 1/2\sqrt{i}$ $(i \geq 1)$. We define: $S_n = \frac{1}{n} \sum\limits_{i=1}^{n} \varepsilon_i^2 T_i$.

   (a) Prove that the sequence $(S_n)_{n \in \mathbb{N}}$ converges in law.

   (b) Let $k \in \mathbb{N}, k > 1$. Prove that the two-dimensional random sequence:

   $$\left(S_{(k-1)n}, S_{kn}\right)$$

   converges in law, as $n \to \infty$, towards $(\alpha T_1, \beta T_1 + \gamma T_2)$, where α, β, γ are three constants which should be expressed in terms of k.

   (c) More generally, prove that, for every $p \in \mathbb{N}, p > 1$, the p-dimensional random sequence

   $$(S_n, S_{2n}, \ldots, S_{pn})$$

   converges in law, as $n \to \infty$, towards an r.v. which may be represented simply in terms of the vector $(T_1, T_2, \ldots, T_p)$.

   (d) Does the sequence (S_n) converge in probability?

3. We now refine the preceding study by introducing continuous time.

   (a) Define $S'_n(t) = \frac{1}{n} \sum\limits_{i=1}^{[nt^2]} \varepsilon_i^2 T_i$, where $[x]$ denotes the integer part of x.

Prove that, as $n \to \infty$, the finite-dimensional distributions of the process $(S'_n(t),\, t \geq 0)$ converge weakly towards those of a process $(T_t,\, t \geq 0)$ which has independent and time-homogeneous increments, that is: if $t_1 < t_2 < \cdots < t_p$, the variables $T_{t_1}, T_{t_2} - T_{t_1}, \ldots, T_{t_p} - T_{t_{p-1}}$ are independent, and the law of $T_{t_{i+1}} - T_{t_i}$ depends only on $(t_{i+1} - t_i)$. The latter distribution should be computed explicitly.

(b) Now, let $X_1, X_2, \ldots, X_n, \ldots$ be a sequence of independent r.v.s, which are centered, have the same distribution, and admit a second moment. Define: $\xi_n(t) = \frac{1}{\sqrt{n}} \sum\limits_{i=1}^{[nt]} X_i$.

Prove that, as $n \to \infty$, the finite-dimensional distributions of the process $(\xi_n(t),\, t \geq 0)$ converge weakly towards those of a process with independent and time homogeneous increments which should be identified.

Comments and references:

(a) If $(B_t,\, t \geq 0)$ is the standard Brownian motion, then the weak limit of the process $(\xi_n(t),\, t \geq 0)$, introduced at the end of the exercise, has the same law as $(\sigma B_t,\, t \geq 0)$, where $\sigma^2 = E[X_i^2]$, in (b) above. In this setting, the process $(T_t,\, t \geq 0)$ may be defined as the first hitting time process of σB, that is, for any $t \geq 0$, $T_t = \inf\{s \geq 0, \sigma B_s \geq t\}$. Since σB has independent and time homogeneous increments, the same properties hold for the increments of $(T_t,\, t \geq 0)$ and the results of questions 3 (a) and 3 (b) follow.

(b) We leave to the reader the project of amplifying this exercise when the T_is are replaced by stable(μ) variables, i.e. variables which satisfy

$$E[\exp(-\lambda T_i)] = \exp(-\lambda^\mu), \quad \lambda \geq 0,$$

for some $0 < \mu < 1$ (the case treated here is $\mu = 1/2$).

** 5.13 The empirical process and the Brownian bridge

Let $U_1, U_2, \ldots$ be independent r.v.s which are uniformly distributed on $[0, 1]$. We define the stochastic process $b^{(n)}$ on the interval $[0,1]$ as follows:

$$b^{(n)}(t) = \sqrt{n} \left(\frac{1}{n} \sum_{k=1}^{n} \mathbb{1}_{[0, t]}(U_k) - t \right), \quad t \in [0, 1]. \tag{5.13.1}$$

1. For any s and t in the interval $[0, 1]$, compute $E[b^{(n)}(t)]$ and $\mathrm{Cov}[b^{(n)}(s), b^{(n)}(t)]$.

2. Prove that, as $n \to \infty$, the finite-dimensional distributions of the process $(b^{(n)}(t),\, t \in [0,1])$ converge weakly towards those of a Gaussian process, $(b(t),\, t \in [0,1])$, whose means and covariances are the same as those of $b^{(n)}$.

Comments and references. The process $b^{(n)}$ is very often involved in mathematical statistics and is usually called the empirical process. For large n, its path properties are very close to those of its weak limit $(b(t), t \in [0, 1])$ which is known as the Brownian bridge. The latter has the law of the Brownian motion on the time interval $[0, 1]$, starting from 0 and conditioned to return to 0 at time 1. Actually, the convergence of $b^{(n)}$ towards b holds in much stronger senses. This is the well-known Glivenko–Cantelli Theorem which may be found in the following references.

G.R. SHORACK: *Probability for Statisticians*. Springer Texts in Statistics, Springer-Verlag, New York, 2000.

G.R. SHORACK AND J.A. WELLNER: *Empirical Processes with Applications to Statistics*. Wiley Series in Probability and Mathematical Statistics: Probability and Mathematical Statistics, John Wiley & Sons, Inc., New York, 1986.

See also Chapter 8, p. 145, of Toulouse [93].

* 5.14 The functional law of large numbers

Let $\{X_n, n \geq 1\}$ be a sequence of i.i.d. real-valued random variables such that $E(|X_1|) < \infty$. Define $S_0 = 0$ and $S_n = \sum_{k=1}^n X_k$, for $k \geq 1$.

Prove that, almost surely, the continuous time stochastic process

$$(n^{-1} S_{[nt]}, t \geq 0)$$

converges uniformly on every compact, towards the drift $(E[X_1] \cdot t, t \geq 0)$, when n tends to $+\infty$. (Here $[x]$ denotes the integer part of the real x.)

Comments and references. This fact is often invoked in discussions about invariance principles for functionals of sequences of converging random walks. It was used, for instance, in the proof of the convergence of the local time at the supremum, of a sequence of converging random walks in:

L. CHAUMONT AND R.A. DONEY: Invariance principles for local times at the maximum of random walks and Lévy processes. *Ann. Probab.*, **38**, no. 4, 1368–1389 (2010).

See Theorem 2 there.

** 5.15 The Poisson process and Brownian motion

Let $c > 0$, and assume that $(N_t^{(c)}, t \geq 0)$ is a Poisson process with parameter c.

Define: $X_t^{(c)} = N_t^{(c)} - ct$.

Show that, for any $n \in \mathbb{N}^*$, and any n-tuple $(t_1, \ldots, t_n) \in \mathbb{R}_+^n$, the random vector $\frac{1}{\sqrt{c}}(X_{t_1}^{(c)}, \ldots, X_{t_n}^{(c)})$ converges in law towards $(\beta_{t_1}, \ldots, \beta_{t_n})$, as $c \to \infty$, where $(\beta_t, t \geq 0)$ is a Gaussian process. Identify this process.

Comments. The processes $(N_t^{(c)}, t \geq 0)$ and $(\beta_t, t \geq 0)$ belong to the important class of stochastic processes with stationary and independent increments, better known as Lévy processes. The Poisson process and Brownian motion may be considered as the two building blocks for Lévy processes.

** 5.16 Brownian bridges converging in law to Brownian motions

Define a standard Brownian bridge $(b(u), 0 \leq u \leq 1)$ as a centered continuous Gaussian process, with covariance function

$$E[b(s)b(t)] = s(1-t), \quad 0 \leq s \leq t \leq 1. \tag{5.16.1}$$

1. (a) Let $(B_t, \leq t \leq 1)$ be a Brownian motion. Prove that

$$b(u) = B_u - uB_1, \quad 0 \leq u \leq 1 \tag{5.16.2}$$

is a standard Brownian bridge, which is independent of B_1. This justifies the following assertion.
Conditionally on $B_1 = 0$, $(B_u, 0 \leq u \leq 1)$ is distributed as $(B_u - uB_1, 0 \leq u \leq 1)$.

   (b) Prove that $(b(1-u), 0 \leq u \leq 1) \overset{(\text{law})}{=} (b(u), 0 \leq u \leq 1)$.

From now on, we shall use the pathwise representation (5.16.2) for the Brownian bridge.

2. Prove that, when $c \to 0$, the family of processes indexed by (u, s, t)

$$\left\{ (b(u), \, u \leq 1); \, \left(\frac{1}{\sqrt{c}} b(cs), \, 0 \leq s \leq \frac{1}{c} \right); \, \left(\frac{1}{\sqrt{c}} b(1-ct), \, 0 \leq t \leq \frac{1}{c} \right) \right\}, \tag{5.16.3}$$

converges in law towards

$$\left\{ (b(u), \, u \leq 1); \, (B_s, \, s \geq 0); \, (B_t', \, t \geq 0) \right\}, \tag{5.16.4}$$

where b, B and B' are independent and B, B' are two Brownian motions.

3. Prove that the family of processes

$$\left(\sqrt{\lambda}\, b(\exp(-t/\lambda)), \, t \geq 0 \right)$$

converges in law, as $\lambda \to +\infty$, towards Brownian motion.

Comments and references. These limits in law may be considered as "classical"; some consequences for the convergence of quadratic functionals of the Brownian bridge are discussed in

G. PECCATI AND M. YOR: Four limit theorems for quadratic functionals of Brownian motion and Brownian bridge. In: *Asymptotic Methods in Stochastics*, Amer. Math. Soc., Comm. Series, in honour of M. Csörgö, **44**, 75–87 (2004).

* 5.17 An almost sure convergence result for sums of stable random variables

Let $X_1, X_2, \ldots, X_n, \ldots$ be i.i.d. stable r.v.s with parameter $\alpha \in (0, 2]$, i.e. for any $n \geq 1$,

$$n^{-\frac{1}{\alpha}} S_n \stackrel{\text{(law)}}{=} X_1, \qquad (5.17.1)$$

where $S_n = \sum_{k=1}^n X_k$. Prove that for any bounded Borel function f,

$$\frac{1}{\log n} \sum_{k=1}^n \frac{1}{k} f\left(k^{-\frac{1}{\alpha}} S_k\right) \to E[f(X_1)], \quad \text{almost surely as } n \to +\infty. \qquad (5.17.2)$$

Hint. Introduce the stable process $(S_t)_{t \geq 0}$ which satisfies $S_1 \stackrel{\text{law}}{=} X_1$, and apply the Ergodic Theorem to the re-scaled process $Z_t = e^{-\frac{t}{\alpha}} S_{e^t}$, and to the shift transformation.

Comments and references. In the article:

P. ERDÖS AND G.A. HUNT: Changes of sign of sums of random variables. *Pac. J. Math.*, **3**, 673–687 (1953)

the authors proved a conjecture of P. Lévy (1937), i.e. for any sequence of partial sums $S_n = X_1 + \cdots + X_n$ of i.i.d. r.v.s (X_n) with continuous symmetric distribution,

$$\lim_{N \to \infty} \frac{1}{\log N} \sum_{k \leq N} \frac{1}{k} \mathbb{I}_{\{S_k > 0\}} = \frac{1}{2}, \quad \text{a.s.}$$

This almost sure convergence result has been later investigated independently in

P. SCHATTE: On strong versions of the central limit theorem. *Math. Nachr.*, **137**, 249–256 (1988)

and

G.A. BROSAMLER: An almost everywhere Central Limit Theorem. *Math. Proc. Camb. Philos. Soc.*, **104**, no. 3, 561–574 (1988).

where the authors proved that any sequence of i.i.d. r.v.s (X_n) with $E[X_1] = 0$ and $E(1|X_1|^{2+\delta}) < \infty$, for some $\delta > 0$ satisfies the so-called "Almost Sure Central Limit

Theorem":

$$\lim_{N \to \infty} \frac{1}{\log N} \sum_{k \leq N} \frac{1}{k} \mathbb{I}_{\left\{ \frac{s_k}{\sqrt{k}} < x \right\}} = \phi(x), \quad \text{a.s., for all } x \in \mathbb{R},$$

(ϕ is the standard Gaussian distribution function). It was shown slightly later that the constant δ may be assumed to be zero.

Many further developments have been made on this topic. The later ones concern for instance: r.v.s which are in the domain of attraction of a stable law (our exercise is a particular case), continuous time semi-stable processes, etc. An exhaustive survey on the topic is done in

I. BERKES: Results and problems related to the pointwise central limit theorem. *Asymptotic Methods in Probability and Statistics* (Ottawa, ON, 1997), 59–96, North-Holland, Amsterdam, 1998.

Comments on Exercises 5.2 and 5.8:

(C.2) Question 3 of Exercise **5.2** may be completed as follows:

(i) If a sequence of independent variables (Y_n) satisfies:
$Y_n \stackrel{\text{(law)}}{=} \sigma_n^2 Y$, with $Y \geq 0$, $E[Y] = 1$, then $\Sigma_n \sigma_n^2 < \infty$ if and only if $\Sigma_n Y_n < \infty$, a.s.

(ii) Exhibit a sequence (Y_n) of independent R_+-valued r.v.s such that $E[Y_n] = 1$ (hence: $\Sigma_m E[Y_n] = \infty$), but $P(\Sigma_n Y_n < \infty) = 1$
Solution: $P\left(Y_n = \frac{1}{n^2 - 1}\right) = 1 - \frac{1}{n^2}$; $P(Y_n = n^2 - 1) = \frac{1}{n^2}$.

(C.8) Question 3 of Exercise **5.8** may be completed as follows:

Let $\alpha > 1$, and let $X_1, X_2, \ldots$, be a sequence of independent r.v.s such that: $P(X_i \in dy) = c \frac{dy}{|y|^\alpha} 1\{|y| \geq 1\}$, for a certain constant c. Give a necessary and sufficient condition on a deterministic sequence $(\varphi(n), n \in \mathbb{N})$ such that: $\varphi(n) \to \infty$, as $n \to \infty$, which implies:

$$\frac{X_1 + \cdots + X_n}{\varphi(n)} \xrightarrow[n \to \infty]{\text{(law)}} Z,$$

where Z is β-stable, for some $\beta \in (0, 2)$.

Solution: $\alpha \in (1, 3)$; take $\varphi(n) = n^{1/\alpha - 1}$, $\beta = \alpha - 1$;

$\alpha = 3$; $\varphi(n) = (n \log n)^{1/2}$, $\beta = 2$;

$\alpha > 3$; $\varphi(n) = \sqrt{n}$, $\beta = 2$.

Solutions for Chapter 5

Solution to Exercise 5.1

1. Write

$$
\begin{aligned}
E[|X_n - Y|] &= E[(X_n - Y)^+] + E[(X_n - Y)^-] \\
&= E[X_n - Y] + 2E[(X_n - Y)^-] \\
&= E[X_n - Y] + 2E[(Y - X_n)^+].
\end{aligned}
$$

From Lebesgue's theorem of dominated convergence, the last term converges towards $2E[(Y - X)^+]$, which allows us to conclude since, by the assumption, the term $E[X_n - Y]$ converges to $E[X - Y] = E[|X - Y|] - 2E[(Y - X)^+]$.

2. This follows from question 1 by taking $Y = X$.

Solution to Exercise 5.2

1. The sum of r.v.s $\sum_i X_i^2$ converges in L^1 if and only if:

$$
\sum_i E[X_i^2] = \sum_i \left(\sigma_i^2 + \mu_i^2\right) < \infty.
$$

2. We show that under assumption (5.2.1), $\sum_i X_i^2$ belongs to L^p. First observe that we can write X_i as $\sigma_i Y_i + \mu_i$, where Y_i is a sequence of independent, centered Gaussian variables with variance 1. Therefore we have:

$$
\begin{aligned}
\left\| \sum_i X_i^2 \right\|_p &\leq \left\| \sum_i \sigma_i^2 Y_i^2 \right\|_p + \left\| \sum_i 2\sigma_i \mu_i Y_i \right\|_p + \sum_i \mu_i^2 \\
&\leq \|Y_1^2\|_p \sum_i \sigma_i^2 + \|Y_1\|_p \sum_i 2|\sigma_i \mu_i| + \sum_i \mu_i^2 ,
\end{aligned}
$$

which is finite thanks to the inequality $2|\sigma_i \mu_i| \leq \sigma_i^2 + \mu_i^2$. Thus, $\sum_{i=1}^n X_i^2$ converges in L^p towards $\sum_{i=1}^\infty X_i^2$, by Lebesgue's dominated convergence.

3. Let Y be a centered Gaussian variable with variance 1. For every n, we have:

$$E\left[e^{-\sum_{i=1}^{n} X_i^2}\right] = \prod_{i=1}^{n} E\left[e^{-\sigma_i^2 Y^2}\right] = \prod_{i=1}^{n}(1 + 2\sigma_i^2)^{-\frac{1}{2}}.$$

Since $\sum_{i=1}^{n} 2\sigma_i^2 \leq \prod_{i=1}^{n}(1 + 2\sigma_i^2)$, when $n \to \infty$, $E\left[e^{-\sum_{i=1}^{n} X_i^2}\right]$ converges towards 0. This proves that $P(\lim_{n \to +\infty} \sum_{i=1}^{n} X_i^2 = +\infty) > 0$. But the event $\{\lim_{n \to +\infty} \sum_{i=1}^{n} X_i^2 = +\infty\}$ belongs to the tail σ-field generated by $(X_n, n \in \mathbb{N})$, which is trivial, so we have $P(\lim_{n \to +\infty} \sum_{i=1}^{n} X_i^2 = +\infty) = 1$.

Solution to Exercise 5.3

1. Since $\frac{1}{k+1} = \int_0^1 du\, u^k$, it follows that $E[f(X_n)]$ converges towards $\int_0^1 du\, f(u)$ for any polynomial function f defined on $[0,1]$. Then, the Weierstrass Theorem allows us to extend this convergence to any continuous function on $[0,1]$. So $(X_n, n \geq 0)$ converges in law towards the uniform law on $[0,1]$.

2. We may use the same argument as in question 1. Another means is to consider the Laplace transform ψ_{X_n} of X_n. For every $\lambda \geq 0$, $\psi_{X_n}(\lambda)$ converges towards

$$\sum_{k \geq 0} \frac{(-\lambda)^k}{k!} \frac{a}{a+k} = \sum_{k \geq 0} \frac{(-\lambda)^k}{k!} E[X^k] = E[e^{-\lambda X}],$$

where X is an r.v. whose law has density: $a x^{a-1} \mathbb{1}_{\{x \in [0,1]\}}$. So $(X_n, n \geq 0)$ converges in law towards the distribution whose density is as above.

Solution to Exercise 5.4

1. It suffices to show that for any pair of bounded continuous functions f and g,

$$\lim_{n \to \infty} E[f(\varphi(X_n))g(Y_n)] = E[f(\varphi(X))g(Y)].$$

Let g_n be a uniformly bounded sequence of Borel functions which satisfies:

$$E[g(Y_n) \,|\, X_n] = g_n(X_n).$$

From the hypothesis, X_n converges in law towards X. Since the law of X_n does not depend on n, this law is the law of X. So, we can write:

$$E[f(\varphi(X_n))g_n(X_n)] = E[f(\varphi(X))g_n(X)].$$

Set $h = f \circ \varphi$ and let (h_n) be a sequence of continuous functions with compact support which converges towards h in the space $L^1(\mathbb{R}, \mathcal{B}_{\mathbb{R}}, \nu_X)$, where ν_X is the law of X. For any k and n, we have the inequalities:

$$\begin{aligned}
|E[h(X)g_n(X) - h(X)g(Y)]| &\leq E[|h(X)g_n(X) - h_k(X)g_n(X)|] \\
&\quad + |E[h_k(X)g_n(X)] - E[h_k(X)g(Y)]| \\
&\quad + E[|h_k(X)g(Y) - h(X)g(Y)|] \\
&\leq BE[|h(X) - h_k(X)|] \\
&\quad + |E[h_k(X)g_n(X)] - E[h_k(X)g(Y)]| \\
&\quad + BE[|h_k(X) - h(X)|],
\end{aligned}$$

where B is a uniform bound for both g and g_n, that is $|g| \leq B$ and $|g_n| \leq B$, for all n. Since h_k is bounded and continuous, and from the hypotheses of convergence in law, we have for every k:

$$\lim_{n \to \infty} |E[h_k(X)g_n(X)] - E[h_k(X)g(Y)]| = \lim_{n \to \infty} |E[h_k(X_n)g(Y_n)] - E[h_k(X)g(Y)]| = 0.$$

Finally, since $h_k(X)$ converges in $L^1(\Omega, \mathcal{F}, P)$ towards $h(X)$, the term $E[|h_k(X) - h(X)|]$ converges towards 0 as k goes to ∞.

2. To construct a counter-example, we do not need to rely on a bivariate sequence of r.v.s (X_n, Y_n). It is sufficient to consider a sequence (X_n) which converges in law towards X and a Borel discontinuous function φ, such that $\varphi(X_n)$ does not converge in law towards $\varphi(X)$.

A very simple counter-example is the following: set $X_n = a + \frac{1}{n}$, with $a \in \mathbb{R}$ and $n \geq 1$. Define also $\varphi(x) = \mathbb{1}_{\{x \leq a\}}$, $x \in \mathbb{R}$, then X_n converges surely, hence in law, towards a, although $\varphi(X_n) = 0$, for all $n \geq 1$. So, it cannot converge in law towards $\varphi(X) = \varphi(a) = 1$.

Solution to Exercise 5.5

Let F_n be the distribution function of $\frac{1}{n}\left(\sup_{i \leq n} X_i\right)$. For every $x \in \mathbb{R}$, F_n is given by:

$$\begin{aligned}
F_n(x) &= P\left(\frac{1}{n}\left(\sup_{i \leq n} X_i\right) \leq x\right) = P(X_1 \leq nx)^n \\
&= \left(1 - \int_{nx}^{\infty} \frac{a\,dy}{\pi(a^2 + y^2)}\right)^n.
\end{aligned}$$

If $x < 0$, then since $P(X_1 \leq nx)$ decreases to 0 as $n \to \infty$, $F_n(x)$ converges to 0. Suppose $x > 0$. In that case, $F_n(x) \sim \exp\left(-n\int_{nx}^{\infty} \frac{a\,dy}{\pi(a^2+y^2)}\right)$, as $n \to +\infty$, and since

$\int_{nx}^{\infty} \frac{a\, dy}{\pi(a^2+y^2)} = \int_x^{\infty} \frac{na\, dz}{\pi(a^2+n^2 z^2)}$, we have

$$\lim_{n\to+\infty} F_n(x) = \exp\left(-\int_x^{\infty} \frac{a\, dz}{\pi z^2}\right) = \exp\left(-\frac{a}{\pi x}\right).$$

To conclude, we check that the function F given by $F(x) = \lim_{n\to\infty} F_n(x) = \exp\left(-\frac{a}{\pi x}\right) \mathbb{1}_{(0,\infty)}(x)$, for all $x \geq 0$, is the distribution function of $\frac{1}{T_c}$, where T_c is an exponential r.v. with parameter $c = \frac{a}{\pi}$:

$$\exp\left(-\frac{c}{x}\right) = P\left(T_c > \frac{1}{x}\right) = P\left(\frac{1}{T_c} < x\right).$$

Comments on the solution. The law of $\left(\frac{1}{T_c}\right)$ is a particular case of the Fréchet distributions (with distribution function: $\exp(-ax^{-\alpha})$, $a > 0$, $\alpha > 0$), which are themselves a special class of limit laws for extremes. See, for example:

P. EMBRECHTS, C. KLÜPPELBERG AND T. MIKOSCH: *Modelling Extremal Events. For Insurance and Finance.* Applications of Mathematics, **33**, Springer-Verlag, Berlin, 1997.

Solution to Exercise 5.6

1. Let M be a matrix such that $K = M^t M$ and set

$$f(x) = \max_{1\leq i\leq n}(Mx)_i.$$

Then we can check from Cauchy–Schwarz inequality, that f is Lipschitz with:

$$\|f\|_{\text{Lip}} = \left(\max_{1\leq i\leq n} \sum_{j=1}^{n} M_{i,j}^2\right)^{\frac{1}{2}} = \max_{1\leq i\leq n} E[X_i^2]^{\frac{1}{2}} = \sigma,$$

hence question 3 of Exercise **3.11** implies:

$$P\left(\max_{1\leq i\leq n} X_i \geq E[\max_{1\leq i\leq n} X_i] + \sigma r\right) \leq e^{-\frac{r^2}{2}}.$$

2. Question 1 yields the inequality

$$\limsup_{r\to+\infty} \frac{1}{r^2} \log P(\max_{1\leq i\leq n} X_i \geq r) \leq -\frac{1}{2\sigma^2}.$$

To get the other inequality, it suffices to note that for any $i = 1, \ldots, n$:

$$P\left(\max_{1\leq i\leq n} X_i \geq r\right) \geq P(X_i \geq r) = 1 - \Phi\left(\frac{r}{\sigma_i}\right) \geq \frac{\exp\left(-\frac{r^2}{2\sigma_i^2}\right)}{\sqrt{2\pi}\left(1 + \frac{r}{\sigma_i}\right)},$$

where $\sigma_i = E[X_i^2]^{\frac{1}{2}}$ and Φ is the distribution function of the centered Gaussian law with variance 1. Hence,

$$\liminf_{r \to \infty} \frac{1}{r^2} \log P(\max_{i \le n} X_i \ge r) \ge -\frac{1}{2\sigma_i^2},$$

for all $i \le n$.

Solution to Exercise 5.7

1. Let f be a bounded Borel function defined on $\mathbb{R}_+$. We have

$$E\left[f\left(\frac{1}{n} \cosh X_n\right)\right] = \int_0^\infty f\left(\frac{1}{n} \cosh u\right) \frac{1}{n} \sinh u \exp\left(\frac{1 - \cosh u}{n}\right) du.$$

With the change of variables $x = \frac{1}{n} \cosh u$, we obtain $du = \frac{n\,dx}{\sqrt{n^2 x^2 - 1}}$, and

$$E\left[f\left(\frac{1}{n} \cosh X_n\right)\right] = \int_{\frac{1}{n}}^\infty f(x) \exp\left(\frac{1 - nx}{n}\right) dx,$$

which shows that the law of $\frac{1}{n} \cosh X_n$ has density $\exp\left(\frac{1}{n} - x\right) \mathbb{1}_{\{x \ge \frac{1}{n}\}}$.

2. From above, the r.v. $\frac{1}{n}(\cosh X_n - 1)$ follows the exponential law with parameter 1, thus $\frac{\cosh X_n}{n}$ converges in law towards the exponential law with parameter 1. Consequently, $\log \frac{\cosh X_n}{n}$ converges in law towards $Y = \log X$, where X follows the exponential law with parameter 1, that is

$$E[f(\log X)] = \int_0^\infty e^{-x} f(\log x)\,dx = \int_{-\infty}^\infty e^{-e^y} e^y f(y)\,dy.$$

3. Since $\log(\cosh X_n) - \log n$ converges in law towards $Y < \infty$, a.s., the ratio $\frac{\log(\cosh X_n)}{\log n} - 1$ converges in law (and thus in probability) towards 0.

Write: $\frac{\log(\cosh X_n)}{\log n} = \frac{X_n}{\log n} + \frac{\log(1 + e^{-2X_n}) - \log 2}{\log n}$. Since X_n is a.s. positive, $\log(1 + e^{-2X_n})$ is a.s. bounded by $\log 2$ and $\frac{\log(1 + e^{-2X_n}) - \log 2}{\log n}$ converges a.s. towards 0. We conclude from above that: $\frac{X_n}{\log n} \xrightarrow[n \to \infty]{(P)} 1$.

4. From the preceding question, we know that

$$\log\left(e^{X_n}\left(\frac{1 + e^{-2X_n}}{2}\right)\right) - \log n \xrightarrow[n \to \infty]{(\text{law})} Y,$$

from which we deduce:
$$X_n - (\log n) - \log 2 \xrightarrow{(\text{law})} Y.$$

A direct proof of this result may be obtained by writing:

$$E[f(X_n - \log n)] = \frac{1}{n} \int_{-(\log n)}^{\infty} (\sinh(v + \log n)) e^{\frac{1}{n}(1 - \cosh(v + \log n))} f(v) dv,$$

for a bounded function f on $\mathbb{R}$, and letting n tend to $+\infty$

Solution to Exercise 5.8

1. For any bounded Borel function f, we have:

$$E\left[f\left(\frac{\varepsilon}{\sqrt{U}}\right)\right] = \frac{1}{2}\left(E\left[f\left(\frac{1}{\sqrt{U}}\right)\right] + E\left[f\left(\frac{-1}{\sqrt{U}}\right)\right]\right)$$

$$= \frac{1}{2}\int_0^1 \left(f\left(\frac{1}{\sqrt{u}}\right) + f\left(\frac{-1}{\sqrt{u}}\right)\right) du = \int_1^{\infty} (f(x) + f(-x)) \frac{dx}{x^3},$$

hence, the law of X has density: $\frac{1}{|x|^3}\mathbb{1}_{\{|x|\geq 1\}}$.

2. The characteristic function of X is given by:

$$\varphi_X(t) = \frac{1}{2}\left(E\left[\exp\left(\frac{it}{\sqrt{U}}\right)\right] + E\left[\exp\left(\frac{-it}{\sqrt{U}}\right)\right]\right)$$

$$= E\left[\cos\frac{t}{\sqrt{U}}\right] = 2t^2 \int_t^{\infty} \cos x \frac{dx}{x^3}, \quad t \in \mathbb{R}.$$

When t goes towards 0, $\varphi_X(t) - 1$ is equivalent to $t^2 \log t$, so when n goes towards ∞, the characteristic function of $\frac{X_1 + \cdots + X_n}{\sqrt{n \log n}}$, that is $\left(\varphi_X\left(\frac{t}{\sqrt{n \log n}}\right)\right)^n$, $t \in \mathbb{R}$ is equivalent to

$$\left(1 + \frac{t^2}{n \log n}\left(\log t - \frac{1}{2}\log n - \frac{1}{2}\log\log n\right)\right)^n, \quad t \in \mathbb{R}.$$

When n goes to ∞, the latter expression converges towards $e^{-\frac{t^2}{2}}$, for any $t \in \mathbb{R}$, which proves the result.

3. Let ε and U be as in question 1. We easily check that the law of $X = \varepsilon U^{-\frac{1}{2+\alpha}}$ has the density which is given in the statement of question 3. Its characteristic function is $E[e^{itX}] = (2 + \alpha)t^{2+\alpha} \int_t^{\infty} \cos x \frac{dx}{x^{2+\alpha+1}}$, and is equivalent to $1 - \frac{2+\alpha}{2\alpha}t^2$, when t goes to 0. Therefore, if $\varphi^2(n) \to +\infty$, as $n \to \infty$, then the characteristic function of $\frac{X_1 + \cdots + X_n}{\varphi(n)}$ is asymptotically equivalent to $\left(1 - \frac{2+\alpha}{2\alpha}\frac{t^2}{\varphi^2(n)}\right)^n$, for any $t \in \mathbb{R}$, as $n \to \infty$. This expression converges towards $e^{-\frac{t^2}{2}}$ if $\varphi^2(n)$ is equivalent to $\frac{2+\alpha}{\alpha}n$, as $n \to \infty$.

Note the important difference between the setups of questions 1 and 3; in the latter, the classical CLT assumptions are satisfied.

Solution to Exercise 5.9

1. First we show that $(|U_n - X|, n \geq 1)$ is uniformly integrable, which follows from the fact that this sequence is bounded in L^2 (see the discussion following Exercise **1.3**). Here are the details. Indeed, for every $n \geq 1$:

$$E[|U_n - X|^2] \leq \left(E[U_n^2]^{\frac{1}{2}} + E[X^2]^{\frac{1}{2}}\right)^2.$$

But $E[U_n^2] = \frac{1}{n}\sum_{k=1}^n E[Y_k^2] = 1$, so that, $E[|U_n - X|^2] \leq \left(1 + E[X^2]^{\frac{1}{2}}\right)^2$. Now, from the Cauchy–Schwarz and Bienaymé–Tchebychev inequalities, we have for any $a > 0$: $a\,E[|U_n - X|\mathbb{1}_{\{|U_n-X|>a\}}] \leq E[(U_n - X)^2]$. Hence

$$E[|U_n - X|\mathbb{1}_{\{|U_n-X|>a\}}] \leq \frac{1}{a}\left(1 + E[X^2]^{\frac{1}{2}}\right)^2,$$

so $\lim_{a\to+\infty}\sup_{n\in\mathbb{N}} E[|U_n - X|\mathbb{1}_{\{|U_n-X|>a\}}] = 0$ and $(|U_n - X|, n \geq 1)$ is uniformly integrable. Moreover, applying the Central Limit Theorem, we obtain that $(|U_n - X|, n \geq 1)$ converges in law towards $|Y - X|$, where Y is an independent copy of X, (recall that (U_n) is independent from X). From the result of Exercise **1.3**, we conclude that:

$$\lim_{n\to\infty} E[|U_n - X|] = E[|Y - X|] = \frac{2}{\sqrt{\pi}}. \tag{5.9.a}$$

The last equality follows from the fact that $X - Y$ is a centered Gaussian variable whose variance equals 2 (see Exercise **3.8**).

2. From the independence between the r.v.s Y_n, we can write, for any bounded continuous function, f, defined on $\mathbb{R}^{p+1}$:

$$E[f(Y_1, \ldots, Y_p, U_n)]$$
$$= \int_{\mathbb{R}^p} E\left[f\left(y_1, \ldots, y_p, \frac{1}{\sqrt{n}}(y_1 + \cdots + y_p) + \frac{1}{\sqrt{n}}(Y_{p+1} + \cdots + Y_n)\right)\right]$$
$$\times P(Y_1 \in dy_1, \ldots, Y_p \in dy_p).$$

The Central Limit Theorem ensures that for any $(y_1, \ldots, y_p) \in \mathbb{R}^p$,

$$\lim_{n\to\infty} E\left[f\left(y_1, \ldots, y_p, \frac{1}{\sqrt{n}}(y_1 + \cdots + y_p) + \frac{1}{\sqrt{n}}(Y_{p+1} + \cdots + Y_n)\right)\right]$$
$$= E[f(y_1, \ldots, y_p, X)],$$

where X is a centered reduced Gaussian variable. Finally, from above and since f is bounded,

$$\lim_{n\to\infty} E[f(Y_1, \ldots, Y_p, U_n)] = E[f(Y_1, \ldots, Y_p, X)].$$

Hence, when n goes to ∞, $(Y_1, \ldots, Y_p, U_n)$ converges in law towards $(Y_1, \ldots, Y_p, X)$, where X is a centered reduced Gaussian variable which is independent of $(Y_1, \ldots, Y_p)$.

3. If U_n would converge in probability, then it would necessarily converge towards an r.v. such as X defined in the previous question. Let us assume that when n goes to ∞, $U_n \xrightarrow{(P)} X$, where X is a centered Gaussian variable with variance 1, independent of the sequence $(Y_n, n \geq 1)$. Since from question 1, $|U_n - X|$, $n \geq 1$ is uniformly integrable, the convergence in probability implies that $U_n \xrightarrow{L^1} X$. But, from (5.9.a) $E[|U_n - X|]$ converges towards $\frac{2}{\sqrt{\pi}}$, which contradicts our hypothesis. So U_n cannot converge in probability.

4. Set $\mathcal{G}_\infty = \cap_p \sigma\{Y_p, Y_{p+1}, \ldots\}$. It follows from the hypothesis that for any $n \geq 1$ and $m \geq 1$ such that $n < m$, $\sigma\{Y_1, \ldots, Y_n\}$ is independent of $\sigma\{Y_k : k \geq m\}$. Since for any $m \geq 1$, $\mathcal{G}_\infty \subset \sigma\{Y_k : k \geq m\}$, the sigma-fields $\sigma\{Y_1, \ldots, Y_n\}$ and $\mathcal{G}_\infty$ are independent for any $n \geq 1$. An application of the Monotone Class Theorem shows that $\sigma\{Y_1, \ldots, Y_n, \ldots\}$ and $\mathcal{G}_\infty$ are independent. Since $\mathcal{G}_\infty \subset \sigma\{Y_1, \ldots, Y_n, \ldots\}$, the sigma-field $\mathcal{G}_\infty$ is independent of itself, so it is trivial.

Now assume that U_n converges in probability. It is well known that hence the limit would be measurable with respect to $\mathcal{G}_\infty$. According to what we just proved, this limit must be constant, but this would contradict the Central Limit Theorem.

Comment on the solution. Actually, the property proved in solving question 4, is valid for any sequence of independent random variables and is known as Kolmogorov's $0-1$ law.

Solution to Exercise 5.10

Set $\overline{S}_n = \frac{1}{\sigma\sqrt{n}} \sum_{i=1}^n (X_i - m)$ and let N be a Gaussian, centered r.v., with variance 1 (defined on an auxiliary probability space). Our goal is to show that for any bounded continuous function f with compact support on $\mathbb{R}$,

$$E_Q[f(\overline{S}_n)] \to E[f(N)], \quad \text{as } n \to \infty. \tag{5.10.a}$$

For $k \geq 1$, put $D_k = E_P[D \mid \mathcal{G}_k]$, with $\mathcal{G}_k = \sigma\{X_1, \ldots, X_k\}$. Since we deal only with variables $X_1, X_2, \ldots, X_n, \ldots$, we may assume, without loss of generality, that D is measurable with respect to $\mathcal{G}_\infty = \lim_{k \uparrow \infty} \mathcal{G}_k$. We then write:

$$E_Q[f(\overline{S}_n)] = E_P[(D - D_k)f(\overline{S}_n)] + E_P[D_k f(\overline{S}_n)]. \tag{5.10.b}$$

Applying the result of Exercise **1.6**, together with the inequality:

$$|E_P[(D - D_k)f(\overline{S}_n)]| \leq E_P[|D - D_k|] \, \|f\|_\infty,$$

we see that the first term of (5.10.b) converges towards 0, as k goes to ∞, uniformly in n. It remains to show that for any $k \geq 1$,

$$E_P[D_k f(\overline{S}_n)] \to E[f(N)], \quad \text{as } n \to \infty. \tag{5.10.c}$$

To this aim, put $\overline{S}_{k,n} = \frac{1}{\sigma\sqrt{n}} \sum_{i=k}^{n} (X_i - m)$ and note that, since $\overline{S}_{k,n}$ is independent of $\mathcal{G}_k$ and $E_P[D_k] = 1$, one has

$$E_P[D_k f(\overline{S}_{k,n})] = E_P[f(\overline{S}_{k,n})] \to E[f(N)], \quad \text{as } n \to \infty. \tag{5.10.d}$$

Since f is uniformly continuous and $|\overline{S}_{k,n} - \overline{S}_n| \to 0$ a.s., as $n \to +\infty$,

$$|E_P[D_k f(\overline{S}_{k,n})] - E_P[D_k f(\overline{S}_n)]| \to 0, \quad \text{as } n \to +\infty, \tag{5.10.e}$$

by Lebesgue's Theorem of dominated convergence. This finally leads to (5.10.c), via (5.10.d) and (5.10.e).

Solution to Exercise 5.11

1. This follows easily from the Mellin transform formula:

$$E[(T_\mu)^{\mu s}] = \frac{\Gamma(1 - s)}{\Gamma(1 - \mu s)}, \quad \text{for } s < 1,$$

which was derived in Exercise **4.19** (cf. formula (4.19.4)). Hence, $\lim_{\mu \to 0} E[(T_\mu)^{\mu s}] = \Gamma(1 - s)$, which proves the result.

2. This follows from question 1, since:

$$\log\left(\frac{T_\mu}{T'_\mu}\right)^\mu = \mu \log T_\mu - \mu \log T'_\mu$$

converges in law towards $\log Z - \log Z'$.

3. The identity in law: $\frac{Z}{Z'} \overset{\text{(law)}}{=} \frac{1}{U} - 1$ follows, e.g., from the beta–gamma algebra, since $(Z, Z') \overset{\text{(law)}}{=} (U, 1 - U)(Z + Z')$ (use question 1 of Exercise **4.2**). Passing to the limit in formula (4.23.3), as $\mu \to 0$, one obtains, as $\mu \to 0$:

$$\left(\frac{T_\mu}{T'_\mu}\right)^\mu \overset{\text{(law)}}{\longrightarrow} \frac{dy}{(y + 1)^2}.$$

and the right hand side is the law of $\frac{1}{U} - 1$.

Solution to Exercise 5.12

1. Question 3 of Exercise **4.17** allows us to write

$$E\left[\exp\left(-\frac{\lambda^2}{2n}\sum_{i=1}^n \varepsilon_i^2 T_i\right)\right] = \prod_{i=1}^n E\left[\exp\left(-\frac{\lambda^2}{2n}\varepsilon_i^2 T_i\right)\right] = \exp\left(-\frac{|\lambda|}{\sqrt{n}}\sum_{i=1}^n \varepsilon_i\right),$$

for every $\lambda \in \mathbb{R}$. This shows that the sequence $\frac{1}{n}\sum_{i=1}^n \varepsilon_i^2 T_i$ converges in law if and only if $\frac{1}{\sqrt{n}}\sum_{i=1}^n \varepsilon_i$ converges towards a finite real value, say ε. In this case, the Laplace transform of the limit in law is: $e^{-\sqrt{2\lambda}\,\varepsilon}$, $\lambda \geq 0$. We deduce from question 3 of Exercise **4.17** that this limit is the law of $\varepsilon^2 T$.

2. (a) The series $\sum_{i=1}^n \frac{1}{2\sqrt{i}}$ may be compared with the integral $\int_1^n \frac{dx}{2\sqrt{x}}$, so it is equivalent to $\sqrt{n}$ and in this case, $\frac{1}{\sqrt{n}}\sum_{i=1}^n \varepsilon_i$ converges to 1. Applying question 1, we see that S_n converges in law to μ.

(b) First write $S_{kn} = \frac{k-1}{k}S_{(k-1)n} + \frac{1}{kn}\sum_{i=(k-1)n+1}^{kn} \varepsilon_i^2 T_i$. It is clear from question 1 that the first term on the right hand side converges in law to $\frac{k-1}{k}T_1$. The Laplace transform of the second term is

$$E\left[\exp\left(-\frac{\lambda^2}{2}\frac{1}{kn}\sum_{i=(k-1)n+1}^{kn} \varepsilon_i^2 T_i\right)\right] = \exp\left(-\frac{\lambda}{2\sqrt{kn}}\sum_{i=(k-1)n+1}^{kn}\frac{1}{\sqrt{i}}\right).$$

It converges to $\exp\left(-\frac{\lambda}{(k+\sqrt{k}\sqrt{k-1})}\right)$, as n goes to ∞, so the sequence $\frac{1}{kn}\sum_{i=(k-1)n+1}^{kn} \varepsilon_i^2 T_i$ converges in law to $\frac{1}{(k+\sqrt{k}\sqrt{k-1})}T_2$, as n goes to ∞. Since both terms in the above decomposition of S_{kn} are independent, we can say that $(S_{(k-1)n}, S_{kn})$ converges in law to $(T_1, \frac{k-1}{k}T_1 + \frac{1}{(k+\sqrt{k}\sqrt{k-1})}T_2)$, as n goes to ∞.

(c) The p-dimensional sequence $(S_n, S_{2n}, \ldots, S_{pn})$ may be written as *-1pt

$$\left(\frac{1}{n}\sum_{i=1}^n \varepsilon_i^2 T_i, \frac{1}{2n}\sum_{i=1}^n \varepsilon_i^2 T_i + \frac{1}{2n}\sum_{i=n+1}^{2n} \varepsilon_i^2 T_i, \ldots, \frac{1}{pn}\sum_{i=1}^n \varepsilon_i^2 T_i + \cdots + \frac{1}{pn}\sum_{i=(p-1)n+1}^{pn} \varepsilon_i^2 T_i\right).$$

Since for each $k \leq p$, the variables $\frac{1}{kn}\sum_{i=(j-1)n+1}^{jn} \varepsilon_i^2 T_i$, $j = 1, 2, \ldots, k$ are independent and converge respectively in law to $\frac{j}{k}\frac{1}{(j+\sqrt{j}\sqrt{j-1})}T_j$, the limit in law of $(S_n, S_{2n}, \ldots, S_{pn})$ is

$$\left(T_1, \frac{1}{2}T_1 + \frac{1}{2+\sqrt{2}}T_2, \ldots, \sum_{j=1}^k \frac{j}{k}\frac{1}{(j+\sqrt{j}\sqrt{j-1})}T_j, \ldots, \sum_{j=1}^p \frac{j}{p}\frac{1}{(j+\sqrt{j}\sqrt{j-1})}T_j\right).$$

(d) From above, we see that when n goes to ∞, $S_{2n} - S_n$ converges in law towards $\frac{1}{2+\sqrt{2}}T_2 - \frac{1}{2}T_1$ which is non-degenerate. This result prevents S_n from converging in probability.

3. (a) The p-dimensional sequence $(S'_{t_1}, S'_{t_2}, \ldots, S'_{t_p})$ may be written as

$$\left(\frac{1}{n} \sum_{i=1}^{[t_1^2 n]} \varepsilon_i^2 T_i, \frac{1}{n} \sum_{i=1}^{[t_1^2 n]} \varepsilon_i^2 T_i + \frac{1}{n} \sum_{i=[t_1^2 n]+1}^{[t_2^2 n]} \varepsilon_i^2 T_i, \ldots, \frac{1}{n} \sum_{i=1}^{[t_1^2 n]} \varepsilon_i^2 T_i + \cdots + \frac{1}{n} \sum_{i=[t_{p-1}^2 n]+1}^{[t_p^2 n]} \varepsilon_i^2 T_i \right).$$

For each $k \leq p$, the Laplace transform

$$E[\exp(-\frac{\lambda^2}{2}(S'_{t_k} - S'_{t'_{k-1}}))] = E\left[\exp\left(-\frac{\lambda^2}{2n} \sum_{i=[t_{k-1}^2 n]+1}^{[t_k^2 n]} \varepsilon_i^2 T_i \right) \right]$$

$$= \exp\left(-\frac{\lambda}{2\sqrt{n}} \sum_{i=[t_{k-1}^2 n]+1}^{[t_k^2 n]} \frac{1}{\sqrt{i}} \right),$$

$\lambda \geq 0$, converges towards $E[\exp(-\frac{\lambda^2}{2}(T_{t_k} - T_{t_{k-1}}))] = \exp(-\lambda(t_k - t_{k-1}))$. It follows from the independence hypothesis that the r.v.s $(S'_{t_k} - S'_{t_{k-1}})_{1 \leq k \leq p}$ are independent, hence so are the increments $(T_{t_k} - T_{t_{k-1}})_{1 \leq k \leq p}$.

In conclusion, the finite-dimensional distributions of the process $(S'_n(t), t \geq 0)$ converge weakly towards those of a process with independent and time homogeneous increments. The increments, $T_{t_k} - T_{t_{k-1}}$, of the limit process are such that $(t_k - t_{k-1})^{-2}(T_{t_k} - T_{t_{k-1}})$ has law μ. This process is the so-called stable (1/2) subordinator.

(b) The increments of the process $(\xi_n(t), t \geq 0)$ at times $t_1 < \cdots < t_p$ are

$$\left(\frac{1}{\sqrt{n}} \sum_{i=1}^{[nt_1]} X_i, \frac{1}{\sqrt{n}} \sum_{i=[nt_1]+1}^{[nt_2]} X_i, \ldots, \frac{1}{\sqrt{n}} \sum_{i=[nt_{p-1}]+1}^{[nt_p]} X_i \right).$$

Put $t_0 = 0$ and call σ^2 the variance of the X_is and ψ their characteristic function. For each $k = 1, \ldots, p$, we can check that $\frac{1}{\sqrt{n}} \sum_{i=[nt_{k-1}]+1}^{[nt_k]} X_i$ converges in law towards a centered Gaussian variable with variance $\sigma^2(t_k - t_{k-1})$ as n goes to $+\infty$. Indeed, the characteristic function of the r.v. $\frac{1}{\sqrt{n}} \sum_{i=[nt_{k-1}]+1}^{[nt_k]} X_i$ is equal to:

$$\varphi_n(t) = \psi\left(\frac{t}{\sqrt{n}} \right)^{[nt_k]-[nt_{k-1}]} = \left(1 - \frac{\sigma^2}{2n} t^2 + o\left(\frac{1}{n} \right) \right)^{[nt_k]-[nt_{k-1}]}, \quad t \in \mathbb{R}.$$

When n tends to $+\infty$, $\varphi_n(t)$ converges towards $\exp\left(-\frac{t^2 \sigma^2}{2}(t_k - t_{k-1}) \right)$. Finally, note that the increments of the process $(\xi_n(t), t \geq 0)$ at times $t_1 < \cdots < t_p$ are independent. So, we conclude as follows.

The finite-dimensional distributions of the process $(\xi_n(t), t \geq 0)$ converge weakly towards those of a process with independent and time homogeneous increments. The increments of the limit process at times $t_1 < \cdots < t_p$ are centered Gaussian variables with variance $\sigma^2(t_k - t_{k-1})$, $k = 2, \ldots, p$. Hence, the limit process is $(\sigma B_t, t \geq 0)$, where $(B_t, t \geq 0)$ is a Brownian motion.

Solution to Exercise 5.13

1. Let s and t belong to the interval $[0, 1]$, and write $b^{(n)}(t) = \frac{1}{\sqrt{n}} \sum_{k=1}^{n} (\mathbb{1}_{[0, t]}(U_k) - t)$. We have

$$E[b^{(n)}(t)] = \frac{1}{\sqrt{n}} \sum_{k=1}^{n} (P(U_k \leq t) - t) = 0.$$

Suppose that $s \leq t$; from the independence between the r.v.s U_k, we obtain:

$$\begin{aligned} \mathrm{Cov}[b^{(n)}(s), b^{(n)}(t)] &= \frac{1}{n} \sum_{k=1}^{n} \mathrm{Cov}(\mathbb{1}_{[0, t]}(U_k), \mathbb{1}_{[0, s]}(U_k)) \\ &= s - st, \end{aligned}$$

so that for any s and t,

$$\mathrm{Cov}[b^{(n)}(s), b^{(n)}(t)] = s \wedge t - st. \tag{5.13.a}$$

2. Let $\lambda_i > 0$ and $t_i \in [0, 1]$, $i = 1, \ldots, j$, for any integer $j \geq 1$ and write:

$$\sum_{i=1}^{j} \lambda_i b^{(n)}(t_i) = \frac{1}{\sqrt{n}} \sum_{k=1}^{n} \left(\sum_{i=1}^{j} \lambda_i \mathbb{1}_{[0, t_i]}(U_k) - \sum_{i=1}^{j} \lambda_i t_i \right). \tag{5.13.b}$$

From (5.13.b) and the Central Limit Theorem, we see that $\sum_{i=1}^{n} \lambda_i b^{(n)}(t_i)$ converges in law towards a centered Gaussian variable with variance:

$$\mathrm{Var}\left(\sum_{i=1}^{j} \lambda_i \mathbb{1}_{[0, t_i]}(U_k) \right) = \sum_{i=1}^{j} \lambda_i^2 (t_i - t_i^2) + \sum_{1 \leq i \neq i' \leq j} \lambda_i \lambda_{i'}(t_i \wedge t_{i'} - t_i t_{i'}).$$

It means that $(b^{(n)}(t_1), \ldots, b^{(n)}(t_j))$ converges in law towards a centered Gaussian vector $(b(t_1), \ldots, b(t_j))$ whose covariance matrix is given by:

$$\mathrm{Cov}(b(t_i), b(t_{i'})) = t_i \wedge t_{i'} - t_i t_{i'} ., \quad i, i' = 1, \ldots, j,$$

which allows us to reach our conclusion.

Solution to Exercise 5.14

First from the strong law of large numbers,

$$\lim_{n \to \infty} \frac{S_n}{n} = E[X_1], \quad \text{a.s.}$$

Therefore, almost surely, for all real $t \geq 0$,

$$\lim_{n \to \infty} \frac{S_{[nt]}}{n} = E[X_1] \cdot t.$$

Then since $t \mapsto E[X_1] \cdot t$ is continuous and increasing and $t \mapsto \frac{S_{[nt]}}{n}$ is right continuous, the result follows from a Dini lemma type argument for deterministic functions.

Solution to Exercise 5.15

The Poisson process with parameter c, $(N_t^{(c)}, t \geq 0)$, is the unique (in law) increasing right continuous process with independent and stationary (or time homogeneous) increments, such that for each $t > 0$, $N_t^{(c)}$ has a Poisson distribution with parameter ct. It follows that the process $(X_t^{(c)}, t \geq 0)$ also has stationary and independent increments. Set $t_0 = 0$ and let $(t_1, \dots, t_n) \in \mathbb{R}_+^n$ be such that $t_1 < t_2 < \cdots < t_n$, then the r.v.s $(X_{t_1}^{(c)}, X_{t_2}^{(c)} - X_{t_1}^{(c)}, \dots, X_{t_n}^{(c)} - X_{t_{n-1}}^{(c)})$ are independent and for each $k = 1, \dots, n$, $X_{t_k}^{(c)} - X_{t_{k-1}}^{(c)}$ has the same distribution as $X_{t_k - t_{k-1}}^{(c)}$. Let us compute the characteristic function of $\frac{1}{\sqrt{c}} X_{t_k - t_{k-1}}^{(c)}$. For any $\lambda \in \mathbb{R}$, we have:

$$E\left[\exp\left(\frac{i\lambda}{\sqrt{c}} X_{t_k - t_{k-1}}^{(c)} \right) \right]$$

$$= \exp\left(-i\lambda(t_k - t_{k-1})\sqrt{c} - c(t_k - t_{k-1}) \right) \sum_{n=0}^{\infty} \frac{(c(t_k - t_{k-1}))^n}{n!} e^{\frac{i\lambda}{\sqrt{c}} n}$$

$$= \exp\left(-c(t_k - t_{k-1})(1 - e^{\frac{i\lambda}{\sqrt{c}}}) - i\lambda(t_k - t_{k-1})\sqrt{c} \right).$$

When $c \to \infty$, the latter expression converges towards $\exp\left(-\frac{\lambda^2}{2}(t_k - t_{k-1}) \right)$, which is the characteristic function of a centered Gaussian variable with variance $t_k - t_{k-1}$. This shows that, when $c \to \infty$, the random vector $\frac{1}{\sqrt{c}}(X_{t_1}^{(c)}, X_{t_2}^{(c)} - X_{t_1}^{(c)}, \dots, X_{t_n}^{(c)} - X_{t_{n-1}}^{(c)})$ converges in law towards a random vector whose coordinates are independent centered Gaussian r.v.s with variances $t_k - t_{k-1}$, $k = 1, \dots, n$. So, we conclude that the random vector $\frac{1}{\sqrt{c}}(X_{t_1}^{(c)}, \dots, X_{t_n}^{(c)})$ converges in law towards $(\beta_{t_1}, \dots, \beta_{t_n})$, as $c \to \infty$, where $(\beta_t, t \geq 0)$ is a centered Gaussian process which has stationary and independent increments. The latter property entirely characterizes the process $(\beta_t, t \geq 0)$, which is known as Brownian motion and has already been encountered in question 3 (b) of Exercise **5.8**. From above, one may easily check that the covariance of Brownian motion is given by:

$$E[\beta_s \beta_t] = s \wedge t, \quad s, t \geq 0.$$

Comments on the solution. As an additional exercise, one could check that the process $(N_t^{(c)}, t \geq 0)$ may be constructed as follows.

Let $(X_k)_{k\geq 1}$ be a sequence of independent exponential r.v.s with parameter c. Set $S_n = \sum_{k=1}^{n} X_k$, $n \geq 1$, then

$$(N_t^{(c)}, t \geq 0) \stackrel{\text{(law)}}{=} \left(\sum_{n=1}^{\infty} \mathbb{1}_{\{S_n \leq t\}}, t \geq 0 \right).$$

Solution to Exercise 5.16

1. (a) This follows immediately from the Gaussian character of Brownian motion, and the fact that

$$E[b(u)B_1] = u - u = 0.$$

(b) It suffices to write $b(1-u) = B_{1-u} - (1-u)B_1 = uB_1 - (B_1 - B_{1-u})$, and to use the fact that $(-(B_1 - B_{1-u}), 0 \leq u \leq 1)$ is a Brownian motion.

2. We first write $\left(\frac{1}{\sqrt{c}} b(cu), 0 \leq u \leq \frac{1}{c} \right)$ and $\left(\frac{1}{\sqrt{c}} b(1-cu), 0 \leq u \leq \frac{1}{c} \right)$ in terms of B

$$\frac{1}{\sqrt{c}} b(cu) = \frac{1}{\sqrt{c}} B_{cu} - \sqrt{c} u B_1,$$

$$\frac{1}{\sqrt{c}} b(1-cu) = \frac{1}{\sqrt{c}} \left(B_{1-cu} - (1-cu)B_1 \right),$$

and it follows that this question may be reduced to showing the result: for fixed $U, S, T > 0$, the process

$$\left[(B_u, 0 \leq u \leq U); \left(\frac{1}{\sqrt{c}} B_{cu}, 0 \leq u \leq S \right); \left(\frac{1}{\sqrt{c}} (B_{1-cu} - B_1), 0 \leq u \leq T \right) \right]$$

(5.16.a)

converges in law towards (B, β, β'), a three-dimensional Brownian motion. This follows from the fact that the mixed covariance function $E\left[B_u \frac{1}{\sqrt{c}} B_{cs} \right]$ converges towards 0 as c goes to 0, and that each of the three families which constitute (5.16.a) is constant in law (i.e. its law does not depend on c).

3. By writing $\sqrt{\lambda} b(e^{-\frac{t}{\lambda}}) = \sqrt{\lambda} b(1 - (1 - e^{-\frac{t}{\lambda}}))$ and using the time reversal property of the Brownian bridge, it suffices to show that $\sqrt{\lambda} b(1 - e^{-\frac{t}{\lambda}})$ converges in law towards Brownian motion. Now, using the representation of b as $b(u) = B_u - uB_1$, $u \leq 1$, we can replace b in $\sqrt{\lambda} b(1 - e^{-\frac{t}{\lambda}})$ by B, a Brownian motion.

Finally, we can restrict ourselves to $t \leq T$, for T fixed and then with the help of the $\frac{1}{2} - \varepsilon$ Hölder property of the Brownian trajectories $(B_u, u \leq T)$, we have

$$\sup_{t \leq T} \sqrt{\lambda} \left| B\left(1 - e^{-\frac{t}{\lambda}} \right) - B\left(\frac{t}{\lambda} \right) \right| \stackrel{\text{(a.s.)}}{\longrightarrow} 0,$$

as $\lambda \to \infty$. Now the result follows from the scaling property.

Solution to Exercise 5.17

Let $(S_t)_{t \geq 0}$ be the stable process which satisfies $S_1 \overset{(\text{law})}{=} X_1$, that is a continuous time càdlàg (i.e. right continuous and left limited) process with independent and stationary increments satisfying the scaling property

$$t^{-\frac{1}{\alpha}} S_t \overset{(\text{law})}{=} S_1 \overset{(\text{law})}{=} X_1. \tag{5.17.a}$$

The process Z defined from S by the Lamperti transformation $Z_t = e^{-\frac{t}{\alpha}} S_{e^t}$, $t \geq 0$ is stationary, i.e.

$$\text{for any fixed } s, \quad (Z_{t+s}, t \geq 0) = \left(e^{-\frac{t+s}{\alpha}} S_{e^{t+s}}, t \geq 0 \right) \overset{(\text{law})}{=} (Z_t, t \geq 0),$$

has independent increments. Hence the tail σ-field $\cap_t \sigma\{Z_{t+s}, s \geq t\}$ is trivial and from the Ergodic Theorem,

$$\frac{1}{t} \int_0^t f(Z_u) \, du \to E[f(Z_1)], \quad \text{a.s., as } t \to \infty.$$

From the change of variables $v = e^u$ and the scaling property of S, we obtain:

$$\frac{1}{\log t} \int_0^t \frac{1}{v} f(v^{-\frac{1}{\alpha}} S_v) \, dv \to E[f(S_1)], \quad \text{a.s., as } t \to \infty.$$

Finally, the result in discrete time easily follows from the above.

Comments on the solution. Lamperti's transformation is a one-to-one transformation between the class of Lévy processes (i.e. processes with stationary and independent increments) and semistable Markov processes (i.e. non-negative Markov processes satisfying a scaling property such as (5.17.a)). This transformation (and more on semistable processes) is presented in:

P. EMBRECHTS AND M. MAEJIMA: *Self-similar Processes.* Princeton University Press, Princeton, NJ, 2002.

Chapter 6

Random processes

> What is a random process ?

In this chapter, we shall consider on a probability space $(\Omega, \mathcal{F}, P)$, *random (or stochastic) processes* $(X_t, t \geq 0)$ indexed by $t \in \mathbb{R}_+$, that is applications $(t, \omega) \mapsto X_t(\omega)$ assumed to be measurable with respect to $\mathcal{B}_{\mathbb{R}_+} \otimes \mathcal{F}$, and valued in $\mathbb{R}$ (or $\mathbb{C}$, or $\mathbb{R}^d$).

In the early stages of developments of stochastic processes (e.g. the 1950s, see Doob's book [24]), it was a major question to decide whether, given finite-dimensional marginals $\{\mu_{t_1, \ldots, t_k}\}$, there existed a (measurable) process $(X_t, t \geq 0)$ with these marginals. Kolmogorov's Existence Theorem (for a projective family of marginals; see, e.g., Neveu [62]) together with Kolmogorov's continuity criterion

$$E[|X_t - X_s|^p] \leq C|t - s|^{1+\varepsilon}, \quad t, s \geq 0,$$

for some $p > 0$, $\varepsilon > 0$, and $C < \infty$, which bears only on the two-dimensional marginals, often allows us to obtain at once continuous paths realizations. It has also been established (Meyer [60]) that on a filtered probability space $(\Omega, \mathcal{F}, (\mathcal{F}_t)_{t \geq 0}, P)$, with $(\mathcal{F}_t)_{t \geq 0}$ right continuous and $(\mathcal{F}, P)$ complete, a supermartingale admits a "càdlàg" (i.e. right continuous, with left limits) version. As a consequence, many processes admit a càdlàg version, and their laws may be constructed on the corresponding canonical space of càdlàg functions (often endowed with the Skorokhod topology). Once the (fundamental) question of an adequate realization solved, then the probabilistic study of a random process often bears upon its Markov (or lack of Markov!) property, as well as upon all kinds of behavior (either local or global), and also upon various transformations of one process into another.

Basic facts about Brownian motion

In this chapter, we draw more directly on research papers on Brownian motion, (and more generally stochastic processes). Thus, the reader will see how some of the random objects (random variables, etc.) used in the previous chapters appear naturally in the Brownian context. Also, a few times, we have given less details in the proofs, and have simply referred the reader to the original papers.

Among continuous time stochastic processes, i.e. families of random variables (X_t) indexed by $t \in \mathbb{R}_+$, Brownian motion – often denoted $(B_t, t \geq 0)$ – is one of the most remarkable:

(i) It is a *Gaussian process*: for any k-uple $t_1 < t_2 < \ldots < t_k$, the random vector $(B_{t_1}, B_{t_2}, \ldots, B_{t_k})$ is Gaussian, centered, with:

$$E[B_{t_i} B_{t_j}] = t_i \wedge t_j .$$

(ii) It has *independent* and *stationary increments*: for every $t_1 < t_2 < \ldots < t_k$, $B_{t_1}, B_{t_2} - B_{t_1}, \ldots, B_{t_k} - B_{t_{k-1}}$ are independent, and $B_{t_i} - B_{t_{i-1}}$ is distributed as $B_{(t_i - t_{i-1})}$.

(iii) Almost surely, the path: $t \to B_t$ is continuous; more precisely, it is *locally Hölder* of order $(\frac{1}{2} - \varepsilon)$, for every $\varepsilon \in (0, \frac{1}{2})$.

(iv) It is a *self-similar process* of order $\frac{1}{2}$, i.e.

$$\text{for any } c > 0, \ (B_{ct}, t \geq 0) \overset{(\text{law})}{=} (\sqrt{c} B_t, t \geq 0) .$$

(Note that the previous chapter just ended with a reference to self-similar processes.)

(v) It is the unique *continuous martingale* $(M_t, t \geq 0)$ such that:

$$\text{for any } s < t, \ E[(M_t - M_s)^2 \,|\, \mathcal{F}_s] = t - s ,$$

a famous result due to Lévy. (For clarity, assume that $(\mathcal{F}_t)$ is the natural filtration associated with (M_t).)

Note also that property (i) on the one hand, and properties (ii) and (iii) on the other hand, characterize Brownian motion.

These facts are presented in every book dealing with Brownian motion, e.g. Karatzas and Shreve [46], Revuz and Yor [75] and Rogers and Williams [77], etc.

A discussion of random processes with respect to filtrations

The basic notions of a constant (respectively increasing, decreasing, affine, differentiable, convex, etc.) function f (which we assume here to be defined on $\mathbb{R}_+$, and taking values in $\mathbb{R}$) have very interesting analogues which apply to stochastic processes $(X_t)_{t \geq 0}$ adapted to an increasing family of σ-fields $(\mathcal{F}_t)_{t \geq 0}$:

• an $(\mathcal{F}_t)$ conditionally constant process (for $s \leq t$, $E[X_t \mid \mathcal{F}_s] = X_s$) is a $(\mathcal{F}_t)$ martingale;

• an $(\mathcal{F}_t)$ conditionally increasing process (for $s < t$, $E[X_t \mid \mathcal{F}_s] \geq X_s$) is a $(\mathcal{F}_t)$ submartingale;

• conditionally affine processes are studied in Exercise **6.24**; their probabilistic name is harnesses;

• conditionally convex processes have been shown to play some interesting role in mathematical finance. See:

P. BANK AND N. EL KAROUI: A stochastic representation theorem with applications to optimization and obstacle problems. *Ann. of Prob.*, **32**, no. 1B, 1030–1067 (2004).

A fundamental result of Doob–Meyer is that every conditionally increasing process (i.e. submartingale) is the sum of a conditionally constant process (i.e. martingale) and an increasing process.

---

* 6.1 Jeulin's lemma deals with the absolute convergence of integrals of random processes

1. Let $(R_t, \, t \geq 0)$ be an $\mathbb{R}_+$-valued process such that

   (i) for every $t \geq 0$, the law of R_t does not depend on t;
   (ii) $0 < E[R_1] < \infty$.

   The aim of this question is to show that if $\mu(dt)$ is a random, nonnegative, σ-finite measure on $(0, 1]$, (which may be infinite), then there is the equivalence:

   $$\int_{(0,1]} \mu(dt)\, R_t < \infty, \quad \text{a.s.} \quad \Leftrightarrow \quad \int_{(0,1]} \mu(dt) < \infty.$$

   (a) First show the implication

   $$\int_{(0,1]} \mu(dt) < \infty \quad \Rightarrow \quad \int_{(0,1]} \mu(dt)\, R_t < \infty, \quad \text{a.s.}$$

(b) Show that, for any measurable set B,

$$E\left[\mathbb{1}_B \int_{(0,1]} R_t \mu(dt)\right] \geq \int_{(0,1]} \mu(dt) \int_0^\infty du\,(E[\mathbb{1}_B] - P(R_t \leq u))^+ .$$

Hint. Use the elementary inequality $P(B \cap C) \geq (P(B) - P(C^c))^+$ where C^c denotes the complement of C in Ω.

(c) Set

$$B_N = \left\{\int_{(0,1]} R_t\,\mu(dt) \leq N\right\},$$

then with the help of the assumption (i), show that

$$N \geq \mu(0,1] \int_0^\infty du\,(E[\mathbb{1}_{B_N}] - P(R_1 \leq u))^+ .$$

(d) Assume that $\int_{(0,1]} \mu(dt)\,R_t < \infty$, a.s. and show that one can choose N, so that

$$\int_0^\infty du(E[\mathbb{1}_{B_N}] - P(R_1 \leq u))^+ > 0 .$$

Then conclude that $\mu(0,1] < \infty$.

2. Let $(B_t, t \geq 0)$ denote a one-dimensional Brownian motion issued from 0.

(a) Prove that the process $\beta_t = B_t - \int_0^t \frac{ds}{s} B_s$, $t \geq 0$, is a one-dimensional Brownian motion.

(b) Prove that $\mathcal{F}_t^B = \mathcal{F}_t^\beta \vee \sigma(B_t)$.

3. Let $f \in L^2([0,1])$.

(a) Prove that $\lim_{\varepsilon \to 0} \int_\varepsilon^1 \frac{dt}{t} f(t) B_t$ exists a.s.

(b) Apply question 1 to show the following:

$$\int_0^1 \frac{dt}{t}|f(t)|\,|B_t| < \infty \Leftrightarrow \int_0^1 \frac{dt|f(t)|}{\sqrt{t}} < \infty .$$

Give an example of a function $f \in L^2([0,1])$ such that:

$$\int_0^1 \frac{dt}{t}|f(t)|\,|B_t| = \infty , \quad \text{a.s.}$$

Comments and references. Jeulin's Lemma is presented in detail in:

T. JEULIN: Sur la convergence absolue de certaines intégrales. *Séminaire de Probabilités XVI, Lecture Notes in Mathematics*, **920**, Springer, Berlin-New York, 1982.

T. JEULIN AND M. YOR: Inégalité de Hardy, semimartingales, et faux-amis. *Séminaire de Probabilités XIII*, 332–359, *Lecture Notes in Mathematics*, **721**, Springer, Berlin, 1979.

M. CSÖRGÖ, L. HORVÁTH AND Q.M. SHAO: Convergence of integrals of uniform empirical and quantile processes. *Stochastic Process. Appl.*, **45**, no. 2, 283–294 (1993).

Jeulin's Lemma has been quite useful in a number of discussions about finiteness of integrals of square of Brownian motion as in (i), (ii) and (iii) below.

(i) J. PITMAN AND M. YOR: A decomposition of Bessel bridges. *Z. Wahrsch. Verw. Gebiete*, **59**, no. 4, 425–457 (1982).

(ii) J. PITMAN AND M. YOR: Some divergent integrals of Brownian motion. *Adv. in Appl. Probab.* suppl., 109–116 (1986).

(iii) P. SALMINEN AND M. YOR: Properties of perpetual integral functionals of Brownian motion with drift. *Ann. Inst. H. Poincaré Probab. Statist.*, **41**, no. 3, 335–347 (2005).

X.X. XUE: A zero-one law for integral functionals of the Bessel process. *Séminaire de Probabilités XXIV*, 137–153, *Lecture Notes in Mathematics*, **1426**, Springer, Berlin, 1990.

H.J. ENGLEBERT AND W. SCHMIDT: 0-1 Gesetze für die Konvergenz von Integral funktionen gewisser Semi-martingale. *Math. Nachrichten*, **128**, 177–185 (1985).

D. KHOSHNEVISAN, P. SALMINEN AND M. YOR: A note on a.s. finiteness of perpetual integral functionals of diffusions. *Electron. Comm. Probab.*, **11**, 108–117 (2006).

Finally our latest reference on this subject is:

A. MATSUMOTO AND K. YANO: On a zero-one law for the norm process of transient random walks. *Séminaire de Probabilités XLIII*, *Lecture Notes in Mathematics*, **2006**, Springer, 2011.

* 6.2 Functions of Brownian motion as solutions to SDEs; the example of $\varphi(x) = \sinh(x)$

Let $(B_t,\ t \geq 0)$ denote a one-dimensional Brownian motion starting from 0.

1. Prove that the SDE

$$X_t = x + \int_0^t \sqrt{1 + X_s^2}\, dB_s + \frac{1}{2} \int_0^t X_s\, ds \qquad (6.2.1)$$

admits a unique strong solution, for every $x \in \mathbb{R}$.

2. Now fix $x \in \mathbb{R}$ and let (β_t) and (γ_t) denote two independent Brownian motions

starting from 0. Prove that

$$Y_t = \exp(\beta_t) \left(x + \int_0^t \exp(-\beta_s) \, d\gamma_s \right) , \quad t \geq 0$$

solves equation (6.2.1) for a certain Brownian motion B. Conclude that

$$(Y_t, \, t \geq 0) \stackrel{(\text{law})}{=} (\sinh(a + B_t), \, t \geq 0),$$

where $a = \arg \sinh x$.

3. We now put the preceding in a more general context.

   (a) Show that if $\varphi : \mathbb{R} \to \mathbb{R}$ is a $\mathcal{C}^2$ homeomorphism of $\mathbb{R}$, then $\Phi_t \equiv \varphi(B_t)$ solves

   $$\Phi_t = \varphi(0) + \int_0^t \sigma\left(\Phi_s\right) dB_s + \int_0^t b\left(\Phi_s\right) ds , \qquad (6.2.2)$$

   with $\sigma(x) = (\varphi' \circ \varphi^{-1})(x)$ and $b(x) = \frac{1}{2}(\varphi'' \circ \varphi^{-1})(x)$.

   (b) Conversely, if $\sigma, b : \mathbb{R} \to \mathbb{R}$ are Lipschitz functions, then there is only one solution to (6.2.2). Under which condition on (σ, b) can we solve the system

   $$\varphi'(y) = \sigma(\varphi(y)), \quad \text{and} \quad \frac{1}{2}\varphi''(y) = b(\varphi(y)), \qquad (6.2.3)$$

   so that the solution to (6.2.2) is $\Phi_t = \varphi(B_t)$?

Comments and references. This exercise has been partly inspired by the book:

P.E. KLOEDEN AND E. PLATEN: *Numerical Solutions of Stochastic Differential Equations. Applications of Mathematics*, **23**, Springer, 1992,

and even more so by:

H. DOSS: Liens entre équations différentielles stochastiques et ordinaires. *Ann. Inst. H. Poincaré, Sect. B (N.S.)*, **13**, no. 2, 99–125 (1977).

*6.3 Bougerol's identity and some Bessel variants

We keep the same notation as in the previous Exercise 6.2.

1. (a) Prove Bougerol's identity in law, i.e. for fixed t

   $$\sinh(B_t) \stackrel{(\text{law})}{=} \int_0^t \exp(\beta_s) \, d\gamma_s \stackrel{(\text{law})}{=} \tilde{\gamma}_{\int_0^t \exp(2\beta_s) \, ds} , \qquad (6.3.1)$$

   where $\tilde{\gamma}$ is a Brownian motion independent from β.

(b) Prove that the two *processes* featured on both sides of (6.3.1) are not identical in law.

2. Let $M_t = \sup_{s \le t} B_s$, and let $(L_t, t \ge 0)$ denote the local time at 0, for $(B_t, t \ge 0)$. Likewise, let $\mu_t = \sup_{s \le t} \gamma_s$ and let $(\lambda_t, t \ge 0)$ denote the local time at 0 of $(\gamma_t, t \ge 0)$. Prove the following identities in law: for fixed $t > 0$,

$$\sinh(M_t) \overset{\text{(law)}}{=} \mu_{\int_0^t du\, \exp(2\beta_u)}$$

$$\sinh(L_t) \overset{\text{(law)}}{=} \lambda_{\int_0^t du\, \exp(2\beta_u)}\,.$$

3. Prove the particular case of Lamperti's representation applied to $\exp(\beta_t)$, $t \ge 0$:

$$\exp(\beta_t) = R_{\int_0^t du\, \exp(2\beta_u)}\,, \quad t \ge 0\,,$$

where $(R_u, u \ge 0)$ denotes a two-dimensional Bessel process issued from 1.

4. Putting the previous results together, prove that, for fixed l,

$$H_{\tau_l} \overset{\text{(law)}}{=} \tau_{a(l)}\,, \tag{6.3.2}$$

where $(\tau_v, v \ge 0)$ denotes a stable $1/2$-subordinator independent from $H_u \equiv \int_0^u \frac{ds}{R_s^2}$, $u \ge 0$, and $a(l) = \text{argsh}(l)$.

Comments and references. (a) The result (6.3.2) has been exploited systematically by S. Vakeroudis in his PhD thesis (Paris VI, April 2011).

(b) It follows from

J. BERTOIN AND M. YOR: Retrieving information from subordination. To appear in a Festschrift volume for Y. Prokhorov (2012)

that (6.3.2) does not hold at the process (in l) level.

c) Further identities alongside (6.3.2) are developed separately by S. Vakeroudis and also in:

J. BERTOIN, D. DUFRESNE AND M. YOR: Some two-dimensional extensions of Bougerol's identity in law for the exponential functional of linear Brownian motion. Reprint arxiv: 1201.1495.

* 6.4 Doléans–Dade exponentials and the Maruyama–Girsanov–Van Schuppen–Wong theorem revisited

Recall that if $(X_t, t \ge 0)$ is a continuous semimartingale (on a filtered probability space $(\Omega, \mathcal{F}, (\mathcal{F}_t), P)$), then the equation:

$$Z_t = 1 + \int_0^t Z_s\, dX_s \tag{6.4.1}$$

admits a unique solution given by:

$$Z_t = \mathcal{E}(X)_t \stackrel{(\text{def})}{=} \exp(X_t - \frac{1}{2}\langle X \rangle_t), \quad t \geq 0, \tag{6.4.2}$$

which is usually called the Doléans–Dade exponential associated with X (here, we assume $X_0 = 0$).

From now on, in this exercise, all semimartingales will be assumed to be continuous.

1. Prove the multiplicativity formula:

$$\mathcal{E}(X)\mathcal{E}(Y) = \mathcal{E}(X + Y + \langle X, Y \rangle) \tag{6.4.3}$$

for every pair X, Y of semimartingales. Extend formula (6.4.3) by expressing $\Pi_{i=1}^{n}\mathcal{E}(X_i)$, for $X_1, \ldots, X_n$, n semimartingales, as $\mathcal{E}(Z)$, for a certain semi-martingale Z.

2. Deduce from (6.4.3) that

$$\mathcal{E}(Y)^{-1} = \mathcal{E}(-Y + \langle Y \rangle). \tag{6.4.4}$$

3. (a) Prove that the application $X \mapsto \mathcal{E}(X)$ is injective.

   (b) Let Y and Z be two (given) semimartingales. Solve the equation:

   $$\mathcal{E}(X)\mathcal{E}(Y) = \mathcal{E}(Z).$$

4. Let $\alpha \in \mathbb{R}$ and X be a given semimartingale. Prove that there exists a unique semimartingale, which we denote by $X^{(\alpha)}$, such that $\mathcal{E}(X)^{\alpha} = \mathcal{E}(X^{(\alpha)})$. Give an explicit formula for $X^{(\alpha)}$. Deduce from the multiplicativity formula that $X^{(\alpha)} + X^{(\beta)} = X^{(\alpha+\beta)} - \alpha\beta\langle X \rangle$.

5. (Application to Girsanov's Theorem.)

   (a) Consider M and Y, two continuous local martingales. Find all processes V with bounded variation such that $\mathcal{E}(M + V)\mathcal{E}(Y)$ is a local martingale.

   (b) Assume that Y is such that: $(\mathcal{E}(Y)_t, t \geq 0)$ is a $(P, (\mathcal{F}_t))$ martingale. Define a probability measure Q as follows: $Q_{|\mathcal{F}_t} = \mathcal{E}(Y)_t P_{|\mathcal{F}_t}$.
   Prove that if (M_t) is a $(P, (\mathcal{F}_t))$ local martingale, then $(M_t - \langle M, Y \rangle_t, t \geq 0)$ is a $(Q, (\mathcal{F}_t))$ local martingale.
   Hint: Prove that (X_t) is a $(Q, \mathcal{F}_t)$ local martingale if and only if $\mathcal{E}(X)$ is a $(Q, \mathcal{F}_t)$ local martingale.

Comments and references. The title of our exercise indicates that it is the result of a long evolution. The result of question 5(b) is the general form of Girsanov's theorem, as stated in:

J.H. VAN SCHUPPEN AND E. WONG: Transformation of local martingales under a change of law. *Ann. Probability*, **2**, 879–888 (1974).

Formula (6.4.3) and similar formulae presented in this exercise originate from:

M. YOR: Sur les intégrales stochastiques optionnelles et une suite remarquable de formules exponentielles. *Séminaire de Probabilités X*, 481–500, *Lecture Notes in Mathematics*, **511**, Springer, Berlin, 1976.

A thorough study of Girsanov's Theorem and absolute continuity relationships between the Wiener measure and Brownian motion with "direct" or "indirect" drifts is undertaken in:

A.S. ÜSTÜNEL AND M. ZAKAI: The change of variables formula on Wiener space. *Séminaire de Probabilités XXXI*, 24–39, *Lecture Notes in Mathematics*, **1655**, Springer, Berlin, 1997.

** 6.5 The range process of Brownian motion

Let $(B_t, t \geq 0)$ be a real valued Brownian motion, starting from 0, and $(\mathcal{F}_t, t \geq 0)$ be its natural filtration.

Define: $S_t = \sup_{s \leq t} B_s$, $I_t = \inf_{s \leq t} B_s$, and, for $c > 0$, $\theta_c = \inf\{t : S_t - I_t = c\}$.

1. Show that, for any $\lambda \in \mathbb{R}$,

$$M_t \equiv \cosh\left(\lambda(S_t - B_t)\right) \exp\left(-\frac{\lambda^2 t}{2}\right)$$

is an $(\mathcal{F}_t)$ martingale.

2. Prove the formula: $E\left[\exp\left(-\frac{\lambda^2}{2}\theta_c\right)\right] = \dfrac{2}{1 + \cosh(\lambda c)} \equiv \dfrac{1}{\cosh^2\left(\lambda\frac{c}{2}\right)}$.

Comments and references. The range process $(S_t - I_t,\ t \geq 0)$ of Brownian motion is studied in:

J.P. IMHOF: On the range of Brownian motion and its inverse process. *Ann. of Prob.*, **13**, no. 3, 1011–1017 (1985),

P. VALLOIS: Amplitude du mouvement brownien et juxtaposition des excursions positives et négatives. *Séminaire de Probabilités XXVI*, 361–373, *Lecture Notes in Mathematics*, **1526**, Springer, Berlin, 1992,

P. VALLOIS: Decomposing the Brownian path via the range process. *Stochastic Process. Appl.*, **55**, no. 2, 211–226 (1995).

* 6.6 Symmetric Lévy processes reflected at their minimum and maximum; E. Csáki's formulae for the ratio of Brownian extremes

1. Let $X = (X_t, \, t \geq 0)$ be a symmetric Lévy process and define its past minimum and past maximum processes as $I_t = \inf_{s \leq t} X_s$ and $S_t = \sup_{s \leq t} X_s$. Prove the following identity in law, for a fixed t

$$(S_t - X_t, X_t - I_t, X_t) \overset{\text{(law)}}{=} (-I_t, S_t, X_t) \overset{\text{(law)}}{=} (S_t, -I_t, -X_t). \qquad (6.6.1)$$

Hint. Use the time reversal property of Lévy processes, that is

$$(X_t - X_{(t-u)-}, \, u \leq t) \overset{\text{(law)}}{=} (X_u, \, u \leq t). \qquad (6.6.2)$$

2. As an application, prove that

$$\frac{S_t - X_t}{S_t - I_t} \overset{\text{(law)}}{=} \frac{S_t}{S_t - I_t}.$$

3. In this question, X denotes a standard Brownian motion and T an exponential time with parameter $\frac{\theta^2}{2}$, independent of X. Prove that, for $x, y \geq 0$

$$P(S_T \leq x, \, -I_T \leq y, \, X_T \in da) = \qquad (6.6.3)$$
$$\frac{\theta}{2} \left(e^{-\theta|a|} - \frac{\sinh(\theta y)}{\sinh(\theta(x+y))} e^{-\theta|a-x|} - \frac{\sinh(\theta x)}{\sinh(\theta(x+y))} e^{-|a+y|} \right) da$$
$$P(S_T - X_T \leq x, \, -I_T + X_T \leq y) = P(S_T \leq x, \, -I_T \leq y) \qquad (6.6.4)$$
$$= 1 - \frac{\sinh(\theta x) + \sinh(\theta y)}{\sinh(\theta(x+y))}.$$

As an application, prove that the distribution function of the ratio $\frac{S_1}{S_1 - I_1}$ admits both the following continuous integral and discrete representations:

$$P\left(\frac{S_1}{S_1 - I_1} \leq a \right) = \frac{1-a}{2} \int_0^\infty dx \, \frac{\sinh(ax)}{\cosh^2\left(\frac{x}{2}\right)} \qquad (6.6.5)$$

$$= (1 - a) \left(2a \sum_{n=1}^\infty \frac{(-1)^{n-1}}{n+a} + \frac{\pi a}{\sin(\pi a)} - 1 \right). \qquad (6.6.6)$$

4. Check that these formulae agree with Csáki's formulae for the same quantity:

$$P\left(\frac{S_1}{S_1 - I_1} \leq a \right) = 2a(1-a) \sum_{n=1}^\infty \frac{(-1)^{n-1}n}{n^2 - a^2} \qquad (6.6.7)$$

$$= a(1-a) \left(\psi\left(1 + \frac{a}{2}\right) - \psi\left(\frac{1}{2} + \frac{a}{2}\right) + \frac{\pi}{\sin(\pi a)} - \frac{1}{a} \right), \qquad (6.6.8)$$

where $\psi(x) = \frac{d}{dx}\left(\log \Gamma(x)\right)$ is the digamma function.

Comments and references. Formulae (6.6.7) and (6.6.8) are due to E. Csáki (reference below), who deduced them from classical theta series expansion for the joint law of (S_t, I_t, X_t), (see e.g. Revuz and Yor [75], p.111, Exercise (3.15)). In the same paper, Csáki also obtained variants of these formulae for the Brownian bridge; these have been interpreted via some path decomposition by Pitman as Yor.

E. Csáki: On some distributions concerning maximum and minimum of a Wiener process. In: *Analytic Function Methods in Probability Theory*, 43–52, *Colloq. Math. Soc. János Bolyai*, **21**, North-Holland, Amsterdam-New York (1979).

J. Pitman and M. Yor: Path decompositions of a Brownian bridge related to the ratio of its maximum and amplitude. *Studia Sci. Math. Hungar.*, **35**, no. 3–4, 457–474 (1999).

*6.7 Infinite divisibility with respect to time

1. Let $(X_t, t \geq 0)$ be a real valued Lévy process, with $X_0 = 0$. Prove that, for every $n \in \mathbb{N}$,

$$(X_{nt}, t \geq 0) \overset{(\text{law})}{=} (X_t^{(1)} + X_t^{(2)} + \ldots + X_t^{(n)}, t \geq 0) \qquad (6.7.1)$$

where, on the right hand side of (6.7.1), the processes $X^{(i)}$, $i = 1, 2, \ldots, n$ are independent copies of X.

2. Give some examples of processes $(X_t, t \geq 0)$ which satisfy (6.7.1), but are not Lévy processes.

   Hint. Consider $\hat{X}_t = \int_{at}^{bt} du\,(X_u/u)$, where X is a Lévy process.

From now on, we shall call a process which satisfies (6.7.1) an IDT process (i.e. Infinitely Divisible with respect to Time).

3. Prove that if X is an IDT process, then the random variable $\overline{X} \overset{(\text{def})}{=} (X_t, t \geq 0)$ taking values in path space is infinitely divisible.

4. (a) Show that the Gaussian processes $B_t^f = \int_0^t f(t/u)\, dB_u$ are IDT processes.

   (b) More generally, describe the centered Gaussian processes which are IDT.

5. Give an example of a random variable $X = (X_t, t \geq 0)$ with values in path space (i.e. $\mathcal{C}(\mathbb{R}_+, \mathbb{R})$), which is infinitely divisible, but does not satisfy (6.7.1).

6. Assume that X is an IDT process, where $\langle \delta, X \rangle = \int_0^\infty dt f(t) X_t$. Is it possible to characterize the Lévy measure of X, i.e. the measure N which satisfies

$$E\left(\exp -\langle f, X \rangle\right) = \exp\left(-\int N(d\omega)(1 - e^{-\langle f, \omega \rangle})\right)?$$

More generally, give a form of the Lévy–Khintchine representation of the IDT processes.

7. The aim of this question is to give examples of processes $(X_t, t \geq 0)$ (apart from Lévy processes) such that for all $k \geq 1$ and all $t_1, t_2, \ldots, t_k$:

$$X_{t_1+t_2+\ldots+t_k} \stackrel{\text{(law)}}{=} X_{t_1}^{(1)} + X_{t_2}^{(2)} \ldots + X_{t_k}^{(k)}, \qquad (6.7.2)$$

where, on the right hand side, $X^{(1)}, X^{(2)}, \ldots, X^{(k)}$ are i.i.d. copies of X.

(a) Prove that if $(Y_t, t \geq 0)$ has the same one-dimensional distributions as a Lévy process, then Y satisfies the above property.

(b) Prove that if X satisfies (6.7.2), then for every given t, the law of X_t is infinitely divisible.

(c) Let (ξ_t) be a Lévy process, and $F_t(x) = P(\xi_t \leq x)$. Prove that if U is uniformly distributed on $[0, 1]$, then

$$Y_t = F_t^{-1}(U) \stackrel{\text{(law)}}{=} \xi_t,$$

where $F_t^{-1}(u) = \inf\{x : F_t(x) \geq u\}$. Hence, Y_t satisfies (6.7.2).

8. Is there a process which satisfies $(X_{s+t}, s, t \geq 0) \stackrel{\text{(law)}}{=} (X_s^{(1)} + X_t^{(2)}, s, t \geq 0)$?

Comments and references. IDT processes have been introduced in:

R. MANSUY: On processes which are infinitely divisible with respect to time, arXiv:math/0504408v1 (2005)

and their study is developed in:

K. ES-SEBAIY AND Y. OUKNINE: How rich is the class of processes which are infinitely divisible with respect to time? *Statist. Probab. Lett.*, **78**, no. 5, 537–547 (2008).

** 6.8 A toy example for Westwater's renormalization

Let $(B_t, t \geq 0)$ be a complex valued BM, starting from 0.

1. Consider, for every $\lambda > 0$, and $z \in \mathbb{C} \setminus \{0\}$, the equation

$$Z_t = z + B_t + \lambda \int_0^t \frac{Z_s ds}{|Z_s|^2}. \qquad (6.8.1)$$

Identify the process $(|Z_t|, t \geq 0)$.

Show the representation:

$$Z_t = |Z_t| \exp(i\gamma_{H_t}), \qquad (6.8.2)$$

where $H_t = \int_0^t \frac{ds}{|Z_s|^2}$, and γ is a process which is independent of $|Z|$. Identify γ.

2. Define P_z^λ to be the law on $C(\mathbb{R}_+, \mathbb{C})$ of the process Z, which is the solution of (6.8.1).

Show that the family $\left(P_z^\lambda; z \in \mathbb{C} \setminus \{0\}\right)$ may be extended by weak continuity to the entire complex plane $\mathbb{C}$. We denote by P_0^λ the limit law of P_z^λ, as $z \to 0$. We still denote by (Z_t) the coordinate process on $C(\mathbb{R}_+, \mathbb{C})$, and $\mathcal{Z}_t = \sigma\{Z_s, s \leq t\}$.

Show that, for every $z \neq 0$, and every $t > 0$, P_z^λ and P_z^0 are equivalent on the σ-field $\mathcal{Z}_t$, and identify the Radon–Nikodym density dP_z^λ/dP_z^0.

3. Show that, for every continuous, bounded functional $F : C([0,1], \mathbb{C}) \to \mathbb{R}$, the quantity

$$\frac{1}{\Lambda_z} E_z^0 \left[F(Z) \exp\left(-\frac{\lambda^2}{2} \int_0^1 \frac{ds}{|Z_s|^2} \right) \right]$$

where

$$\Lambda_z = E_z^0 \left[\exp -\frac{\lambda^2}{2} \int_0^1 \frac{ds}{|Z_s|^2} \right]$$

converges, as $z \to 0$, but $z \neq 0$. Describe the limit in terms of P_0^λ.

4. Let $(R_t, t \geq 0)$ be a Bessel process with dimension $d > 2$, starting from 1. Define the process $(\varphi_u, u \geq 0)$ via the equation:

$$\log R_t = \varphi_{H_t}, \quad \text{where } H_t = \int_0^t \frac{ds}{R_s^2} .$$

(a) Identify the process $(\varphi_u, u \geq 0)$.

(b) Prove that $\frac{1}{\log t} H_t$ converges in probability towards a constant which should be computed.

(c) Prove the same result with the help of the Ergodic Theorem by considering: $\int_1^{a^n} \frac{ds}{R_s^2}$, as $n \to \infty$, for some fixed $a > 1$.

(d) Assume now that $(R_t, t \geq 0)$ is starting from 0.

Show that: $\frac{1}{\log \frac{1}{\varepsilon}} \int_\varepsilon^1 \frac{ds}{R_s^2}$ converges in probability as $\varepsilon \to 0$.

(e) Show the convergence in law, as $\varepsilon \to 0$, and identify the limit of: $\frac{1}{\left(\log \frac{1}{\varepsilon}\right)^{1/2}}\theta_{(\varepsilon,1)}$, where $\theta_{(\varepsilon,1)}$ is the increment between ε and 1 of a continuous determination of the angle around 0 made by the process Z, under the law P_0^λ.

Comments and references. This is a "toy example", the aim of which is to provide some insight into (or, at least, some simple analogue for!) *Westwater's renormalization result* which we now discuss briefly.

First, consider a two-dimensional Brownian motion $(B_t, t \geq 0)$ and let $f_n(z) = n^2 f(nz)$, where $f : \mathbb{R}^2 (\simeq \mathbb{C}) \to \mathbb{R}_+$ is a continuous function with compact support, such that

$$\int dx dy f(x,y) = 1.$$

It was proved originally by S. Varadhan (1969) [this result has then been reproved and extended in many ways, by J.F. Le Gall and J. Rosen in particular] that, if we denote $\{X\} = X - E(X)$ for a generic integrable variable X, the sequence:

$$\int_0^1 ds \int_0^1 dt \{f_n(B_t - B_s)\}$$

converges a.s. and in every L^p, as $n \to \infty$ towards what is now called the *renormalized local time of intersection* γ.

Moreover, for any $k \in \mathbb{R}_+$, the sequence of probabilities:

$$W_n^{(k)} \stackrel{\text{def}}{=} c_n \exp\left(-k \int_0^1 ds \int_0^1 dt \{f_n(B_t - B_s)\}\right) \cdot P|_{\sigma(B_s, s \leq 1)}$$

converges weakly to $c \exp(k\gamma) \cdot P|_{\sigma(B_s, s \leq 1)}$.

Westwater's renormalization result is that for the similar problem involving now the three-dimensional Brownian motion $(B_s, s \leq 1)$, the sequence $(W_n^{(k)}, n \geq 1)$ converges weakly towards a probability $W^{(k)}$, which is singular w.r.t. $W^{(0)} \equiv P|_{\sigma(B_s, s \leq 1)}$. In other words, Varadhan's L^p-convergence result does not extend to three-dimensional Brownian motion, although a weak convergence result holds.

Further studies were made by S. Kusuoka, showing that, under $W^{(k)}$, the reference process still has double intersections. Thus, both for $d = 2$ and $d = 3$, the objective (central in Constructive Quantum Field Theory) to construct in this way some kind of self avoiding Brownian motion could not be reached. Intensive research on this subject is presently being done, by G. Lawler, O. Schramm, W. Werner, using completely different ideas (i.e. a stochastic version of Loewner's equation on the complex plane).

** 6.9 Some asymptotic laws of planar Brownian motion

1. Let $(\gamma_t, t \geq 0)$ be a real-valued Brownian motion, starting from 0.

   (a) Let $c \in \mathbb{R}$. Decompose the process $(e^{ic\gamma_u}, u \geq 0)$ as the sum of a continuous martingale, and a continuous process with bounded variation.

   (b) Show that the continuous process:

   $$\left(\gamma_u, \int_0^u d\gamma_s \exp(ic\gamma_s); u \geq 0\right)$$

   which takes its values in $\mathbb{R} \times \mathbb{C}$ converges in law as $c \to \infty$ towards

   $$\left(\gamma_u; \frac{1}{\sqrt{2}}(\gamma_u' + i\gamma_u''); u \geq 0\right),$$

   where $\gamma, \gamma', \gamma''$ are three real-valued independent Brownian motions, starting from 0.

2. Let $(Z_t, t \geq 0)$ be a complex-valued Brownian motion, starting from $Z_0 = 1$. Show that, as $t \to \infty$,

   $$\frac{1}{\log t} \int_0^t ds \frac{Z_s}{|Z_s|^3}$$

   converges in law towards a limit, the law of which will be described.

Comments and references. A much more complete picture of limit laws for planar Brownian motion is given in:

J.W. PITMAN AND M. YOR: Further asymptotic laws of planar Brownian motion. *Ann. Prob.*, **17**, (3), 965–1011 (1989).

See also D. Revuz and M. Yor [75], Chapter XIII and

Y. HU AND M. YOR: Asymptotic studies of Brownian functionals. In: *Random Walks*, 187–217, *Bolyai Soc. Math. Stud.*, **9**, Janos Bolyai Math. Soc., Budapest, 1999.

The following articles on asymptotic distributions and strong approximations for diffusions are also recommended:

A. FÖLDES: Asymptotic independence and strong approximation. A survey. In: *Endre Csáki 65. Period. Math. Hungar.*, **41**, no. 1–2, 121–147 (2000)

E. Csaki, A. Földes and Y. Hu: Strong approximations of additive functionals of a planar Brownian motion. *Prépublication* 779, Laboratoire de Probabilités et Modèles Aléatoires (2002).

** 6.10 Windings of the three-dimensional Brownian motion around a line

Let $B = (X, Y, Z)$ be a Brownian motion in $\mathbb{R}^3$, such that $B_0 \notin D \equiv \{x = y = 0\}$.

Let $(\theta_t, t \geq 0)$ denote a continuous determination of the winding of B around D, which may be defined by taking a continuous determination of the argument of the planar Brownian motion $(X_u + iY_u, u \leq t)$ around $0 = 0 + i0$. To every Borel function $f : \mathbb{R}_+ \to \mathbb{R}_+$, we associate the volume of revolution

$$\Gamma^f = \left\{ (x, y, z) : (x^2 + y^2)^{1/2} \leq f(|z|) \right\} \subset \mathbb{R}^3 .$$

Denote $\theta_t^f = \int\limits_0^t d\theta_s 1_{(B_s \in \Gamma^f)}$.

1. Show that,

$$\text{if} \quad \frac{\log f(\lambda)}{\log \lambda} \xrightarrow[\lambda \to \infty]{} a, \quad \text{then} \quad \frac{2\theta_t^f}{\log t} \xrightarrow[t \to \infty]{\text{(law)}} \int\limits_0^\sigma d\gamma_u 1_{(\beta_u \leq aS_u)}$$

where (β, γ) is a Brownian motion in $\mathbb{R}^2$, starting from 0,

$$S_u = \sup_{s \leq u} \beta_s , \quad \text{and} \quad \sigma = \inf\{u : \beta_u = 1\} .$$

2. Show, by a simple extension of the preceding result, that, if $a > 1$, then:

$$\frac{1}{\log t}(\theta_t - \theta_t^f) \xrightarrow[(t \to \infty)]{(P)} 0 .$$

Show, under the same hypothesis, that, in fact:

$$\theta_t - \theta_t^f \quad \text{converges a.s., as } t \to \infty.$$

Comments and references. The aim of this exercise is to explain how the wanderings of B in different surfaces of revolution in $\mathbb{R}^3$, around D, contribute to the asymptotic windings: $\frac{2\theta_t}{\log t} \xrightarrow{\text{(law)}} \gamma_\sigma$, as $t \to \infty$. For a full proof and motivations, see:

J.F. Le Gall and M. Yor: Enlacements du mouvement Brownien autour des courbes de l'espace. *Trans. Amer. Math. Soc.*, **317**, 687–722 (1990).

** 6.11 Cyclic exchangeability property and uniform law related to the Brownian bridge

Let $\{b_t, \ 0 \le t \le 1\}$ be the standard Brownian bridge and $\mathcal{F}$ the σ-field it generates. Define the family of transformations Θ_u, $u \in [0,1]$, acting on the paths of b as follows:

$$\Theta_u(b)_t = \begin{cases} b_{t+u} - b_u, & \text{if } t < 1 - u, \\ b_{t-(1-u)} - b_u, & \text{if } 1 - u \le t \le 1 \end{cases} .$$

This transformation consists in re-ordering the paths $\{b_t, \ 0 \le t \le u\}$ and $\{b_t, \ u \le t \le 1\}$ in such a way that the new process is continuous and vanishes at times 0 and 1. For a better understanding of the sequel, it is worth drawing a picture.

1. Prove the cyclic exchangeability property for the Brownian bridge, that is:

$$\Theta_u(b) \stackrel{(\text{law})}{=} b, \quad \text{for any } u \in [0,1] .$$

Let $\mathcal{I}(\subset \mathcal{F})$ be the sub-σ-field of the events invariant under the transformations Θ_u, $u \in [0,1]$, that is:

for any given $u \in [0,1]$, $\Lambda \in \mathcal{I}$, if and only if $\mathbb{1}_\Lambda(b) = \mathbb{1}_\Lambda \circ \Theta_u(b)$, $P-$a.s.

An example of a non trivial $\mathcal{I}$–measurable r.v. is given by the amplitude of the bridge b, i.e. $\sup_{0 \le u \le 1} b_u - \inf_{0 \le u \le 1} b_u$.

2. Prove that for any functional $F(b) \in L^1(P)$,

$$E[F(b) \,|\, \mathcal{I}] = \int_0^1 du \, F \circ \Theta_u(b) . \qquad (6.11.1)$$

   Hint. Prove that for every $u, v \in [0,1]$, $\Theta_u \Theta_v = \Theta_{\{u+v\}}$, where $\{x\}$ is the fractional part of x.

3. Let m be the time at which b reaches its absolute minimum. It can be shown that m is almost surely unique. Hence, $m = \inf\{t : b_t = \inf_{s \in [0,1]} b_s\}$. Prove that m is uniformly distributed and is independent of the invariant σ-field $\mathcal{I}$.

   Hint. First note that $\{m \circ \Theta_u + u\} = m$, a.s.

4. Let $A_0 = \int_0^1 du \, \mathbb{1}_{\{b_u \le 0\}}$ be the time that b spends under the level 0. Prove that A_0 is uniformly distributed and independent of the invariant σ-field $\mathcal{I}$.

Comments and references. The *uniform law* for the minimum time and the time spent in $\mathbb{R}_+$ (or $\mathbb{R}_-$) by the Brownian bridge has about the same history as the *arcsine law* for the same functionals of Brownian motion and goes back to

P. LÉVY: Sur certains processus stochastiques homogènes. *Compositio Math.*, **7**, 283–339 (1939).

By now, many proofs and extensions of this uniform law are known; those which are presented in this exercise are drawn from:

L. CHAUMONT, D.G. HOBSON AND M. YOR: Some consequences of the cyclic exchangeability property for exponential functionals of Lévy processes. *Séminaire de Probabilités XXXV*, 334–347, *Lecture Notes in Mathematics,* **1755**, Springer, Berlin, 2001.

The identity in law between the time at which the process reaches its absolute minimum and the time it spends under the level 0 is actually satisfied by any process with exchangeable increments, as shown in

F.B. KNIGHT: The uniform law for exchangeable and Lévy process bridges. Hommage à P.A. Meyer et J. Neveu. *Astérisque,* **236**, 171–188 (1996)

L. CHAUMONT: A path transformation and its applications to fluctuation theory. *J. London Math. Soc.*, (2), **59**, no. 2, 729–741 (1999).

This property admits an analogous version in discrete time and may be obtained as a consequence of fluctuation identities as first noticed in

E. SPARRE-ANDERSEN: On sums of symmetrically dependent random variables. *Scand. Aktuar. Tidskr.*, **26**, 123–138 (1953).

** 6.12 Local time and hitting time distributions for the Brownian bridge

Let P_a be the law of the real valued Brownian motion starting from $a \in \mathbb{R}$ and $P_{a \to x}^{(t)}$ be the law of the Brownian bridge with length t, starting from a and ending at $x \in \mathbb{R}$ at time t. Let $(X_u, u \geq 0)$ be the canonical coordinate process on $C(\mathbb{R}_+, \mathbb{R})$. The density of the Brownian semigroup will be denoted by p_t, i.e.

$$p_t(a,b) = \frac{1}{\sqrt{2\pi t}} \exp\left(-\frac{(a-b)^2}{2t}\right). \tag{6.12.1}$$

1. Prove that $P_{a \to x}^{(t)}$ may be realized as the law of the process

$$\left(a + \left(B_u - \frac{u}{t}B_t\right) + \frac{u}{t}x,\, u \leq t\right).$$

2. We put, for any $y \in \mathbb{R}$, $T_y = \inf\{t : X_t = y\}$. Prove the reflection principle for Brownian motion, that is, under P_0, the process X^y defined by

$$X_t^y = X_t, \quad \text{on } \{t \leq T_y\}, \quad X_t^y = 2y - X_t, \quad \text{on } \{t > T_y\},$$

has the same law as X. Let $S_t = \sup_{s \le t} X_s$. For $x \le y$, $y > 0$, prove that

$$P_0(S_t > y, X_t < x) = P_0(X_t < x - 2y), \qquad (6.12.2)$$

and compute the law of the pair (X_t, S_t) under P_0.

3. Deduce from question 1 that for any $a, y \in \mathbb{R}$,

$$P_a(T_y \in dt) = \left| \frac{\partial}{\partial y} p_t(y, a) \right| dt = \frac{|y - a|}{\sqrt{2\pi t^3}} \exp\left(-\frac{(y - a)^2}{2t} \right) dt, \quad t > 0.$$

$$(6.12.3)$$

(Note that when $a = y$, $P_a(T_a - 0) - 1$.)

4. Let $(L_u, u \le 1)$ be the local time of X at level 0. Using Lévy's identity:

$$(S - X, S) \overset{\text{(law)}}{=} (|X|, L), \quad \text{under } P_0,$$

prove that under P_0, (X_t, L_t) has density:

$$\left(\frac{1}{2\pi t^3} \right)^{\frac{1}{2}} (|x| + y) \exp\left(-\frac{(|x| + y)^2}{2t} \right), \quad x \in \mathbb{R}, y \ge 0. \qquad (6.12.4)$$

Deduce from the preceding formula an explicit expression for the joint law of (X_t, L_t), for a fixed $t < 1$, under the law $P_{0 \to 0}^{(1)}$ of the standard Brownian bridge.

Comments and references. A detailed discussion of the (very classical!) result stated in the first question is found in:

D. LAMBERTON AND B. LAPEYRE: *Introduction to Stochastic Calculus Applied to Finance.* Chapman & Hall, London, 1996. French second edition: Ellipses, Paris, 1997.

We may also cite D. Freedman's book [32] which uses the reflection principle an infinite number of times to deduce the joint law of the maximum and minimum of Brownian motion, a result which goes back to Bachelier!

L. BACHELIER: Probabilités des oscillations maxima. *C. R. Acad. Sci. Paris,* **212,** 836–838 (1941).

Further properties of the local times of the Brownian bridges are discussed in

J.W. PITMAN: The distribution of local times of a Brownian bridge. *Séminaire de Probabilités XXXIII,* 388–394, *Lecture Notes in Mathematics,* **1709,** Springer, Berlin, 1999.

** 6.13 Partial absolute continuity of the Brownian bridge distribution with respect to the Brownian distribution

We keep the same notation as in Exercise **6.12**.

1. Prove that, under the Wiener measure P, the process

$$\left\{ a\left(1 - \frac{u}{t}\right) + \frac{u}{t}x + uX_{(\frac{1}{u} - \frac{1}{t})},\, u \leq t \right\}$$

has law $P_{a \to x}^{(t)}$. In particular, the family $\{P_{a \to x}^{(t)}; a \in \mathbb{R}, x \in \mathbb{R}\}$ depends continuously on a and x.

Hint. Use the Markov property and the invariance of the Brownian law by time inversion.

Deduce therefrom the law of T_0 under $P_{a \to x}^{(t)}$, when $a \neq 0$.

Hint. Take for granted the following result for a nice transient one-dimensional diffusion (X_t)
$$P_a(L_y \in dt) = C\, p_t(a, y)\, dt\,,$$

where L_y is the last passage time at $y \in \mathbb{R}$ by (X_t) (i.e. $L_y = \sup\{t \geq 0 : X_t = y\}$), and $p_t(a, y)$ denotes the density of the semigroup of (X_t) with respect to Lebesgue measure dy. See the reference in the comments below.

2. Prove the following absolute continuity relation, for every $s < t$:

$$E_{a \to x}^{(t)}[F(X_u, u \leq s)] = E_a\left[F(X_u, u \leq s)\frac{p_{t-s}(X_s, x)}{p_t(a, x)} \right],   \tag{6.13.1}$$

where $F : C([0, s], \mathbb{R}) \to \mathbb{R}_+$ is any bounded Borel functional.

3. More generally, show that for every stopping time S and every $s < t$,

$$E_{a \to x}^{(t)}[\mathbb{1}_{\{S < s\}} F(X_u, u \leq S)] = E_a\left[\mathbb{1}_{\{S < s\}} F(X_u, u \leq S)\frac{p_{t-s}(X_S, x)}{p_t(a, x)} \right].$$
$$\tag{6.13.2}$$

4. Prove the equality: for every $s \leq t$,

$$P_{a \to x}^{(t)}(T_y < s) = E_a\left[\mathbb{1}_{\{T_y < s\}}\frac{p_{t-T_y}(y, x)}{p_t(a, x)} \right].   \tag{6.13.3}$$

Then, using question 2 of Exercise **6.12**, prove that for every $s < t$,

$$P_{a \to x}^{(t)}(T_y \in ds) = ds \left| \frac{\partial}{\partial y} p_s(y, a) \right| \frac{p_{t-s}(y, x)}{p_t(a, x)}\,,   \tag{6.13.4}$$

and check that when $y = 0$, it agrees with the result found in question 1.

5. Using the symmetry property of Brownian motion, prove that:

$$P^{(t)}_{a \to x}(T_0 < s) = \frac{p_t(a, -x)}{p_t(a, x)} P^{(t)}_{a \to -x}(T_0 < s), \quad x, a > 0, \qquad (6.13.5)$$

$$P^{(t)}_{a \to x}(T_0 < t) = \frac{p_t(a, -x)}{p_t(a, x)}, \quad x, a > 0. \qquad (6.13.6)$$

Comments and references. We have just seen how the time inversion property of Brownian motion (which is shared by only a few diffusions!) allows us to express a Brownian bridge in terms of Brownian motion. This is further exploited in Exercise **6.18**. See also related discussions in:

S. WATANABE: On time inversion of one-dimensional diffusion processes. *Z. Wahrscheinlichkeitstheorie und Verw. Gebiete,* **31**, 115–124 (1974/75)

J. PITMAN AND M. YOR: Bessel processes and infinitely divisible laws. In: *Stochastic integrals* (Proc. Sympos., Univ. Durham, Durham, 1980), 285–370, *Lecture Notes in Mathematics,* **851**, Springer, Berlin, 1981.

* 6.14 A Brownian interpretation of the duplication formula for the gamma function

Recall (see (4.5.2)) that the duplication formula is:

$$\Gamma(2z) = \frac{1}{\sqrt{2\pi}} 2^{2z - \frac{1}{2}} \Gamma(z) \Gamma\left(z + \frac{1}{2}\right).$$

1. Prove that (4.5.2) is equivalent to the identity in law

$$Z_1 \overset{(\text{law})}{=} 2 \sqrt{Z_1 Z_{1/2}} \qquad (6.14.1)$$

where Z_a denotes the standard gamma variable with parameter a, and Z_1 and $Z_{1/2}$ are assumed to be independent.

Remark that (6.14.1) may also be written as

$$Z_1 \overset{(\text{law})}{=} \sqrt{2Z_1} |N|, \qquad (6.14.2)$$

where N denotes a standard Gaussian variable with variance 1, independent of Z_1. (The identity (6.14.2) is, in fact, the identity (4.8.1).)

2. Let $(B_t, t \geq 0)$ denote standard Brownian motion starting from 0, and let $x \geq 0$. Define $T_x = \inf\{t \geq 0 : B_t = x\}$. Prove the following formulae:

$$P(T_x < Z_1) = \exp(-\sqrt{2}x), \qquad (6.14.3)$$

$$P(T_x < Z_1) = P(x < \sqrt{Z_1} |N|) \qquad (6.14.4)$$

where, on the left hand sides, T_x and Z_1 are assumed to be independent and on the right hand side, Z_1 and N are independent.

3. Deduce from (6.14.3) and (6.14.4) that (6.14.2), hence (6.14.1), are satisfied.

** 6.15 Some deterministic time-changes of Brownian motion

Consider a linear Brownian motion $(B_t, t \geq 0)$ starting from 0, and two regular functions $u, v \colon \mathbb{R}_+ \to \mathbb{R}$. We assume that u satisfies $u(0) = 0$, and is strictly increasing, and is C^1. Assume also that $v(t) \neq 0$, for every $t > 0$, and that v has bounded variations.

1. Prove that the process $X_t = v(t)B_{u(t)}$, $t \geq 0$ is a semimartingale (in its own filtration) and that its martingale part is:

$$\int_0^t v(s)dB_{u(s)} , \quad t \geq 0.$$

2. Prove that the martingale part of X is a Brownian motion if, and only if:

$$v^2(s)u'(s) \equiv 1.$$

3. We denote by (L_t^X) and (L_t^B) the respective local times at 0 of X and B. Prove the formula:

$$L_t^X = \int_0^{u(t)} v\left(u^{-1}(s)\right) dL_s^B. \tag{6.15.1}$$

4. If B is a Brownian motion, and $\beta \in \mathbb{R}$, $\beta \neq 0$, the formula:

$$U_t = e^{\beta t}B\left(\frac{1 - e^{-2\beta t}}{2\beta}\right), \quad t \geq 0$$

defines an Ornstein–Uhlenbeck process with parameter β, i.e. (U_t) solves the SDE:

$$dU_t = d\gamma_t + \beta U_t \, dt,$$

where γ is a Brownian motion.

Apply the formula in the preceding question to relate the local times at 0 of U and B.

5. If $(b(t), t \leq 1)$ is a standard Brownian bridge, it follows from question 1 of Exercise **6.13** that the process

$$B_t = (1 + t)b\left(\frac{t}{1 + t}\right), \quad t \geq 0,$$

is a Brownian motion.

Prove that, if $(L_t, t \geq 0)$, resp. $(\ell_u, u \leq 1)$, is the local time of B, resp. b, at 0, then:

$$\int_0^t \frac{dL_v}{1+v} = \ell\left(\frac{t}{1+t}\right) \qquad (t \geq 0). \tag{6.15.2}$$

Give an explicit expression for the law of $\int_0^t \frac{dL_v}{1+v}$, for fixed t.

Hint. Use the result obtained in Exercise **6.12**, question 3.

*6.16 A new path construction of Brownian and Bessel bridges

Let X be any selfsimilar process of index $\alpha > 0$ with $X_0 = 0$, i.e. X satisfies the scaling property

$$(X_s, s \geq 0) \overset{\text{(law)}}{=} (k^{-\alpha} X_{ks}, s \geq 0), \quad \text{for any } k > 0.$$

1. Show that if for any $a \geq 0$, $P(X_1 > a) > 0$ and if the σ-field $\cap_{t>0}\sigma\{X_s, s \leq t\}$ is trivial, then
    $$\limsup_{t \downarrow 0} \frac{X_t}{t^\alpha} = +\infty, \quad \text{almost surely}.$$

2. Let B be standard Brownian motion. For $a \geq 0$, we define the last hitting time of the curve $t \mapsto a\sqrt{t}$ by B before time 1, i.e. $g_a = \sup\{t \leq 1 : B_t = a\sqrt{t}\}$. Prove that the process
    $$b^{0 \to a} \overset{\text{(def)}}{=} (g_a^{-1/2} B_{g_a t}, 0 \leq t \leq 1)$$
    is the standard bridge of Brownian motion from 0 to a.
    Hint. Think of the time inversion property of Brownian motion and use Exercises **6.12** and **6.13**.

Comments and references. Actually, using the time inversion property is not necessary to prove question 2. In fact the result presented here is valid for any selfsimilar Markov process, with index $\alpha > 0$, provided it attains the curve $t \mapsto at^\alpha$, almost surely, infinitely often as $t \to 0$. In particular, it provides a construction of bridges of Bessel processes of any dimension. See:

L. CHAUMONT AND G. URIBE: Markovian Bridges: weak continuity and pathwise constructions. *Ann. Probab.*, **39**, no. 2, 609–647 (2011).

Other relevant references are:

L. ALILI AND P. PATIE: On the first crossing times of a Brownian motion and a family of continuous curves. *C. R. Math. Acad. Sci. Paris*, **340**, no. 3, 225–228 (2005).

L. BREIMAN: First exit times from a square root boundary. *1967 Proc. Fifth Berkeley Sympos. Math. Statist. and Probability* (Berkeley, California 1965/66), Vol. II: Contributions to Probability Theory, Part 2, 9–16, University of California Press, Berkeley, California.

L.A. SHEPP: A first passage problem for the Wiener process. *Ann. Math. Statist.*, **38**, 1912–1914 (1967).

* 6.17 Random scaling of the Brownian bridge

Let $(B_t, t \geq 0)$ be a real valued Brownian motion, starting from 0, and define $\Lambda = \sup\{t > 0 : B_t - t = 0\}$.

1. Prove that

$$\sqrt{\Lambda} \overset{\text{(law)}}{=} |N|,\tag{6.17.1}$$

where N is Gaussian, centered, with variance 1.

2. Prove that:

$$(B_t - t, t \leq \Lambda) \overset{\text{(law)}}{=} \left(\sqrt{\Lambda} b\left(\frac{t}{\Lambda}\right), t \leq \Lambda\right),\tag{6.17.2}$$

where, on the right hand side, Λ is independent of the standard Brownian bridge $(b(u), u \leq 1)$.

3. Let $\mu \in \mathbb{R}$. Prove the following extension of the two previous questions:

(i)

$$\Lambda_\mu \overset{\text{(law)}}{=} \frac{N^2}{\mu^2}.\tag{6.17.3}$$

(ii)

$$(B_t - \mu t, t \leq \Lambda_\mu) \overset{\text{(law)}}{=} \left(\sqrt{\Lambda_\mu} b\left(\frac{t}{\Lambda_\mu}\right), t \leq \Lambda_\mu\right),\tag{6.17.4}$$

where, on the right hand side, $\Lambda_\mu = \sup\{t > 0 : B_t - \mu t = 0\}$ is independent of the standard Brownian bridge $(b(u), u \leq 1)$.

Deduce from the identity (6.17.4) that conditionally on $\Lambda_\mu = \lambda$, the process $(B_t - \mu t, t \leq \lambda)$ is distributed as a Brownian bridge of length λ.

Comments and references:

(a) One can also prove that for any $\mu \in \mathbb{R}$, the law of the process $(B_u - \mu u, u \leq t)$, given $B_t - \mu t = 0$ does not depend on μ and is distributed as a Brownian bridge with length t. This invariance property follows from the Cameron–Martin absolute continuity relation between $(B_u - \mu u, u \leq t)$ and $(B_u, u \leq t)$. There are in fact other diffusions than Brownian motion with drifts which have the same bridges as Brownian motion. See:

I. BENJAMINI AND S. LEE: Conditioned diffusions which are Brownian bridges. *J. Theoret. Probab.*, **10**, no. 3, 733–736 (1997),

P. FITZSIMMONS: Markov processes with identical bridges. *Electron. J. Probab.*, **3**, no. 12, 12 pp. (1998).

(b) It is often interesting, in order to study a particular property of Brownian motion, to consider simultaneously its extensions for Brownian motion with drift. Many examples appear in:

D. WILLIAMS: Path decomposition and continuity of local time for one-dimensional diffusions I. *Proc. London Math. Soc.*, **28**, (3), 738–768 (1974).

*6.18 Time-inversion and quadratic functionals of Brownian motion; Lévy's stochastic area formula

Let $P^{(m)}_{a \to 0}$ denote the law of the Brownian bridge with duration m, starting at a, and ending at 0.

1. Prove that, if $(B_t, t \geq 0)$ denotes Brownian motion starting at 0, then the law of the process

$$\left\{ (1+t)B\left(\frac{1}{1+t} - \frac{1}{1+m} \right) \; ; \quad 0 \leq t \leq m \right\}$$

conditioned by $\left\{ B\left(1 - \frac{1}{1+m} \right) = a \right\}$ is $P^{(m)}_{a \to 0}$.

Hint. Use the Markov property and the invariance of the Brownian law by time inversion.

2. Let $m > 0$, and consider $\varphi : [0, m] \to \mathbb{R}$ a bounded, Borel function. Prove that, if $(B_t, t \geq 0)$ denotes Brownian motion starting at 0, then the law of $\int_0^m dt\, \varphi(t) B_t^2$, conditioned by $\{ B_m = a \}$ is the same as the law of

$$(1+m)^2 \int_0^m \frac{ds\, B_s^2}{(1+s)^4} \varphi\left(\frac{(1+m)s}{1+s} \right),$$

conditioned by $\left\{ B_m = a\sqrt{1+m} \right\}$.

Show that, consequently, one has:

$$E\left[\exp -\lambda \int_0^m dt\, B_t^2 \mid B_m = a \right]$$

$$= E\left[\exp -\lambda(1+m)^2 \int_0^m \frac{ds\, B_s^2}{(1+s)^4} \Big| B_m = a\sqrt{1+m} \right]$$

for every $\lambda \geq 0$.

Set $\lambda = \frac{b^2}{2}$. Prove that this common quantity equals, in terms of a, b and m:

$$\left(\frac{bm}{\sinh(bm)}\right)^{\frac{1}{2}} \exp\left(-\frac{a^2}{2m}(bm\coth(bm) - 1)\right). \qquad (6.18.1)$$

Comments and references. This exercise is strongly inspired by:

F. B. KNIGHT: Inverse local times, positive sojourns, and maxima for Brownian motion. *Colloque Paul Lévy, Astérisque*, **157–158**, 233–247 (1988).

The expression (6.18.1) of $E\left[\exp\left(-\frac{b^2}{2}\int\limits_0^m dt\, B_t^2\right) \mid B_m = a\right]$ is due to P. Lévy (1951). It has since been the subject of many studies; here we simply refer to

M. YOR: *Some Aspects of Brownian Motion. Part I. Some Special Functionals.* Lectures in Mathematics ETH Zürich. Birkhäuser Verlag, Basel, 1992, p. 18, formula (2.5).

Note that in Exercises **2.14** and **2.15**, we saw another instance of this formula in another disguise involving time spent by Brownian motion up to its first hit of 1.

** 6.19 Quadratic variation and local time of semimartingales

Consider, on a filtered probability space, a continuous semimartingale $X_t = M_t + V_t$ for $t \geq 0$, such that: $X_0 = M_0 = V_0 = 0$, and

$$E\left[\langle M\rangle_\infty + \left(\int\limits_0^\infty |dV_s|\right)^2\right] < \infty. \qquad (6.19.1)$$

(Recall that $\langle X\rangle$, the quadratic variation of X does not depend on V, i.e. it is equal to $\langle M\rangle$.)

1. Prove that (X_t) satisfies:

$$E\left[(X_T)^2\right] = E\left[\langle X\rangle_T\right] \qquad (6.19.2)$$

   for every stopping time T if, and only if,

$$1_{(X_s \neq 0)}|dV_s| = 0, \quad \text{a.s.} \qquad (6.19.3)$$

2. Prove that (X_t) satisfies (6.19.2) if and only if (X_t^+) and (X_t^-) satisfy (6.19.2).

3. (i) Show that if (M_t) is a square integrable martingale, with $M_0 = 0$, then $X_t = -M_t + S_t^M$, where $S_t^M = \sup_{s \leq t} M_s$, satisfies (6.19.2).

(ii) Show that if (M_t) is a square integrable martingale, with $M_0 = 0$, then $X_t = M_t + L_t^M$ satisfies (6.19.2) if and only if $L_t^M \equiv 0$, which is equivalent to $M_t \equiv 0$.

4. Prove that if (L_t) denotes the local time of (X_t) at 0, then (X_t) satisfies:

$$E\left[|X_T|\right] = E[L_T] \tag{6.19.4}$$

for every stopping time T if, and only if, (X_t) is a martingale.

Comments. This exercise constitutes a warning that the well-known identities (6.19.2) and (6.19.4), which are valid for square integrable martingales, do not extend to general semimartingales.

** 6.20 Geometric Brownian motion

Let $(B_t, t \geq 0)$ and $(W_t, t \geq 0)$ be two independent real-valued Brownian motions.

Prove that the two-dimensional process:

$$(X_t, Y_t; t \geq 0) \overset{\text{def}}{=} \left(\exp(B_t), \int_0^t \exp(B_s) dW_s; t \geq 0 \right) \tag{6.20.1}$$

which takes values in $\mathbb{R}^2$ converges to ∞ as t tends to ∞, (i.e. $X_t^2 + Y_t^2 \to \infty$, as $t \to \infty$).

Hint. Consider the time change obtained by inverting:

$$A_t = \int_0^t ds \exp(2B_s) \ .$$

Comments and references. The process (6.20.1) occurs very naturally in relation with hyperbolic Brownian motion, i.e a diffusion in the plane with the infinitesimal generator $\frac{y^2}{2} \left(\frac{\partial^2}{\partial x^2} + \frac{\partial^2}{\partial y^2} \right)$.

J.C. GRUET: Semi-groupe du mouvement brownien hyperbolique. *Stochastics and Stochastic Rep.*, **56**, no. 1–2, 53–61 (1996).

For some computations related to this exercise, see e.g.

M. YOR: On some exponential functionals of Brownian motion. *Adv. in Appl. Probab.*, **24**, no. 3, 509–531 (1992).

The following figure presents the graphs of the densities of the distributions of A_t for some values of t. We thank K. Ishiyama (Nagoya University) for kindly providing us with this picture.

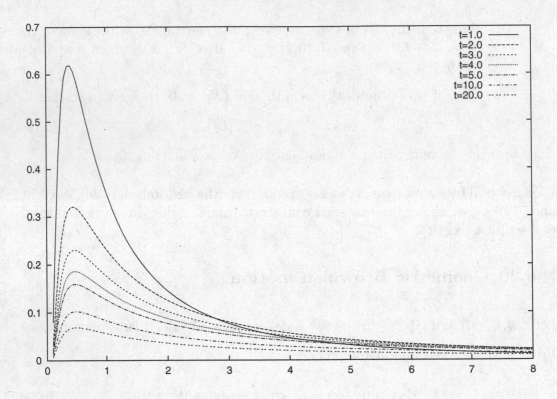

* 6.21 0-self similar processes and conditional expectation

Let $(X_u, u \geq 0)$ be 0-self similar, i.e.

$$(X_{cu}, u \geq 0) \overset{(\text{law})}{=} (X_u, u \geq 0), \quad \text{for any } c > 0 \qquad (6.21.1)$$

and assume that $E[|X_1|] < \infty$.

Prove that, for every $t > 0$,

$$E[X_t \,|\, \overline{X}_t] = \overline{X}_t, \qquad (6.21.2)$$

where $\overline{X}_t = \frac{1}{t} \int_0^t ds \, X_s$.

Comments and references:

(a) Let C be a cone in $\mathbb{R}^n$, with vertex at 0 and define:

$$X_t = \mathbb{1}_{\{B_t \in C\}},$$

where (B_t) denotes the n-dimensional Brownian motion. Then formula (6.21.2) yields:

$$E\left[\mathbb{1}_{\{B_t \in C\}} \,|\, \frac{1}{t} \int_0^t ds \, \mathbb{1}_{\{B_s \in C\}} = a\right] = a,$$

a remarkable result since the law of $\int_0^t ds\, \mathbb{1}_{\{B_s \in C\}}$ is unknown except in the very particular case where $C = \{x \in \mathbb{R}^n : x_1 > 0\}$. See the following papers for a number of developments:

R. PEMANTLE, Y. PERES, J. PITMAN AND M. YOR: Where did the Brownian particle go? *Electron. J. Probab.*, **6**, no. 10, 22 pp. (2001)

J.W. PITMAN AND M. YOR: Quelques identités en loi pour les processus de Bessel. Hommage à P.A. Meyer et J. Neveu. *Astérisque*, **236**, 249–276 (1996).

(b) In the paper:

N.H. BINGHAM AND R.A. DONEY: On higher-dimensional analogues of the arc-sine law. *J. Appl. Probab.*, **25**, no. 1, 120–131 (1988)

the authors show that in general, time spent in a cone (up to 1) by n-dimensional Brownian motion (e.g. in the first quadrant, by two-dimensional Brownian motion) is not beta distributed, which *a priori* was a reasonable guess, given Lévy's arc-sine law for the time spent in $\mathbb{R}_+$ by linear Brownian motion.

** 6.22 A Taylor formula for semimartingales; Markov martingales and iterated infinitesimal generators

Let $(\Omega, \mathcal{F}, (\mathcal{F}_t), P)$ be a filtered probability space. We say that a right continuous process, $(X_t, t \geq 0)$ which is $(\mathcal{F}_t)$ adapted is $(\mathcal{F}_t)$ differentiable if there exists another right continuous process, which we denote as (X_t'), such that:

(i) $M_t^X \stackrel{\text{def}}{=} X_t - \int_0^t ds\, X_s'$ is an $(\mathcal{F}_t)$-martingale.
(ii) $E[|X_t|] < \infty$ and $E\left[\int_0^t ds\, |X_s'|\right] < \infty$.

1. Prove that if (X_t) is $(\mathcal{F}_t)$ differentiable, then its $(\mathcal{F}_t)$ derivative, (X_t'), is unique, up to indistinguability.

2. Assume that (X_t) is $(\mathcal{F}_t)$ differentiable up to order $(n+1)$ and use the notation: $X'' = (X')'$ and by recurrence $X^{(n)} = (X^{(n-1)})'$, $n \geq 3$. Then prove the formula:

$$X_t - tX_t' + \frac{t^2}{2}X_t'' + \cdots + (-1)^n \frac{t^n}{n!} X_t^{(n)} \tag{6.22.1}$$
$$= \int_0^t \frac{s^n}{n!} X_s^{(n+1)}\, ds + M_t^X - \int_0^t s\, dM_s^{X'} + \cdots + (-1)^n \int_0^t \frac{s^n}{n!}\, dM_s^{X^{(n)}}.$$

Hint: Use a recurrence formula and integration by parts.

3. Let $(B_t, t \geq 0)$ denote a real valued Brownian motion. Define the Hermite polynomials $H_k(x,t)$, as:

$$H_k(x,t) = x^k - t\frac{k(k-1)}{2}x^{k-2} \qquad (6.22.2)$$

$$+ \frac{t^2}{2}\frac{k(k-1)(k-2)(k-3)}{2^2}x^{k-4} + \cdots \qquad (6.22.3)$$

$$+ (-1)^{[k/2]}\frac{t^{[k/2]}}{[k/2]!}\frac{k(k-1)\cdots(k-2[k/2]+1)}{2^{[k/2]}}x^{k-2[k/2]}.$$

Prove that $(H_k(B_t,t), t \geq 0)$ is a martingale.

4. Give another proof of this martingale property using the classical generating function expansion:

$$\exp\left(ax - \frac{a^2 t}{2}\right) = \sum_{k=0}^{\infty}\frac{a^k}{k!}H_k(x,t), \quad a \in \mathbb{R}, (x,t) \in \mathbb{R} \times \mathbb{R}_+.$$

Comments and references. The sequence of Hermite polynomials is that of orthogonal polynomials with respect to the Gaussian distribution; hence, it is not astonishing that the Hermite polynomials may be related to Brownian motion. For orthogonal polynomials associated to other Markov processes, see:

W. SCHOUTENS: *Stochastic Processes and Orthogonal Polynomials. Lecture Notes in Statistics*, **146**, Springer-Verlag, New York, 2000.

In particular, the Laguerre polynomials, resp. Charlier polynomials are associated respectively to Bessel processes, and to the Poisson and Gamma processes.

** 6.23 A remark of D. Williams: the optional stopping theorem may hold for certain "non-stopping times"

Let $W_0 = 0, W_1, W_2, \ldots, W_n, \ldots$ be a standard coin-tossing random walk, i.e. the variables $W_1, W_2 - W_1, \ldots, W_k - W_{k-1}, \ldots$ are independent Bernoulli variables such that $P(W_k - W_{k-1} = \pm 1) = \frac{1}{2}, k = 1, 2, \ldots$. For $p \geq 1$, define:

$$\sigma(p) = \inf\{k \geq 0 : W_k = p\}, \qquad \eta = \sup\{k \leq \sigma(p) : W_k = 0\},$$
$$m = \sup\{W_k, k \leq \eta\}, \qquad \gamma = \inf\{k \geq 0 : W_k = m\}.$$

It is known (see the reference below) that:

(i) m is uniformly distributed on $\{0, 1, \ldots, p-1\}$;

(ii) Conditionally on $\{m = j\}$, the family $\{W_k, k \leq \gamma\}$ is distributed as $\{W_k, k \leq \sigma(j)\}$.

Take these two results for granted. For $n \in \mathbb{N}$, denote by $\mathcal{F}_n$ the σ-field generated by $W_0, W_1, \ldots, W_n$.

Prove that for every bounded $(\mathcal{F}_n)_{n \geq 0}$ martingale $(M_n)_{n \geq 0}$, and every $j \in \{0, 1, \ldots, p-1\}$, one has:

$$E\left[M_\gamma \mathbb{1}_{\{m=j\}}\right] = E[M_0]P(m = j).$$

In particular, one has: $E[M_\gamma] = E[M_0]$.

Comments and references:

(a) (i) and (ii), and more generally the analogue for a standard random walk of D. Williams's path decomposition of Brownian motion $\{B_t, t \leq T_p\}$, where $T_p = \inf\{t : B_t = p\}$ were obtained by J.F. Le Gall in:

J.F. LE GALL: Une approche élémentaire des théorèmes de décomposition de Williams. *Séminaire de Probabilités XX*, 1984/85, 447–464, *Lecture Notes in Mathematics*, **1204**, Springer, Berlin, 1986.

Donsker's Theorem then allows us to deduce D. Williams' decomposition results from those for the random walk skeletons.

(b) In turn, D. Williams (2002) showed the analogue for Brownian motion of the optional stopping result presented in the above exercise.

Our exercise is strongly inspired from D. Williams' result, in:

D. WILLIAMS: A "non-stopping" time with the optional stopping property. *Bull. London Math. Soc.*, **34**, 610–612 (2002).

(c) It would be very interesting to characterize the random times, which we might call "pseudo-stopping times", for which the optional stopping theorem is still valid.

* 6.24 Stochastic affine processes, also known as "Harnesses"

Note that given three reals $s < t < T$, the only reals A and B which satisfy the equality

$$f(t) = Af(s) + Bf(T)$$

for every affine function $f(u) = \alpha u + \beta$, $(\alpha, \beta \in \mathbb{R})$ are:

$$A = \frac{T - t}{T - s}, \quad \text{and} \quad B = \frac{t - s}{T - s}.$$

Now assume that on a probability space $(\Omega, \mathcal{F}, P)$, a process $(\Phi_t, t \geq 0)$ is given such that: $E[|\Phi_t|] < \infty$, for each $t \geq 0$. We define the past–future filtration associated with Φ as $\mathcal{F}_{s,T} = \sigma\{\Phi_u : u \in [0,s] \cup [T,\infty)\}$. We shall call Φ an affine process if it satisfies

$$E[\Phi_t \mid \mathcal{F}_{s,T}] = \frac{T-t}{T-s}\Phi_s + \frac{(t-s)}{T-s}\Phi_T, \qquad s < t < T. \qquad (6.24.1)$$

1. Check that (6.24.1) is equivalent to the property that for $s \leq t < t' \leq u$, the quantity

$$E\left[\frac{\Phi_t - \Phi_{t'}}{t - t'} \mid \mathcal{F}_{s,u}\right], \qquad (6.24.2)$$

   does not depend on the pair (t,t'), hence is equal to $\frac{\Phi_u - \Phi_s}{u-s}$.

2. Prove that Brownian motion is an affine process.

3. Let $(X_t, t \geq 0)$ be a centered Gaussian process, which is Markovian (possibly inhomogeneous). Prove that there exist constants $\alpha_{s,t,T}$ and $\beta_{s,t,T}$ such that

$$E[X_t \mid \mathcal{F}_{s,T}] = \alpha_{s,t,T} X_s + \beta_{s,t,T} X_T.$$

   Compute the functions α and β in terms of the covariance function $K(s,t) = E[X_s X_t]$. Which among such processes (X_t) are affine processes?

4. Let $(X_t, t \geq 0)$ be a Lévy process, with $E[|X_t|] < \infty$, for every t. Prove that (X_t) is an affine process.

   Hint. Prove that for $a < b$, and for every $\lambda \in \mathbb{R}$:

$$E[a^{-1} X_a \exp(i\lambda X_b)] = E[b^{-1} X_b \exp(i\lambda X_b)].$$

5. Prove that, although a subordinator $(T_t, t \geq 0)$ may not satisfy $E[T_t] < \infty$, nonetheless the conditional property (6.24.1) is satisfied.

   Hint. Prove that, for $a < b$, and every $\lambda > 0$,

$$E\left[\frac{T_a}{a} \exp(-\lambda T_b)\right] = E\left[\frac{T_b}{b} \exp(-\lambda T_b)\right].$$

6. Prove that if an affine process (X_t) is L^1-differentiable, i.e.

$$\frac{X_{t+h} - X_t}{h} \xrightarrow{L^1} Y_t, \quad \text{as } h \to 0,$$

   then for every s, t, $P(Y_s = Y_t) = 1$, and (X_t) is affine in the strong sense, i.e. $X_t = \alpha t + \beta$, for some random variables α and β which are $\mathcal{F}_0 \vee \mathcal{G}_\infty$-measurable.

7. We now assume only that $(X_t, t \geq 0)$ is a process with exchangeable increments, which is continuous in L^1. Prove that it is an affine process.

Comments and references:

(a) The general notion of a *Harness* is due to J. Hammersley. The particular case we are studying here is called a *simple Harness* by D. Williams (1980), who proved that essentially the only continuous Harnesses are Brownian motions with drifts. In the present exercise, we preferred the term *affine process*. In his book [98], D. Williams studies Harnesses indexed by $\mathbb{Z}$, and proves that they are only affine processes in the strong sense.

(b) Without knowing (or mentioning) the terminology, Jacod and Protter (1988) showed that every integrable Lévy process is a Harness. In fact, their proof only uses the exchangeability of increments, as we do to solve question 7. In relation with that question, we should recall that any process with exchangeable increments is a mixture of Lévy processes. (See the comments in Exercise **2.6**, and the reference to Aldous' St-Flour course, Proposition 10.5, given there).

(c) As could be expected, Lévy was first on the scene! (See Lévy's papers referred to below.) Indeed, he remarked that, for Φ a Brownian motion, the property (6.24.1) holds because

$$\Phi_t - \left(\frac{T-t}{T-s}\Phi_s + \frac{t-s}{T-s}\Phi_T \right)$$

is independent from $\mathcal{F}_{s,T}$.

(d) If ($\Phi_t = \gamma_t, t \geq 0$) is the gamma subordinator, then with the notation in (6.24.2), $\frac{\Phi_t - \Phi_{t'}}{\Phi_u - \Phi_s}$ is independent from $\mathcal{F}_{s,u}$, which implies *a fortiori* that (Φ_t) is a Harness.

(e) The arguments used in questions 4 and 5 also appear in Bertoin's book ([67], p. 85), although no reference to Harnesses is made there.

P. Lévy: Un théorème d'invariance projective relatif au mouvement Brownien. *Comment. Math. Helv.*, **16**, 242–248 (1943).

P. Lévy: Une propriété d'invariance projective dans le mouvement Brownien. *C. R. Acad. Sci., Paris*, **219**, 377–379 (1944).

J.M. Hammersley: Harnesses. *Proc. Fifth Berkeley Sympos. Mathematical Statistics and Probability, Vol. III: Physical Sciences*, 89–117, Univ. California Press, Berkeley, CA, 1967.

J. Jacod and P. Protter: Time reversal on Lévy processes. *Ann. Probab.*, **16**, no. 2, 620–641 (1988).

D. Williams: *Brownian motion as a Harness*. Unpublished manuscript, (1980).

D. Williams: Some basic theorems on Harnesses. In: *Stochastic Analysis*, (a tribute to the memory of Rollo Davidson), eds: D. Kendall, H. Harding, 349–363, Wiley, London, 1973.

* 6.25 More on Harnesses

Recall from Exercise 6.24 that every integrable Lévy process (X_t) is a Harness, with respect to its past–future filtration $\mathcal{F}_{s,u} = \sigma\{X_h, \, h \leq s, \, X_k, \, k \geq u\}, \, s \leq u$, i.e.

$$E\left[\frac{X_d - X_c}{d - c} \mid \mathcal{F}_{a,b}\right] = \frac{X_b - X_a}{b - a}. \tag{6.25.1}$$

We now assume only that we are dealing with a Harness (it is not necessarily a Lévy process).

1. Let $T > 0$ be given. Prove that there exists $(M_t^{(T)}, \, t \leq T)$, a $(\mathcal{F}_{t,T}, \, t < T)$ martingale such that:

$$X_t = M_t^{(T)} + \int_0^t ds \, \frac{X_T - X_s}{T - s}. \tag{6.25.2}$$

2. Conversely, assume that the property (6.25.2) is satisfied for every t, T, with $t < T$. Prove that $(X_t, \, t \geq 0)$ is a Harness.

Comments and references. Formula (6.25.2) has been obtained by Jacod–Protter for Lévy processes, see e.g. Protter's book [71]. Here we show that this formula is equivalent to the Harness property. For a developed discussion, see:

R. MANSUY AND M. YOR: Harnesses, Lévy bridges and Monsieur Jourdain. *Stochastic Process. Appl.*, **115**, no. 2, 329–338 (2005).

* 6.26 A martingale "in the mean over time" is a martingale

Consider a process $(\alpha(s), \, s \geq 0)$ that is adapted to the filtration $(\mathcal{F}_s)_{s \geq 0}$, and satisfies: for any $t > 0$, $E[|\alpha(t)|] < \infty$ and $E\left[\int_0^t du \, |\alpha(u)|\right] < \infty$. Prove that the two following conditions are equivalent:

(i) $(\alpha(u), \, u \geq 0)$ is a $(\mathcal{F}_u)$ martingale,

(ii) for every $t > s$, $E\left[\frac{1}{t-s} \int_s^t dh \, \alpha(h) \mid \mathcal{F}_s\right] = \alpha(s)$.

Comments and references. Quantities such as on the left hand side of (ii) appear very naturally when one discusses whether a process is a semimartingale with respect to a given filtration by using the method of "Laplaciens approchés" due to P.A. Meyer. For precise statements, see

C. STRICKER: Une caractérisation des quasimartingales. *Séminaire de Probabilités IX*, 420–424, *Lecture Notes in Mathematics*, **465**, Springer, Berlin, 1975.

*6.27 A reinforcement of Exercise 6.26

Consider a process $(H(s),\ s \geq 0)$ not necessarily adapted to the filtration $(\mathcal{F}_s)_{s \geq 0}$, which satisfies:

(i) for any $s > 0$, $E[\|H(s)\|] < \infty$.

(ii) for any s, the process $t \mapsto E\left[\frac{H_t - H_s}{t - s} \mid \mathcal{F}_s\right]$, $(t > s)$ does not depend on t.

We call the common value $\alpha(s)$.

1. Prove that $(\alpha(s),\ s \geq 0)$ is an $(\mathcal{F}_s)$-martingale.

2. Assume that $(H(s),\ s \geq 0)$ is $(\mathcal{F}_s)$ adapted and that $(\alpha(s),\ s \geq 0)$ is measurable. Prove that $(H_t,\ t \geq 0)$ is of the form

$$H_t = M_t + \int_0^t \alpha(u)\, du\,,$$

where $(M_t,\ t \geq 0)$ is a $(\mathcal{F}_t)$-martingale.

Comments. For a general discussion on Exercises **6.22** to **6.27**, see the comments at the beginning of this chapter.

*6.28 Some past-and-future Brownian martingales

For $0 \leq a < b$, let: $\mathcal{F}_{a,b} \overset{(\text{def})}{=} \sigma\{B_u : u \leq a\} \vee \sigma\{B_v : v \geq b\}$. The aim of this exercise is to find as many $(\mathcal{F}_{a,b})_{a<b}$ martingales as possible in the Brownian setup.

1. Let $\hat{\varphi} = (\varphi_-, \varphi_+) \in (L(\mathbb{R}_+, ds))^2$ and consider the two parameter process:

$$M_{a,b} = \int_0^a \varphi_-(s)\, dB_s + \int_b^\infty \varphi_+(s)\, dB_s + C_{a,b}(B_b - B_a),\quad 0 < a \leq b < \infty\,,$$

where $C_{a,b}$ is a family of reals. Find conditions on this family so that $(M_{a,b})$ is a $(\mathcal{F}_{a,b})$ martingale.

2. Let $\varphi \in L^2(\mathbb{R}_+, ds)$; show that

$$E\left[\exp\left(\int_0^\infty \varphi(s)\, dB_s\right) \mid \mathcal{F}_{a,b}\right]$$
$$= \exp\left(\int_0^a \varphi(s)\, dB_s + \int_b^\infty \varphi(s)\, dB_s + C_{a,b}^\varphi(B_b - B_a) + D_{a,b}^\varphi\right)\,.$$

Give explicit formulae for $C_{a,b}^\varphi$ and $D_{a,b}^\varphi$.

3. Let $\hat{\varphi} = (\varphi_-, \varphi_+)$ be as in question 1. Do there exist two families of reals, $C_{a,b}^{\hat{\varphi}}$ and $D_{a,b}^{\hat{\varphi}}$, such that:

$$\mathcal{E}_{a,b} \stackrel{\text{(def)}}{=} \exp\left(\int_0^a \varphi_-(s)\, dB_s + \int_b^\infty \varphi_+(s)\, dB_s + C_{a,b}^{\hat{\varphi}}(B_b - B_a) + D_{a,b}^{\hat{\varphi}} \right)$$

is a $(\mathcal{F}_{a,b})$ martingale? If yes, then compute $C_{a,b}^{\hat{\varphi}}$ and $D_{a,b}^{\hat{\varphi}}$.

Comments and references. Past-and-future (sub)-martingales are presented in greater generality in section 5.2 of:

CH. PROFETA, B. ROYNETTE AND M. YOR: Option prices as probabilities. *A New Look at Generalized Black–Scholes Formulae.* Springer Finance, Springer-Verlag, Berlin, 2010.

* 6.29 Additive and multiplicative martingale decompositions of Brownian motion

Let $(B_t,\, t \geq 0)$ denote a one-dimensional Brownian motion starting from 0. Let $k > 1$, $k \in \mathbb{N}$.

1. (a) Prove that k independent martingales $M^{(1)}, \ldots, M^{(k)}$ such that $M_0^{(1)} = \ldots = M_0^{(k)} = 0$ satisfy the identity in law

    $$M^{(1)} + M^{(2)} + \cdots + M^{(k)} \stackrel{\text{(law)}}{=} B, \qquad (6.29.1)$$

    if and only if their joint law may be expressed as:

    $$(M^{(1)}, \ldots, M^{(k)}) \stackrel{\text{(law)}}{=} \left(\int_0^\cdot m_s^{(1)}\, dB_s^{(1)}, \ldots, \int_0^\cdot m_s^{(k)}\, dB_s^{(k)} \right),$$

    where $(B^{(1)}, \ldots, B^{(k)})$ are k independent Brownian motions, and $m^{(1)}, \ldots, m^{(k)}$ are k deterministic functions such that:

    $$\sum_{i=1}^k (m_s^{(i)})^2 = 1, \quad ds - a.e.$$

    (b) Note that a particular case may be obtained from a partition $\Pi = (\Pi^{(1)}, \ldots, \Pi^{(k)})$ of $\mathbb{R}_+$, as one considers the k-dimensional martingale:

    $$M_t^{(\Pi)} = \left(\int_0^t \mathbb{1}_{\Pi^{(i)}}(s)\, dB_s,\ i = 1, 2, \ldots, k \right), \quad t \geq 0.$$

2. The vector-valued martingale $(M_t^{(\Pi)})$ presented in question 1(b) obviously satisfies (with the notations introduced in question 1(a)):

$$m_s^{(i)} m_s^{(j)} = 0, \quad \text{for } i \neq j. \tag{6.29.2}$$

(a) Give an example of martingales $(M^{(1)}, M^{(2)})$ satisfying (6.29.1), but not (6.29.2).

Hint. Consider $M^{(1)} = (W^{(1)} + W^{(2)})/2$ and $M^{(2)} = (W^{(1)} - W^{(2)})/2$, where $W^{(1)}$ and $W^{(2)}$ are two independent real-valued Brownian motions.

(b) In order to understand better, in the general framework of question 1, when (6.29.2) is satisfied, let us consider the Kunita–Watanabe orthogonal decompositions:

$$M_t^{(i)} = \int_0^t \mu_s^{(i)} \, dB_s + N_t^{(i)}, \quad i = 1, 2, \ldots, k,$$

where $B_t \stackrel{\text{(def)}}{=} \sum_{i=1}^k M_t^{(i)}$, and $\langle B, N^{(i)} \rangle = 0$. Then compute $\langle N^{(i)}, N^{(j)} \rangle$ in terms of $\mu^{(i)}, \mu^{(j)}$, for $i \neq j$.

3. We now consider the strictly positive martingale which is geometric Brownian motion: $\exp\left(B_t - \frac{t}{2}\right)$. Characterize the strictly positive martingales $(L_t^{(1)}, \ldots, L_t^{(k)})$ such that:

(i) $L_0^{(i)} = 1, \quad i = 1, 2, \ldots, k,$
(ii) $L^{(1)}, \ldots, L^{(k)}$ are independent,   (6.29.3)
(iii) $\exp\left(B_t - \frac{t}{2}\right) = \prod_{i=1}^k L^{(i)}(t).$

∘ 4. Are there other pairs of continuous independent martingales $\left(\Phi_t^{(1)}, \Phi_t^{(2)}\right)$ such that

$$B_t = \Phi_t^{(1)} \Phi_t^{(2)} \tag{6.29.4}$$

than "the obvious solutions"

$$\Phi_t^{(1)} = C, \quad \Phi_t^{(2)} = \frac{1}{C} B_t, \quad t \geq 0,$$

for any $C \neq 0$. We make the conjecture that the answer is NO!

Comments and references. (a) Had we assumed $M^{(i)}, i = 1, \ldots, k$ to be only semimartingales, then a very large family of k-tuples is available. Indeed, there are many examples of semimartingales $(\beta_t, t \geq 0)$ in a given filtration $(\mathcal{F}_t)$ that turn out to be Brownian motions. These are the so-called non-canonical representations of Brownian motion, e.g. $\beta_t = B_t - \int_0^t \frac{ds}{s} B_s$, which we already encountered in Exercise 6.1.

(b) We may ask the same problem when B is replaced by the compensated Poisson process: $(N_t - t,\ t \geq 0)$. This seems to be a natural question since, as the reader shall see, we begin our proof of (6.29.1) with the help of Cramer's theorem for Gaussian variables. Now there is also the analogous Raikov's theorem for Poisson variables, see Stoyanov [86], Section [12]. See also Lukacs [55].

c) So far, we have not completely solved (this is a euphemism!!) question 4, but we conjecture that the only solutions are:

$$M_t^{(1)} = C \quad \text{and} \quad M_t^{(2)} = \frac{1}{C} B_t,$$

for some $C \neq 0$ (and, of course, vice versa, with the roles of $M^{(1)}$ and $M^{(2)}$ inter-verted!).

Solutions for Chapter 6

In the solutions developed below, $(\mathcal{F}_t)$ will often denote the "obvious" filtration involved in the question, and $F = F(X_u, u \leq t)$ a functional on the canonical path space $C(\mathbb{R}_+, \mathbb{R}^d)$, which is measurable with respect to the past up to time t.

Solution to Exercise 6.1

1. (a) By assumption, $(R_t, t \geq 0)$ is a positive process, so it follows from Fubini theorem that

$$E\left[\int_{(0,1]} R_t \mu(dt)\right] = \int_{(0,1]} \mu(dt) E\left[R_t\right] .$$

Then we derive from the assumptions (i) and (ii) that $E\left[R_t\right] = E\left[R_1\right] < \infty$, so that if $\int_{(0,1]} \mu(dt) < \infty$, then $E\left[\int_{(0,1]} R_t \mu(dt)\right] < \infty$, and hence $\int_{(0,1]} R_t \mu(dt) < \infty$, a.s.

(b) Applying Fubini theorem, we obtain that for any measurable set B,

$$
\begin{aligned}
E\left[\mathbb{1}_B \int_{(0,1]} R_t \mu(dt)\right] &= \int_{(0,1]} \mu(dt) E\left[\mathbb{1}_B R_t\right] \\
&= \int_{(0,1]} \mu(dt) E\left[\mathbb{1}_B \int_0^\infty du \mathbb{1}_{\{R_t \geq u\}}\right] \\
&= \int_{(0,1]} \mu(dt) \int_0^\infty du P\left(B \cap \{R_t \geq u\}\right) .
\end{aligned}
$$

Then the inequality follows from the hint.

(c) By definition of the set B_N, we have $N \geq E\left[\mathbb{1}_{B_N} \int_{(0,1]} R_t \mu(dt)\right]$. Moreover, the assumption (i) implies that

$$\int_{(0,1]} \mu(dt) \int_0^\infty du \left(E[\mathbb{1}_{B_N}] - P(R_t \leq u)\right)^+ = \mu(0,1] \int_0^\infty du \left(E[\mathbb{1}_{B_N}] - P(R_1 \leq u)\right)^+ .$$

Then we conclude thanks to the inequality of question 1(a).

(d) If $\int_{(0,1]} \mu(dt)\, R_t < \infty$, a.s., then $\lim_{N \to \infty} P(B_N) = 1$. Let us exclude the trivial case where $R_1 = 0$, a.s. So we can choose N sufficiently large and $\varepsilon > 0$ sufficiently small to have $E[\mathbb{1}_{B_N}] - P(R_1 \leq u) > 0$, for all $u \in (0, \varepsilon]$, from which we derive that

$$\int_0^\infty du (E[\mathbb{1}_{B_N}] - P(R_1 \leq u))^+ > 0 .$$

Then we deduce that $\mu(0,1] < \infty$ from the inequality of question 1 (c).

2. (a) Let us compute $E(\beta_t \beta_s) = E\left[\left(B_t - \int_0^t \frac{dh}{h} B_h\right)\left(B_s - \int_0^s \frac{dh}{h} B_h\right)\right]$, for $s < t$. First we note that

$$\int_0^t \frac{dh}{h} B_h = \int_0^t \frac{dh}{h} \int_0^h dB_s = \int_0^t dB_s \left(\int_s^t \frac{dh}{h}\right) = \int_0^t dB_s \log(t/s) ,$$

so that $\beta_t = \int_0^t dB_s\,(1 - \log(t/s))$, and

$$E(\beta_s \beta_t) = \int_0^t dh\,(1 - \log(s/h))\,(1 - \log(t/h)) = s .$$

The process β is clearly a continuous centered Gaussian process. Moreover, as we have just proved, its covariance function is given by $E(\beta_t \beta_s) = s \wedge t$, $s, t \geq 0$, hence β is a standard Brownian motion

(b) Fix $t > 0$ and set $\hat{B}_h = B_t - B_{t-h}$ and $\hat{\beta}_h = \beta_t - \beta_{t-h}$, for $h \leq t$ and let us write $\beta_t - \beta_{t-u} = (B_t - B_{t-u}) - \int_0^u \frac{dh}{t-h} B_{t-h}$. This identity may be expressed as

$$\hat{B}_u = \hat{\beta}_u + \int_0^u \frac{dh}{t-h}(B_t - \hat{B}_h) .$$

Then considering the above equation as a linear differential equation whose solution exists and is $(\hat{B}_u,\, u < t)$, we see that $\hat{B}$ can be written as

$$\hat{B}_u = \Phi(\hat{\beta}_h, h \leq u, B_t) ,$$

for some measurable functional Φ, hence $\mathcal{F}_t^B \subseteq \mathcal{F}_t^\beta \vee \sigma(B_t)$.

3. (a) From the expression of β defined in the previous question, we may write for all $\varepsilon \in (0,1)$,

$$\int_\varepsilon^1 f(s)\, dB_s = \int_\varepsilon^1 f(s)\, d\beta_s + \int_\varepsilon^1 \frac{ds}{s} f(s)\, B_s .$$

Since the integrals of f with respect to B and β are well defined a.s. for all $f \in L^2([0,1])$, the result follows by making ε tend to 0 in the above identity.

(b) The process $R_t = |B_t|/\sqrt{t}$ clearly satisfies the assumptions (i) and (ii) of question 1. By setting $\mu(dt) = |f(t)| dt/\sqrt{t}$, we see that the equivalence

$$\int_0^1 \frac{dt}{t} |f(t)||B_t| < \infty \Leftrightarrow \int_0^1 \frac{dt|f(t)|}{\sqrt{t}} < \infty$$

follows from this question. Using the process $R_t = |B_t|/\sqrt{t}$ and applying question 1 again, we see that $\int_0^1 \frac{dt}{t} |f(t)||B_t| = \infty$, a.s. for all $f \in L^2([0,1])$ such that $\int_0^1 \frac{dt|f(t)|}{\sqrt{t}} = \infty$. For instance, the function

$$f(t) = \frac{(-\ln t)^{-\alpha}}{\sqrt{t}} \mathbb{1}_{\{t \in (0,1/2)\}}$$

satisfies these (non) integrability properties for $L \in (1/2, 1)$.

Solution to Exercise 6.2

1. The coefficients of the SDE (6.2.1) satisfy the Lipschitz condition: $|\sqrt{1+y^2} - \sqrt{1+z^2}| + \frac{1}{2}|y - z| \le \frac{3}{2}|y - z|$, so from the general theory of stochastic differential equations, this equation admits a unique strong solution, for every $x \in \mathbb{R}$.

2. Itô's formula yields

$$
\begin{aligned}
Y_t &= x + \int_0^t \left(e^{\beta_s} d\beta_s + \frac{1}{2} e^{\beta_s} ds \right) e^{-\beta_s} Y_s + \int_0^t e^{\beta_s} e^{-\beta_s} d\gamma_s \\
&= x + \gamma_t + \int_0^t Y_s d\beta_s + \frac{1}{2} \int_0^t Y_s ds \\
&= x + \int_0^t \sqrt{1 + Y_s^2}\, dB_s + \frac{1}{2} \int_0^t Y_s ds,
\end{aligned}
$$

for a certain Brownian motion $(B_t, t \ge 0)$. This proves that Y satisfies equation (6.2.1). From question 1, we have $(Y_t, t \ge 0) \overset{(\text{law})}{=} (\sinh(a + B_t), t \ge 0)$.

3. (a) This follows immediately from Itô's formula.

(b) Considering the system (6.2.3), we differentiate the first identity, which yields:

$$\varphi''(y) = \sigma'(\varphi(y))\varphi'(y) = (\sigma'\sigma)(\varphi(y)).$$

Hence, the pair (σ, b) must satisfy: $\frac{1}{2}(\sigma'\sigma) = b$, i.e. $\sigma^2(x) = \int^x dy\, b(y)$. Now, if b is Lipschitz and bounded, we associate with it the function $\sigma(x) = \sqrt{\int^x dy\, b(y)}$, which satisfies

$$|\sigma(x) - \sigma(x')| \le C\sqrt{|x - x'|},$$

and there is only one strong solution.

Solution to Exercise 6.3

1. (a) We take $x = a = 0$, and we may write: $Y_t = \int_0^t \exp(\beta_t - \beta_s)\, d\gamma_s$. Then, using the time reversal $s = t - u$ on $[0, t]$ and the fact that $\beta_t - \beta_{t-u}$ and $\gamma_t - \gamma_{t-u}$ are two

independent Brownian motions, we obtain, for fixed t:

$$Y_t \overset{\text{(law)}}{=} \int_0^t \exp(\beta_u)\, d\gamma_u \,.$$

The second identity in law follows from the Dambis–Dubins–Schwarz representation of the martingale $\left(\int_0^t \exp(\beta_u)\, d\gamma_u \right)$.

(b) It suffices to note that the right hand side of (6.3.1) is a martingale, whereas the left hand side is not.

As another proof, note that the quadratic variations of both sides (as processes) are, respectively: $\int_0^t [\cosh(\beta_s)]^2\, ds$ and $\int_0^t ds\, \exp(2\beta_s)$. These two processes have different laws. Indeed, $[\cosh(\beta_s)]^2$ and $\exp(2\beta_s)$ have different expectations, since, on the one hand

$$E[\cosh^2(\beta_s)] = \frac{1}{4} E[e^{2\beta_s} + e^{-2\beta_s} + 2] = \frac{1}{4}\left(2 + 2E[e^{2\beta_s}]\right) = \frac{1}{2}(1 + e^{2s})$$

and, on the other hand, $E[e^{2\beta_s}] = e^{2s}$.

2. We easily deduce from (6.3.1) that, for fixed t,

$$\sinh(|B_t|) \overset{\text{(law)}}{=} |\gamma|_{\int_0^t du\, \exp(2\beta_u)} \,.$$

Next, since $M_t \overset{\text{(law)}}{=} |B_t|$ and $|\gamma_h| \overset{\text{(law)}}{=} \mu_h \overset{\text{(law)}}{=} \lambda_h$, for fixed t and fixed h (by the symmetry principle), we obtain the desired result.

3. This is a consequence of the use of Itô's formula and time changing for the process $(\exp(\beta_t),\ t \geq 0)$.

4. We proceed by passing to the inverse process, i.e.

$$
\begin{aligned}
(H_{\tau_l} < \alpha) &= (\tau_l < A_\alpha) \\
&= (l < \lambda_{A_\alpha}) \\
&\overset{\text{(law)}}{=} (l < \sinh(L_\alpha)) \\
&= (a(l) < L_\alpha) \\
&= \left(\tau_{a(l)} < \alpha \right) .
\end{aligned}
$$

Solution to Exercise 6.4

1. The covariation application: $(X, Y) \mapsto \langle X, Y \rangle$ is a symmetric and bilinear form, hence: $\langle X + Y \rangle = \langle X \rangle + \langle Y \rangle + 2 \langle X, Y \rangle$ and the formula follows. Now we can prove the second part of the question. For $i = 3$, one has

$$
\begin{aligned}
\mathcal{E}(X_1)\mathcal{E}(X_2)\mathcal{E}(X_3) &= \mathcal{E}(X_1 + X_2 + \langle X_1, X_2 \rangle)\mathcal{E}(X_3) \\
&= \mathcal{E}(X_1 + X_2 + X_3 + \langle X_1, X_2 \rangle + \langle X_1, X_3 \rangle + \langle X_2, X_3 \rangle).
\end{aligned}
$$

By iteration, we easily deduce the general case:

$$\Pi_{i=1}^n \mathcal{E}(X_i) = \mathcal{E}\left(\sum_i X_i + \sum_{i<j} \langle X_i, X_j \rangle\right).$$

2. We obtain the result by taking $X = -Y + \langle Y \rangle$ in the formula of question 1.

3. (a) The injectivity follows from the formula:

$$X_t = \int_0^t \frac{d\mathcal{E}(X)_s}{\mathcal{E}(X)_s}.$$

(b) Questions 1 and 2 imply

$$\mathcal{E}(X) = \mathcal{E}(Z)\mathcal{E}(Y)^{-1} = \mathcal{E}(Z)\mathcal{E}(-Y + \langle Y \rangle) = \mathcal{E}(Z - Y - \langle Z - Y, Y \rangle).$$

Then from the injectivity of $X \mapsto \mathcal{E}(X)$, we deduce that $X = Z - Y - \langle Z - Y, Y \rangle = Z - Y - \langle Z, Y \rangle + \langle Y \rangle$.

4. Since $\mathcal{E}(X)_t^\alpha = \exp(\alpha X_t - \frac{\alpha}{2}\langle X \rangle_t)$, we are looking for a semimartingale $(X_t^{(\alpha)})$ such that $\mathcal{E}(X^{(\alpha)})_t = \exp(\alpha X_t - \frac{\alpha}{2}\langle X \rangle_t)$, i.e. $X_t^{(\alpha)} - \frac{1}{2}\left\langle X^{(\alpha)} \right\rangle_t = \alpha X_t - \frac{\alpha}{2}\langle X \rangle_t$. But this last equation gives $\left\langle X^{(\alpha)} \right\rangle = \alpha^2 \langle X \rangle$ and proves the uniqueness of $X_t^{(\alpha)} = \alpha X_t + \frac{\alpha^2 - \alpha}{2}\langle X \rangle_t$.

Then from the multiplicativity formula established in question 1, we obtain:

$$\mathcal{E}(X^{(\alpha+\beta)}) = (\mathcal{E}(X))^{\alpha+\beta} = \mathcal{E}(X^{(\alpha)})\mathcal{E}(X^{(\beta)})$$
$$= \mathcal{E}\left(X^{(\alpha)} + X^{(\beta)} + \langle X^{(\alpha)}, X^{(\beta)} \rangle\right).$$

Hence, $X^{(\alpha+\beta)} = X^{(\alpha)} + X^{(\beta)} + \langle X^{(\alpha)}, X^{(\beta)} \rangle$. But $X^{(\alpha)} = \alpha X + \frac{\alpha^2 - \alpha}{2}\langle X \rangle$. Hence $\langle X^{(\alpha)}, X^{(\beta)} \rangle = \alpha\beta\langle X, X \rangle$, so that:

$$X^{(\alpha+\beta)} = X^{(\alpha)} + X^{(\beta)} + \alpha\beta\langle X, X \rangle.$$

5. (a) First recall that from equation (6.4.1)

$$\mathcal{E}(Y)_t = 1 + \int_0^t \mathcal{E}(Y)_s \, dY_s$$

$$\mathcal{E}(M + V)_t = 1 + \int_0^t \mathcal{E}(M + V)_s \, dM_s + \int_0^t \mathcal{E}(M + V)_s \, dV_s,$$

so that from Itô formula

$$\mathcal{E}(M + V)_t \mathcal{E}(Y)_t = 1 + \int_0^t \mathcal{E}(M + V)_s \mathcal{E}(Y)_s \, dY_s + \int_0^t \mathcal{E}(M + V)_s \mathcal{E}(Y)_s \, dM_s$$

$$+ \int_0^t \mathcal{E}(M + V)_s \mathcal{E}(Y)_s \, dV_s + \int_0^t \mathcal{E}(M + V)_s \mathcal{E}(Y)_s \, d\langle M, Y \rangle_s.$$

Thus $\mathcal{E}(M+V)_t\mathcal{E}(Y)_t$ is a local martingale if and only if $\int_0^t \mathcal{E}(M+V)_s\mathcal{E}(Y)_s\,d(V + \langle M,Y\rangle)_s = 0$, for any t. Hence, the only continuous processes with bounded variation such that $V_0 = 0$ and which satisfy this property are given by $V = c - \langle M,Y\rangle$, where c is any constant.

(b) Note that $\mathcal{E}(M - \langle Y,M\rangle)$ is a $(Q,\mathcal{F}_t)$ local martingale since from question 5(a), $\mathcal{E}(M - \langle Y,M\rangle)\mathcal{E}(Y)$ is a $(P,\mathcal{F}_t)$ local martingale. But it is clear from equation (6.4.1) that $\mathcal{E}(X)$ is a local martingale if and only if X is a local martingale (and this property does not depend on the probability measure). Hence, $M - \langle M,Y\rangle$ is a $(Q,\mathcal{F}_t)$ local martingale.

Solution to Exercise 6.5

1. Let $F : \mathbb{R} \times \mathbb{R}_+^2 \mapsto \mathbb{R}$ be the function defined by $F(x,y,z) = \cosh(\lambda(y - x))\exp\left(-\frac{\lambda^2}{2}z\right)$. This function satisfies:

$$\frac{1}{2}\frac{\partial^2}{\partial x^2}F(x,y,z) + \frac{\partial}{\partial z}F(x,y,z) = 0, \qquad \frac{\partial}{\partial y}F(x,x,z) = 0, \quad (x,y,z) \in \mathbb{R} \times \mathbb{R}_+^2\,.$$
$$(6.5.a)$$

Thanks to the first part of (6.5.a) and the fact that $(S_t,\, t \geq 0)$ is an increasing process, Itô's formula applied to the semimartingale: $(F(B_t, S_t, t),\, t \geq 0)$ gives:

$$F(B_t, S_t, t) = F(0,0,0) + \int_0^t \frac{\partial}{\partial x}F(B_u, S_u, u)\,dB_u + \int_0^t \frac{\partial}{\partial y}F(B_u, S_u, u)\,dS_u\,.$$

Now observe that $B_u = S_u$, for each time u at which S increases. This remark and the second part of (6.5.a) show that the integral $\int_0^t \frac{\partial}{\partial y}F(B_u, S_u, u)\,dS_u$ vanishes, hence

$$F(B_t, S_t, t) = 1 + \int_0^t \frac{\partial}{\partial x}F(B_u, S_u, u)\,dB_u,$$

which proves the result.

2. Since the distribution of Brownian motion is symmetric, we can replace in question 1, $S_t - B_t$ by $B_t - I_t$ and S_t by I_t, so that

$$N_t \equiv \cosh\left(\lambda(B_t - I_t)\right)\exp\left(-\frac{\lambda^2 t}{2}\right)$$

is a $(\mathcal{F}_t)$ martingale. Observe that θ_c is a $(\mathcal{F}_t)$ stopping time and that the martingales $(M_{t\wedge\theta_c})$ and $(N_{t\wedge\theta_c})$ are bounded, therefore, we may apply the optional stopping theorem which yields:

$$E[M_{\theta_c}] = 1, \quad \text{and} \quad E[N_{\theta_c}] = 1. \tag{6.5.b}$$

Note also that almost surely, either $M_{\theta_c} - B_{\theta_c} = c$ and $B_{\theta_c} - I_{\theta_c} = 0$, or $M_{\theta_c} - B_{\theta_c} = 0$ and $B_{\theta_c} - I_{\theta_c} = c$, so that

$$M_{\theta_c} + N_{\theta_c} = \exp\left(-\frac{\lambda^2 \theta_c}{2}\right)\left(\cosh(\lambda c)\mathbb{1}_{\{M_{\theta_c} - B_{\theta_c} = c\}} + \mathbb{1}_{\{M_{\theta_c} - B_{\theta_c} = 0\}}\right.$$

$$\left. + \cosh(\lambda c)\mathbb{1}_{\{B_{\theta_c} - I_{\theta_c} = c\}} + \mathbb{1}_{\{B_{\theta_c} - I_{\theta_c} = 0\}}\right)$$

$$= \exp\left(-\frac{\lambda^2 \theta_c}{2}\right)(\cosh(\lambda c) + 1)\ .$$

The result now follows from equations (6.5.b).

Comments on the solution: The result of this exercise implies that

$$\theta_c \stackrel{\text{(law)}}{=} T_{\frac{c}{2}} + \tilde{T}_{\frac{c}{2}}\ ,$$

where T_a and $\tilde{T}_a$ are two independent copies of $\inf\{t : |B_t| = a\}$. This follows from the fact that $\cosh(\lambda B_t)\exp\left(-\frac{\lambda^2 t}{2}\right)$ is a $(\mathcal{F}_t)$ martingale.

Solution to Exercise 6.6

The solutions to questions 1 and 2 are immediate from the *Hint*; as an example,

$$S_t - X_t \stackrel{\text{(law)}}{=} \sup_{u \le t}(X_{(t-u)-} - X_t) \stackrel{\text{(law)}}{=} -I_t\ .$$

3. First note the following identities for $x \ge 0$ and $y \ge 0$

$$P(S_T \le x, -I_T \le y, X_T \in da) = P(T \le T_x, T \le T_{-y}, X_T \in da)$$

$$= P(T \le T_x \wedge T_{-y}, X_T \in da)$$

$$= E\left[\left(1 - \mathbb{1}_{\{T_x \wedge T_{-y} \le T\}}\right)\mathbb{1}_{\{X_T \in da\}}\right]$$

$$= P(X_T \in da) - \frac{\theta^2}{2}E\left[\int_{T_x \wedge T_{-y}} e^{-\frac{\theta^2}{2}t}\mathbb{1}_{\{X_t \in da\}}\,dt\right]$$

$$= P(X_T \in da)$$

$$- \frac{\theta^2}{2}E\left[e^{-\frac{\theta^2}{2}T_x \wedge T_{-y}}E_{X_{T_x \wedge T_{-y}}}\left(\int_0^\infty e^{-\frac{\theta^2}{2}u}\mathbb{1}_{\{X_u \in da\}}\,du\right)\right].$$

Now, identity (6.6.3) is easily deduced from the following well-known facts.

$$P(X_T \in da) = \frac{\theta}{2}e^{-\theta|a|}\,da$$

$$E\left[e^{-\frac{\theta^2}{2}T_x \wedge T_{-y}}\mathbb{1}_{\{T_x < T_{-y}\}}\right] = \frac{\sinh(\theta y)}{\sinh(\theta(x+y))}$$

$$E\left[e^{-\frac{\theta^2}{2}T_x \wedge T_{-y}}\mathbb{1}_{\{T_{-y} < T_x\}}\right] = \frac{\sinh(\theta x)}{\sinh(\theta(x+y))}$$

$$E_x\left[\int_0^\infty e^{-\frac{\theta^2}{2}u}\mathbb{1}_{\{X_u \in da\}}\,du\right] = \frac{1}{\theta}e^{-\theta|a-x|}\,da\ ,$$

where E_x denotes the expectation with respect to P_x, the law of Brownian motion starting at x, $(P \equiv P_0)$. The identity (6.6.4) follows from both question 1 and integration with respect to the law of X_T in (6.6.3).

Finally, we deduce the results (6.6.5) and (6.6.6) by using the scaling property of Brownian motion so that

$$\frac{S_1}{S_1 - I_1} \overset{\text{(law)}}{=} \frac{S_T}{S_T - I_T}.$$

4. Csáki's formulae (6.6.7) and (6.6.8) can easily be recovered from respectively (6.6.6) and (6.6.5). For the latter, one may use the following integral representation due to Gauss of the digamma function.

$$\psi(x) = \int_0^\infty \left(\frac{e^{-z}}{z} - \frac{e^{-zx}}{1 - e^{-z}} \right) dz,$$

see Theorem 1.6.1, p. 26, in [1]; see also Lebedev [51].

Solution to Exercise 6.7

1. Since both sides of the identity in law (6.7.1) involve Lévy processes, i.e. processes with homogeneous (in time) and independent increments, it suffices to prove the identity for a fixed time t_0. But the latter is true since

$$X_{nt_0} = X_{t_0} + (X_{2t_0} - X_{t_0}) + \ldots + (X_{nt_0} - X_{(n-1)t_0})$$

is identical in law with $X_{t_0}^{(1)} + \ldots + X_{t_0}^{(n)}$.

2. We write $\hat{X}_t = \int_a^b \frac{du}{u} X_{ut}$, hence $\hat{X}_{nt} = \int_a^b \frac{du}{u} X_{unt}$. It now remains for us to apply the identity (6.7.1) for $X_{nut} \overset{\text{(law)}}{=} X_{ut}^{(1)} + X_{ut}^{(2)} + \cdots + X_{ut}^{(n)}$, where the identity holds as processes in ut. We also need to prove that $\hat{X}$ is not a Lévy process. But note that, for $t > 0$, $\hat{X}$ is differentiable on the right, i.e.

$$\lim_{\varepsilon \to 0} \frac{1}{\varepsilon} (\hat{X}_{t+\varepsilon} - \hat{X}_t) = b \frac{X_{bt}}{bt} - a \frac{X_{at}}{at}$$

$$= \frac{1}{t} (X_{bt} - X_{at}),$$

which cannot hold for a continuous Lévy process that differs from a pure drift.

3. This is immediate from identity (6.7.1), since, if X is an IDT process, then $\overline{X} \overset{\text{(law)}}{=} \overline{X}_{\cdot/n}^{(1)} + \overline{X}_{\cdot/n}^{(2)} + \ldots + \overline{X}_{\cdot/n}^{(n)}$, where the $\overline{X}^{(i)}$s are independent copies of $\overline{X}$.

4. (a) It suffices to write

$$B_{nt}^f = \int_0^{nt} f(nt/u) \, dB_u$$

$$= \int_0^t f(t/v) \, dB_{vn},$$

and to use (6.7.1) for B. Note that this construction of IDT processes also holds when replacing B by any Lévy process (recall that any Lévy process is a semimartingale).

(b) If $c(t,s) = E(X_t X_s)$ is the covariance of our Gaussian process, then we must have

$$c(nt, ns) = nc(t, s). \tag{6.7.a}$$

For example, if (X_t) is a fractional Brownian motion with index α, then $(t^{1/2-\alpha}X_t,$ $t \geq 0)$ satisfies the desired property (6.7.a).

5. Squares of Bessel processes are nice sources of examples of processes that are infinitely divisible, but not IDT. For instance, let $X = (R_N^2(t), t \geq 0)$ be a squared Bessel process with dimension N, and $X^{(i)}$, $i = 1, \ldots, n$ be independent copies of X. Then we have the identity in law

$$(X_t^{(1)} + \ldots + X_t^{(n)}, t \geq 0) \stackrel{(\text{law})}{=} (R_{nN}^2(t), t \geq 0).$$

On the other hand, it follows from the scaling property of X that

$$(R_N^2(nt), t \geq 0) \stackrel{(\text{law})}{=} (nR_N^2(t), t \geq 0).$$

We then have to compare the laws of $(nR_N^2(t), t \geq 0)$ and $(R_{nN}^2(t), t \geq 0)$. Let us denote by γ_a a Gamma variable with parameter a. On the one hand, we have $nR_N^2(t) \stackrel{(\text{law})}{=} 2nt\gamma_{N/2}$ and on the other hand we have $R_{nN}^2(t) \stackrel{(\text{law})}{=} 2t\gamma_{nN/2}$. In particular, for $t = 1$ and $n = N = 2$, we have $nR_N^2(1) \stackrel{(\text{law})}{=} 4\gamma_1$ and $R_{nN}^2(1) \stackrel{(\text{law})}{=} 2\gamma_2$ and the laws of $2\gamma_1$ and γ_2 differ.

6. We can write the Lévy–Khintchine representation as follows (for, say, $X \geq 0$):

$$E\left[\exp\left(-\int \mu(dt)\, X_t\right)\right] = \exp\left[-\int_{\mathcal{C}(\mathbb{R}_+, \mathbb{R}_+)} N(d\omega)\left(1 - \exp\left(-\int \mu(dt)\omega(t)\right)\right)\right],$$

where $\mu(dt)$ is a nice positive measure on $\mathbb{R}_+$. Taking $\mu(dt) = f(t)\, dt$ and using the definition of an IDT process, we obtain the identity:

$$E\left[\exp\left(-\int dt f(t) X_{nt}\right)\right] = E\left[\exp\left(-\int \frac{du}{n} f\left(\frac{u}{n}\right) X_u\right)\right]$$

$$= \exp\left[-\int_{\mathcal{C}(\mathbb{R}_+, \mathbb{R}_+)} N(d\omega)\left(1 - \exp\left(-\int \frac{du}{n} f\left(\frac{u}{n}\right)\omega(u)\right)\right)\right].$$

On the other hand, this expression is equal to

$$\exp\left[-\int_{\mathcal{C}(\mathbb{R}_+, \mathbb{R}_+)} nN(d\omega)\left(1 - \exp\left(-\int dt\, f(t)\omega(t)\right)\right)\right].$$

Hence N is the Lévy measure of an IDT process if and only if it satisfies $m_n(N) = nN$, where m_n is the transformation of $\mathcal{C}(\mathbb{R}_+, \mathbb{R})$ which is defined by $m_n(\omega)(t) = \omega(nt)$.

7. (a) First we observe that any Lévy process X satisfies this property. It is a direct consequence of the independence and stationarity properties of the increments of Lévy processes. Then, since the law of each side of (6.7.2) depends only on one-dimensional distributions of the process X, it follows that if Y has the same one-dimensional distributions as a Lévy process, then Y also satisfies this property.

(b) As a consequence of (6.7.2), we have for any integer $n \geq 1$,

$$X_t \overset{\text{(law)}}{=} X_{t/n}^{(1)} + X_{t/n}^{(2)} \ldots + X_{t/n}^{(k)}, \tag{6.7.b}$$

which shows that X_t is an infinitely divisible random variable.

(c) It is a straightforward consequence of the following well-known property: for any real random variable ξ with distribution function F, if U is uniformly distributed on $[0, 1]$ and for $u \in [0, 1]$, $F^{-1}(u) = \inf\{x : F(x) \geq u\}$, then the random variable $F^{-1}(U)$ has the same law as ξ.

8. Suppose that a process X satisfies $(X_{s+t}, s, t \geq 0) \overset{\text{(law)}}{=} (X_s^{(1)} + X_t^{(2)}, s, t \geq 0)$. Then this identity for $s = t = 0$ shows that $X_0 = 0$, a.s. On the other hand, applying this identity to the sum of r.v.s $X_0^{(1)} + X_t^{(2)}$ and then to the sum of r.v.s $X_t^{(1)} + X_0^{(2)}$, we obtain:

$$(X_t, X_t) \overset{\text{(law)}}{=} (X_0^{(1)} + X_t^{(2)}, X_t^{(1)} + X_0^{(2)}) = (X_t^{(2)}, X_t^{(1)}),$$

so that

$$0 \overset{\text{(law)}}{=} X_t^{(2)} - X_t^{(1)}.$$

The last equality, means that the random variable $X_t^{(1)}$ is independent of itself. Hence it is almost surely constant. So, the only measurable processes satisfying the identity of the statement are linear and deterministic.

Solution to Exercise 6.8

1. Applying Itô's formula to $|Z_t|^2 = Z_t \overline{Z}_t$, we obtain:

$$\begin{aligned} d(Z_t \overline{Z}_t) &= Z_t \, d\overline{Z}_t + \overline{Z}_t \, dZ_t + 2 \, dt \\ &= (Z_t \, d\overline{B}_t + \overline{Z}_t \, dB_t) + (2\lambda + 2) \, dt \\ &= 2|Z_t| \, d\beta_t + (2\lambda + 2) \, dt, \end{aligned}$$

where, from Lévy's Characterization Theorem, $d\beta_t = \text{Re}(Z_t \, d\overline{B}_t)/|Z_t|$ defines a real valued Brownian motion. Hence, $|Z_t|^2$ satisfies the equation of the square of a Bessel process with dimension $\delta = 2(\lambda + 1)$. Thus, $(Z_t, \, t \geq 0)$ does not take the value 0, a.s.

Let X and Y be the real part and imaginary parts of Z, respectively. Using Itô's formula again, we can now check that $Z_t = z \exp\left(\int_0^t \frac{dZ_u}{Z_u}\right)$, hence

$$Z_t = z \exp\left(\int_0^t \frac{\overline{Z}_u dZ_u}{|Z_u|^2}\right) = z \exp\left(\int_0^t \frac{\overline{Z}_u \, dB_u}{|Z_u|^2} + \lambda H_t\right)$$

$$= z \exp\left(\int_0^t \frac{d\beta_u}{|Z_u|} + \lambda H_t + i \int_0^t \frac{d\xi_u}{|Z_u|}\right),$$

where, from Lévy's Characterization Theorem, $d\beta_t = \mathrm{Re}(\overline{Z}_t \, dB_t)/|Z_t|$ and $d\xi_t = \mathrm{Im}(\overline{Z}_t \, dB_t)/|Z_t|$ define two independent real-valued Brownian motions. Finally we obtain the representation (6.8.1) by time changing with the inverse of (H_t). Then (γ_u) is a Brownian motion independent of $|Z|$.

2. Clearly, assuming that P_0^λ exists, it should satisfy the following properties:

   (i) under P_0^λ, the law of $(Z_t, \, t \geq 0)$ is left invariant by any deterministic rotation;

   (ii) $(|Z_t|^2, \, t \geq 0)$ is the square of a $\delta = 2(\lambda + 1)$ Bessel process starting from 0.

Following this intuition, we can now show the existence of P_0^λ.

For $\varepsilon > 0$, we set $B_t^{(\varepsilon)} = B_{t+\varepsilon} - B_\varepsilon$. Let P_0^λ be the law under which B is a Brownian motion, and the coordinate process $(Z_t, \, t \geq 0)$ is such that $Z_0 = 0$, a.s. and for every $\varepsilon > 0$, the process $Z^{(\varepsilon)} = (Z_{t+\varepsilon}, \, t \geq 0)$ satisfies the equation

$$Z_t^{(\varepsilon)} = Z_\varepsilon + B_t^{(\varepsilon)} + \lambda \int_0^t \frac{Z_s^{(\varepsilon)} \, ds}{|Z_s^{(\varepsilon)}|^2}. \tag{6.8.a}$$

Moreover, we assume that the law of Z_ε is radial, (i.e. it is invariant by rotation), and that $|Z_\varepsilon|$ is distributed as $\sqrt{\varepsilon}\rho$, where ρ is the value at time 1 of a Bessel process with dimension δ, starting from 0. With the help of Kolmogorov's Consistency Theorem, the probability measure P_0^λ is now well defined and under P_0^λ, the process $(|Z_t|^2, \, t \geq 0)$ is the square of a Bessel process of dimension $\delta = 2(\lambda + 1)$, starting from 0, so Z_ε converges P_0^λ-almost surely towards 0, as $\varepsilon \to 0$. Moreover, from the above equation, under P_0^λ, conditionally on Z_ε, the process $Z^{(\varepsilon)}$ has law P_{Z_ε}. This shows the weak convergence of $(P_z^\lambda, \, z > 0)$ towards P_0^λ.

An application of Girsanov's Theorem shows that for every t, the absolute continuity relationship holds for every $z \neq 0$:

$$P_{z\,|\mathcal{F}_t}^\lambda = \left|\frac{Z_t}{z}\right|^\lambda \exp\left(-\frac{\lambda^2}{2} \int_0^t \frac{ds}{|Z_s|^2}\right) P_{z\,|\mathcal{F}_t}^0. \tag{6.8.b}$$

3. From (6.8.b), we have, for every bounded, continuous functional F which is $\mathcal{Z}_1$ measurable:

$$\frac{1}{\Lambda_z} E_z^0 \left[F(Z) \exp \left(-\frac{\lambda^2}{2} \int_0^1 \frac{ds}{|Z_s|^2} \right) \right] = \frac{1}{E_z^\lambda \left(\frac{1}{|Z_1|^\lambda} \right)} E_z^\lambda \left(F(Z) \frac{1}{|Z_1|^\lambda} \right),$$

and from the weak convergence established in the previous question, together with the nice integrability properties of $\frac{1}{|Z_1|^\lambda}$, under the $\{P_z^0\}$ laws, we obtain that the above expression converge towards $\left(E_0^\lambda \left(\frac{1}{|Z_1|^\lambda} \right) \right)^{-1} E_0^\lambda \left(F(Z) \frac{1}{|Z_1|^\lambda} \right)$, as $z \to 0$.

4. (a) Recall that R satisfies the equation

$$R_t = 1 + B_t + \frac{d-1}{2} \int_0^t \frac{ds}{R_s},$$

where B is a standard Brownian motion, so that since R does not take the value 0, a.s., we obtain, from Itô's formula:

$$\log R_t = \int_0^t \frac{dB_s}{R_s} + \frac{d-2}{2} \int_0^t \frac{ds}{R_s^2}.$$

Changing time in this equation with the inverse H^{-1} of H (which is equal to the quadratic variation of $\int_0^t \frac{dB_s}{R_s}$), we obtain:

$$\log R_{H_u^{-1}} = \beta_u + \frac{d-2}{2} u,$$

where $\beta_u = \int_0^{H_u^{-1}} \frac{dB_s}{R_s}$ is a standard Brownian motion. This shows that $\varphi_u = \beta_u + \frac{d-2}{2} u$.

(b) First, we deduce from the scaling property of R (i.e. $R_t \overset{\text{(law)}}{=} t^{1/2} R_1$) that

$$\frac{1}{\log t} \log R_t \overset{P}{\longrightarrow} \frac{1}{2}, \quad \text{as } t \to \infty.$$

From the scaling property of Brownian motion, we obtain:

$$\frac{1}{\log t} \beta_{H_t} = W \left(\frac{1}{(\log t)^2} H_t \right),$$

where $(W_s = \frac{1}{\log t} \beta((\log t)_s^2), \ s \geq 0)$ is a standard Brownian motion. Then we verify from the scaling property of R, that $\frac{1}{(\log t)^2} E[H_t] \to 0$, as $t \to \infty$, hence $W \left(\frac{1}{(\log t)^2} H_t \right)$ converges towards 0 in probability. Then we deduce from above and the representation of $\log R_t$ established in the previous question that

$$\frac{1}{\log t} H_t \overset{P}{\longrightarrow} \frac{1}{d-2}, \quad \text{as } t \to \infty.$$

(c) For the following argument, it suffices to assume that R starts from 0. Then, the scaling property of R implies that $\int_1^a \frac{ds}{R_s^2} \overset{(\text{law})}{=} \int_{a^{n-1}}^{a^n} \frac{ds}{R_s^2}$, for every positive integer n. Since the scaling transform of R (: $R \mapsto \frac{1}{\sqrt{c}} R_{c\cdot}$, $c \neq 1$) is ergodic, the Ergodic Theorem implies:

$$\lim_{n \to \infty} \frac{1}{n} \sum_{p=1}^n \int_{a^{p-1}}^{a^p} \frac{ds}{R_s^2} \longrightarrow E\left[\int_1^a \frac{ds}{R_s^2}\right] = (\log a) E\left[\left(\frac{1}{R_1^2}\right)\right] = \frac{\log a}{d-2}, \quad \text{a.s.}$$

hence, $\frac{1}{\log a^n} \int_1^{a^n} \frac{ds}{R_s^2}$ converges almost surely towards $\frac{1}{d-2}$, as $n \to \infty$. It is then easily shown that $\frac{1}{\log t} \int_1^t \frac{ds}{R_s^2}$ converges almost surely towards $\frac{1}{d-2}$, as $t \to \infty$.

(d) Again, it follows from the scaling property of R that

$$\frac{1}{\log \frac{1}{\varepsilon}} \int_\varepsilon^1 \frac{ds}{R_s^2} \overset{(\text{law})}{=} \frac{1}{\log \frac{1}{\varepsilon}} \int_1^{\frac{1}{\varepsilon}} \frac{du}{R_u^2},$$

so we deduce from the previous questions that $\frac{1}{\log \frac{1}{\varepsilon}} \int_\varepsilon^1 \frac{ds}{R_s^2}$ converges in probability towards $\frac{1}{d-2}$, as $\varepsilon \to 0$.

(e) The representation proved in question 1 shows that

$$\theta_{(\varepsilon,1)} = \gamma_{H_1} - \gamma_{H_\varepsilon} \overset{(\text{law})}{=} \gamma\left(\int_\varepsilon^1 \frac{ds}{|Z_s|^2}\right),$$

so that from the scaling property of γ:

$$\left(\log \frac{1}{\varepsilon}\right)^{-\frac{1}{2}} \theta_{(\varepsilon,1)} \overset{(\text{law})}{=} \gamma\left(\left(\log \frac{1}{\varepsilon}\right)^{-1} \int_1^{\frac{1}{\varepsilon}} \frac{ds}{|Z_s|^2}\right).$$

We finally deduce from question (d) that the right hand side of the above equality converges in law towards $\gamma_{1/(d-2)}$, as $\varepsilon \to 0$.

Comments on the solution. As an important complement to the asymptotic result

$$\frac{1}{\log t} \int_0^t \frac{ds}{R_s^2} \overset{\text{a.s.}}{\longrightarrow} \frac{1}{d-2}, \quad \text{as } t \to \infty,$$

for R a Bessel process with dimension $d > 2$, we present the following

$$\frac{4}{(\log t)^2} \int_0^t \frac{ds}{R_s^2} \overset{(\text{law})}{\longrightarrow} T = \inf\{t : \beta_t = 1\}, \quad \text{as } t \to \infty, \tag{6.8.c}$$

where R is a two-dimensional Bessel process starting from $R_0 \neq 0$, and T is stable$(1/2)$. See Revuz and Yor ([75], Chapter XIII) for consequences of (6.8.c) for the asymptotics of the winding number of planar Brownian motion.

Solution to Exercise 6.9

1. (a) Itô's formula allows us to write

$$e^{ic\gamma_t} = 1 + ic \int_0^t e^{ic\gamma_u} \, d\gamma_u - \frac{c^2}{2} \int_0^t e^{ic\gamma_u} \, du \,.$$

So the process $(e^{ic\gamma_t}, \, t \geq 0)$ is the sum of the martingale $\left(ic \int_0^t e^{ic\gamma_u} \, d\gamma_u, \, t \geq 0 \right)$ and the process with bounded variation $\left(1 - \frac{c^2}{2} \int_0^t e^{ic\gamma_u} \, du, \, t \geq 0 \right)$.

(b) Define $Z_u^{(c)} = \int\limits_0^u d\gamma_s \exp(ic\gamma_s)$, $u \geq 0$, and consider the $\mathbb{R}^3$-valued martingale $(\gamma, \mathrm{Re}(Z^{(c)}), \mathrm{Im}(Z^{(c)}))$. The different variation and covariation processes of this martingale, e.g. $\left\langle \gamma, \mathrm{Re}(Z^{(c)}) \right\rangle_t = \int_0^t ds \, \cos(c\gamma_s)$, converge a.s., as $c \to \infty$, towards those of $\left(\gamma, \frac{1}{\sqrt{2}}\gamma', \frac{1}{\sqrt{2}}\gamma'' \right)$, (this follows from the occupation time density formula). The desired result may be deduced from this, but we shall not give the details.

2. Recall from question 1 of Exercise **6.8** (with $\lambda = 0$), the following representation of the complex Brownian motion

$$Z_t = |Z_t| \exp\left(i\gamma_{H_t} \right),$$

where $H_t = \int_0^t \frac{du}{|Z_u|^2}$, and γ is a standard Brownian motion which is independent of $|Z|$. (This is known as the *skew-product* representation of complex Brownian motion.) From obvious changes of variables, we obtain:

$$\frac{1}{\log t} \int_0^t ds \, \frac{Z_s}{|Z_s|^3} = \frac{1}{\log t} \int_0^{H_t} dv \, \exp i\gamma_v = \log t \int_0^{\frac{H_t}{\log^2 t}} ds \, \exp\left(i(\log t)\tilde{\gamma}_s \right),$$

where $\tilde{\gamma}_s = \frac{1}{\log t}\gamma_{(\log t)^2 s}$. Now, on the one hand, recall from question 1 (a) and (b) that $\left(\frac{c}{2} \int_0^u ds \, \exp(ic\gamma_s), \, u \geq 0 \right)$, converges in law towards $\left(\frac{1}{\sqrt{2}}(\gamma_u' + i\gamma_u''), \, u \geq 0 \right)$, as $c \to \infty$ (γ' and γ'' are independent Brownian motions). On the other hand, as stated in the *Comments and references* of Exercise **6.8**, $\frac{4}{(\log t)^2} H_t$ converges in law towards a $1/2$ stable unilateral random variable, T, as $t \to \infty$; in fact, from the independence of γ and $|Z|$, we may deduce that when $t \to \infty$,

$$\frac{1}{\log t} \int_0^t ds \, \frac{Z_s}{|Z_s|^3} = \log t \int_0^{\frac{H_t}{\log^2 t}} ds \, \exp\left(i(\log t)\tilde{\gamma}_s \right)$$

converges towards

$$\sqrt{2} \left(\gamma_{\frac{1}{4}T}' + i\gamma_{\frac{1}{4}T}'' \right) \overset{\text{(law)}}{=} \frac{1}{\sqrt{2}} \left(\gamma_T' + i\gamma_T'' \right),$$

where T, γ' and γ'' are independent. This independence and the scaling property of Brownian motion allow us to write:

$$\frac{1}{\sqrt{2}} \left(\gamma_T' + i\gamma_T'' \right) \overset{\text{(law)}}{=} \frac{1}{\sqrt{2}} \frac{1}{\gamma_1} \left(\gamma_1' + i\gamma_1'' \right),$$

with the help of question 1 of Exercise **4.17**. Finally, from question 5 of Exercise **4.17**, we derive the characteristic function of the above variable, which is $\exp\left(-\frac{|\lambda|}{\sqrt{2}}\right)$, $\lambda \in \mathbb{R}^2$.

Solution to Exercise 6.10

1. For this question, we need to use the so-called *skew-product* representation of the planar Brownian motion (X, Y) (which in fact we have already proven in the solution to question 1 of Exercise **6.8**): put $X_t + iY_t = R_t \exp i\theta_t$. Then there exists a planar Brownian motion (β, γ) such that

$$R_t = |B_0| \exp\left(\beta_{H_t}\right), \quad \text{and} \quad \theta_t = \theta_0 + \gamma_{H_t}, \quad \text{with} \quad H_t \overset{\text{def}}{=} \int_0^t \frac{ds}{R_s^2}. \tag{6.10.a}$$

Now, we may write $\theta_t^f = \int_0^t d\gamma_{H_s} \mathbb{1}_{\{R_s \leq f(|Z_s|)\}}$.

To proceed, we shall admit that, up to the division by $\log t$, we may replace in the definition of θ_t^f, the process $f(|Z_s|)$ by $s^{\frac{a}{2}}$.

So, in the sequel, we put $\tilde{\theta}_t^f = \int_0^t d\gamma_{H_s} \mathbb{1}_{\{\beta_{H_s} \leq s^{\frac{a}{2}}\}}$. Since $u \mapsto \int_0^u dv \exp(2\beta_v)$ is the inverse of $s \mapsto H_s$, we have

$$\frac{2}{\log t} \tilde{\theta}_t^f = \frac{2}{\log t} \int_0^{H_t} d\gamma_u \mathbb{1}_{\{\beta_u \leq \frac{a}{2} \log\left(\int_0^u dv \exp(2\beta_v)\right)\}}.$$

Now, put $\lambda = \frac{\log t}{2}$, make the change of variables $u = \lambda^2 v$, and use the fact that $\tilde{\beta}_v = \frac{1}{\lambda}\beta_{\lambda^2 v}$, and $\tilde{\gamma}_v = \frac{1}{\lambda}\gamma_{\lambda^2 v}$, $v \geq 0$, are two independent Brownian motions. We obtain:

$$\frac{1}{\lambda}\tilde{\theta}_t^f = \int_0^{\frac{1}{\lambda^2}H_t} d\tilde{\gamma}_v \mathbb{1}_{\{\tilde{\beta}_v \leq \frac{a}{2\lambda} \log \lambda^2 \int_0^v dh \, \exp(2\lambda\tilde{\beta}_h)\}}.$$

Now observe that the stochastic integral process $\int_0^\cdot d\tilde{\gamma}_v \mathbb{1}_{\{\tilde{\beta}_v \leq \frac{a}{2\lambda} \log \lambda^2 \int_0^v dh \, \exp(2\lambda\tilde{\beta}_h)\}}$ converges in probability (uniformly on any interval of time) towards: $\int_0^\cdot d\tilde{\gamma}_v \mathbb{1}_{\{\tilde{\beta}_v \leq a \sup_{s \leq v} \tilde{\beta}_s\}}$. This convergence holds jointly with that of $\frac{1}{\lambda^2}H_t$ towards $\tilde{T} = \inf\{v : \tilde{\beta}_v = 1\}$, which proves the desired result.

2. The same computation as above leads to

$$\frac{1}{\log t}(\theta_t - \theta_t^f) \xrightarrow[t \to \infty]{\text{(law)}} \int_0^\sigma d\gamma_u \mathbb{1}_{(\beta_u > aS_u)} = 0.$$

So that the convergence holds in probability.

Solution to Exercise 6.11

1. For any $u \in [0,1]$, let Θ_u be the transformation which consists in re-ordering the paths $\{f(t), 0 \le t \le u\}$ and $\{f(t), u \le t \le 1\}$ of any continuous function f on $[0,1]$, such that $f(0) = 0$, in such a way that the new function $\Theta_u(f)$ is continuous and vanishes at 0. Formally, $\Theta_u(f)$ is defined by

$$\Theta_u(f)(t) = \begin{cases} f(t+u) - f(u), & \text{if } t < 1-u, \\ f(t-(1-u)) + f(1) - f(u), & \text{if } 1-u \le t \le 1 \end{cases}.$$

Let B_t, $t \in [0,1]$ be a real valued Brownian motion on $[0,1]$. Since B has independent and time homogeneous increments, for any $u \in [0,1]$, we have

$$\Theta_u(B) \overset{(\text{law})}{=} B. \tag{6.11.a}$$

Now we use the representation of the Brownian bridge as $b_t \overset{(\text{law})}{=} B_t - tB_1$, $t \in [0,1]$, which is proved in question 5 of Exercise **6.13**. For $u \in [0,1]$, put $B' = \Theta_u(B)$, then it is not difficult to check that $\Theta_u(b)_t = B'_t - tB'_1$, $t \in [0,1]$, and the result follows from (6.11.a).

2. First note that $\Theta_v(b) = \{b_{\{t+v\}} - b_v, 0 \le t \le 1\}$. So,

$$\begin{aligned} \Theta_u \circ \Theta_v(b) &= \Theta_u(\{b_{\{t+v\}} - b_v, 0 \le t \le 1\}) \\ &= \{b_{\{\{t+u\}+v\}} - b_{\{u+v\}}, 0 \le t \le 1\} = \Theta_{\{u+v\}}(b). \end{aligned}$$

Let $F \in L^1(P)$; we can check that $\int_0^1 F \circ \Theta_u \, du$ is $\mathcal{I}$-measurable. Indeed, from above we have for any $v \in [0,1]$,

$$\begin{aligned} \left(\int_0^1 F \circ \Theta_u \, du\right) \circ \Theta_v &= \int_0^1 F \circ \Theta_u \circ \Theta_v \, du \\ &= \int_0^1 F \circ \Theta_{\{u+v\}} \, du = \int_0^1 F \circ \Theta_u \, du. \end{aligned}$$

On the other hand, from question 1, for any $A \in \mathcal{I}$, and any $u \in [0,1]$, we have $E[F \mathbb{1}_A] = E[F \circ \Theta_u \mathbb{1}_A]$. Integrating this expression over $[0,1]$ and using Fubini's Theorem, we obtain:

$$E[F \mathbb{1}_A] = E\left[\int_0^1 du \, F \circ \Theta_u \mathbb{1}_A\right],$$

which proves the result.

3. Consider G, a bounded measurable functional; the result follows from the computation:

$$\begin{aligned} E[G(m) \,|\, \mathcal{I}] &= \int_0^1 G(m) \circ \Theta_u \, du \\ &= \int_0^m du \, G(m-u) + \int_m^1 du \, G(1+m-u) \\ &= \int_0^1 ds \, G(s). \end{aligned}$$

The first equality above comes from (6.11.1), and the second one comes from the relation $\{m \circ \Theta_u + u\} = m$ which is straightforward and may easily be seen on a picture.

4. As in question 3, consider G a bounded measurable functional and write

$$E[G(A_0) \,|\, \mathcal{I}] = \int_0^1 G(A_0 \circ \Theta_u)\, du$$
$$= \int_0^1 du\, G\left(\int_0^1 ds\, \mathbb{1}_{\{b_s \leq b_u\}}\right).$$

Let (l_1^x) be the local time of b, at time 1 and level $x \in \mathbb{R}$. Applying the occupation time density formula, we obtain:

$$E[G(A_0) \,|\, \mathcal{I}] = \int_0^1 du\, G\left(\int_0^1 ds\, \mathbb{1}_{\{b_s \leq b_u\}}\right)$$
$$= \int_{-\infty}^{+\infty} l_1^x\, dx\, G\left(\int_0^1 ds\, \mathbb{1}_{\{b_s \leq x\}}\right)$$
$$= \int_{-\infty}^{+\infty} l_1^x\, dx\, G\left(\int_{-\infty}^x l_1^y\, dy\right).$$

Making the change of variables: $z = \int_{-\infty}^x l_1^y\, dy$ in the above equality, we obtain

$$E[G(A_0) \,|\, \mathcal{I}] = \int_0^1 dz\, G(z),$$

which proves the desired result.

Solution to Exercise 6.12

1. This follows immediately from the Gaussian character of Brownian motion, as in question 1.(a) of Exercise **5.16**.

2. Note that $(X_{T_y+t}^y - y, t \geq 0) = (-(X_{T_y+t} - y), t \geq 0)$, hence from the Markov property of Brownian motion at time T_y, under P_0, $(X_{T_y+t}^y - y, t \geq 0)$ is independent of $(X_t^y, t \leq T_y) = (X_t, t \leq T_y)$ and has the same law as $(X_t, t \geq 0)$. This proves that X^y is a standard Brownian motion.

We prove identity (6.12.2) from the following observation:

$$\{S_t > y, X_t < x\} = \{X_t^y > 2y - x\}, \quad P_0 \text{ a.s.}$$

which we easily verify on a picture. Then, differentiating (6.12.2) with respect to x and y, we obtain the density of the law of the pair, (X_t, S_t) under P_0:

$$\left(\frac{2}{\pi t^3}\right)^{\frac{1}{2}} (2y - x)\exp\left(-\frac{(2y-x)^2}{2t}\right), \quad x \leq y, \ y > 0. \tag{6.12.a}$$

3. Put $x = y$ in (6.12.2), we obtain $P_0(S_t > y, X_t < y) = P_0(X_t > y)$. Since $P_0(S_t > y) = P_0(S_t > y, X_t \geq y) + P_0(S_t > y, X_t < y)$, we deduce from the previous identity that

$$P_0(S_t > y) = 2P_0(X_t > y), \quad y > 0. \tag{6.12.b}$$

Now, (6.12.b) together with the fact that $P_0(S_t > y) = P_0(T_y < t)$ imply $P_0(T_y < t) = 2P_0(X_t > y)$. Moreover, the homogeneity of the Brownian law (see 6.12.1) yields the equality $P_0(T_{y-a} < t) = P_a(T_y < t)$, for any $a < y$, which finally gives

$$P_a(T_y < t) = 2P_0(X_t > y - a), \quad y > 0. \tag{6.12.c}$$

Then equation (6.12.3), for any $a < y$, follows from (6.12.c) by differentiating with respect to t. The general case is obtained from the symmetry property of Brownian motion.

4. From Lévy's identity, we have:

$$(S_t - X_t, S_t) \stackrel{\text{(law)}}{=} (|X_t|, L_t),$$

under P_0. This identity in law together with (6.12.a), yield the density of the law of the pair $(|X_t|, L_t)$:

$$\left(\frac{2}{\pi t^3}\right)^{\frac{1}{2}} (x + y) \exp\left(-\frac{(x+y)^2}{2t}\right), \quad x, y \geq 0.$$

To deduce from above the density of (X_t, L_t), first write $(X_t, L_t) = (\text{sgn}(X_t)|X_t|, L_t)$. Now, observe that since L_t is a functional of $(|X_u|, u \geq 0)$ and $\text{sgn}(X_t)$ is independent of $(|X_u|, u \geq 0)$, then $\text{sgn}(X_t)$ is independent of $(|X_t|, L_t)$. Moreover, $\text{sgn}(X_t)$ is a Bernoulli symmetric random variable. Hence, the density of the pair (X_t, L_t) under P_0 is:

$$\left(\frac{1}{2\pi t^3}\right)^{\frac{1}{2}} (|x| + y) \exp\left(-\frac{(|x|+y)^2}{2t}\right), \quad x \in \mathbb{R}, y \geq 0.$$

We may deduce from above and the absolute continuity relation (6.13.1), an explicit expression for the density of the joint law of (X_t, L_t), $t < 1$, under the law of the standard Brownian bridge, $P_{0 \to 0}^{(1)}$:

$$\left(\frac{1}{2\pi (1-t)t^3}\right)^{\frac{1}{2}} (|x| + y) \exp\left(-\frac{(|x|+y)^2}{2t} - \frac{x^2}{2(1-t)}\right), \quad x \in \mathbb{R}, y \geq 0.$$

Solution to Exercise 6.13

1. For any $a, \mu \in \mathbb{R}$, let P_μ^a be the law of the Brownian motion starting at μ with drift a, that is the law of $(\mu + X_s + as, s \geq 0)$ under P. The invariance of the Brownian law by time inversion may be stated as

$$\left[\left(sX_{\frac{1}{s}}, s > 0\right), P_\mu^a\right] = \left[(X_s, s > 0), P_a^\mu\right].$$

In other words, time inversion interchanges the drift and the starting point. From the above identity, we have for any $x \in \mathbb{R}$:

$$\left[\left(sX_{\frac{1}{s}}, s > 0\right), P_\mu^a\left(\cdot \mid X_{\frac{1}{t}} = \frac{x}{t}\right)\right] = \left[(X_s, s > 0), P_a^\mu(\cdot \mid X_t = x)\right].$$

Applying the Markov property at time $s = \frac{1}{t}$ to the process $\left[\left(sX_{\frac{1}{s}}, s \geq 0\right), P_\mu^a\right]$ gives

$$\left[\left(sX_{\frac{1}{s} - \frac{1}{t}}, 0 \leq s \leq t\right), P_{\frac{x}{t}}^a\right] = \left[(X_s, 0 \leq s \leq t), P_a^\mu(\cdot \mid X_t = x)\right],$$

which is the first part of the question. Note that the law of the process on the right hand side of the above identity does not depend on μ.

We now deduce the law of T_0 under $P_{a \to x}^{(t)}$, with the help of the hint, and of the representation of the bridge in terms of Brownian motion with drift. Indeed, it follows from these two arguments that under $P_{a \to x}^{(t)}$, the r.v. $\frac{1}{T_0} - \frac{1}{t}$ is distributed as the last passage time at 0 of the process: $\left(\frac{x}{t} + X_s + as, s \geq 0\right)$, where $(X_s, s \geq 0)$ is a standard Brownian motion. Thus, from the hint, we deduce:

$$P_{a \to x}^{(t)}\left(\frac{1}{T_0} - \frac{1}{t} \in ds\right) = (\text{cst}) \, p_s^{(a)}\left(\frac{x}{t}, 0\right) ds, \tag{6.13.a}$$

where $p_s^{(a)}(x, y)$ is the density of the semigroup of Brownian motion with drift a. It is easily shown that:

$$p_s^{(a)}(x, y) = \frac{1}{\sqrt{2\pi s}} \exp - \left(\frac{1}{2s}(x - y)^2 + a(x - y) + \frac{a^2 s}{2}\right),$$

so that, it follows from (6.13.a) that:

$$P_{a \to x}^{(t)}(T_0 \in ds) = |a| \, ds \sqrt{\frac{t}{2\pi s^3(t - s)}} \exp - \left(\frac{1}{2}\frac{x^2 s}{t(t - s)} + a\frac{x}{s} + \frac{a^2}{2}\left(\frac{1}{s} - \frac{1}{t}\right)\right). \tag{6.13.b}$$

2. Let f be a generic bounded Borel function. It follows from the definition of $P_{a \to x}^{(t)}$ that

$$\int_{\mathbb{R}} E_{a \to x}^{(t)}[F(X_u, u \leq s)] p_t(a, x) f(x) \, dx = E_a[F(X_u, u \leq s) f(X_t)]$$

$$= E_a[F(X_u, u \leq s) E_{X_s}[f(X_{t-s})]]$$

$$= \int_{\mathbb{R}} E_a[F(X_u, u \leq s) p_{t-s}(X_s, x)] f(x) \, dx,$$

where the second equality is obtained by applying the Markov property at time s. We obtain relation (6.13.1) by identifying the first and last terms of these equalities, using the continuity in x for both quantities: $E_{a \to x}^{(t)}[F(X_u, u \leq s)] p_t(a, x)$ and $E_a[F(X_u, u \leq s) p_{t-s}(X_s, x)]$, for F a bounded continuous functional.

3. Relation (6.13.2) follows from (6.13.1) using the fact that $(s \mapsto p_{t-s}(X_s, x)$, $s < t)$ is a P_a-martingale, together with the optional stopping theorem.

4. We obtain (6.13.3) by taking $S = T_y$ and $F = \mathbb{1}_{\{T_y < s\}}$ in (6.13.2).

We obtain (6.13.4) from (6.13.3) by writing:

$$P_{a \to x}^{(t)}(T_y < s) = \int_0^s \frac{p_{t-u}(y, x)}{p_t(a, x)} P_a(T_y \in du).$$

Then, differentiate with respect to s and use (6.12.3).

Finally, it is immediate to recover the density of the law of T_0, given in (6.13.b) from (6.13.3).

5. To prove (6.13.5), observe that $p_u(0, x) = p_u(0, -x)$, for all $u \geq 0$, hence applying (6.13.3), we have:

$$P_{a \to -x}^{(t)}(T_0 < s) = E_a \left[\mathbb{1}_{\{T_0 < s\}} \frac{p_{t-T_0}(0, x)}{p_t(a, -x)} \right].$$

By combining the above equality with (6.13.3), we obtain (6.13.5). In particular, since $P_{a \to -x}^{(t)}(T_0 < t) = 1$, we have (6.13.6).

Solution to Exercise 6.14

1. Recall that Z_1 and $2\sqrt{Z_1 Z_{\frac{1}{2}}}$, have the same law if and only if their real moments coincide (see Exercise 4.3). For every $z \geq 0$, we have $E[Z_1^{2z-1}] = \Gamma(2z)$, on the one hand and

$$E\left[(2\sqrt{Z_1 Z_{\frac{1}{2}}})^{2z-1}\right] = 2^{2z-1} E\left[Z_1^{z-\frac{1}{2}}\right] E\left[Z_{\frac{1}{2}}^{z-\frac{1}{2}}\right]$$
$$= 2^{2z-1} \Gamma(z + \frac{1}{2}) \Gamma(\frac{1}{2})^{-1} \Gamma(z) = \frac{1}{\sqrt{2\pi}} 2^{2z-\frac{1}{2}} \Gamma(z) \Gamma\left(z + \frac{1}{2}\right)$$

on the other hand.

The identity (6.14.2) follows from: $N^2 \overset{(\text{law})}{=} 2Z_{\frac{1}{2}}$.

2. It follows from (6.12.3) that $\frac{T_x}{2x^2}$ has a unilateral stable law with parameter $\frac{1}{2}$. So, we may write

$$P(T_x < Z_1) = \int_0^\infty P(T_x < t) e^{-t} \, dt = E[e^{-T_x}] = \exp(-\sqrt{2}x).$$

The second formula follows from properties of Brownian motion which we now recall. At first, observe that if $\overline{B}$ is the past maximum process of B, defined by

$\overline{B}_t = \sup_{0 \le s \le t} B_s$, then (6.14.4) can be written as $P(T_x < Z_1) = P(x < \overline{B}_{Z_1})$. Secondly, it follows from the so-called reflection principle of Brownian motion that $\overline{B}_t \overset{(\mathrm{law})}{=} |B_t|$, for every $t \ge 0$. Finally, the scaling property of Brownian motion implies that $B_t \overset{(\mathrm{law})}{=} t^{\frac{1}{2}} B_1$ for every deterministic t, so that we have $P(x < \overline{B}_{Z_1}) = P(x < |B_{Z_1}|) = P(x < \sqrt{Z_1}|B_1|) = P(x < \sqrt{Z_1}|N|)$.

3. With (6.14.3) and (6.14.4), we obtain $P(x < \sqrt{Z_1}|N|) = \exp(-\sqrt{2}x)$, which precisely implies identity (6.14.2).

Solution to Exercise 6.15

1. The process $(B_{u(s)}, s \ge 0)$ is a martingale in its own filtration which is also that of X, since $v(t) \ne 0$, for all $t > 0$. Hence, using Itô's formula, we may write:

$$X_t = \int_0^t v(s)\, dB_{u(s)} + \int_0^t B_{u(s)}\, dv(s)\,, \quad t \ge 0\,.$$

The process $(\int_0^t B_{u(s)}\, dv(s), t \ge 0)$ is adapted to the same filtration and has bounded variation. This proves that (X_t) is a semimartingale whose martingale part is $\int_0^t v(s)\, dB_{u(s)}$.

2. According to Lévy's Characterization Theorem, the martingale $(\int_0^t v(s)\, dB_{u(s)}, t \ge 0)$ is a Brownian motion if and only if its quadratic variation is t, that is

$$\int_0^t v^2(s)\, du(s) = t\,, \quad \text{for every } t \ge 0\,.$$

Since u' and v are continuous functions, this is equivalent to $v^2(s)u'(s) \equiv 1$.

3. Tanaka's formula applied to X gives

$$|X_t| = \int_0^t v(s)\mathrm{sgn}(B_{u(s)})\, dB_{u(s)} + \int_0^t |B_{u(s)}|\, dv(s) + L_t^X\,. \tag{6.15.a}$$

Thanks to the same formula, we also have

$$|B_{u(t)}| = \int_0^t \mathrm{sgn}(B_{u(s)})\, dB_{u(s)} + L_{u(t)}^B\,,$$

which leads to

$$\int_0^t v(s)\, d|B_{u(s)}| = \int_0^t v(s)\mathrm{sgn}(B_{u(s)})\, dB_{u(s)} + \int_0^t v(s)\, dL_{u(s)}^B\,.$$

It follows from the above identity and the decomposition $|X_t| = v(t)|B_{u(t)}| = \int_0^t v(s)\, d|B_{u(s)}| + \int_0^t |B_{u(s)}|\, dv(s)$, that

$$|X_t| = \int_0^t v(s)\mathrm{sgn}(B_{u(s)})\, dB_{u(s)} + \int_0^t |B_{u(s)}|\, dv(s) + \int_0^t v(s)\, dL_{u(s)}^B\,. \tag{6.15.b}$$

We obtain the result by comparing (6.15.a) and (6.15.b).

4. With $u(t) = \frac{1-e^{-2\beta t}}{2\beta}$, $u^{-1}(t) = \frac{-1}{2\beta}\log(1-2\beta t)$, and $v(t) = e^{\beta t}$, we have from (6.15.1):

$$L_t^U = \int_0^{\frac{1-e^{-2\beta t}}{2\beta}} \frac{dL_s^B}{\sqrt{1-2\beta s}}.$$

5. We obtain this formula simply by applying (6.15.1) with $u(t) = \frac{t}{1-t}$, hence $u^{-1}(t) = \frac{t}{1+t}$ and $v(t) = t$.

According to question 2 of Exercise **6.12**, the density of the law of l_t is

$$
\begin{aligned}
h_t(y) &= \left(\frac{1}{2\pi(1-t)t^3}\right)^{\frac{1}{2}} \int_{-\infty}^{\infty} (|x|+y)\exp\left(-\frac{(|x|+y)^2}{2t} - \frac{x^2}{2(1-t)}\right) dx \\
&= \left(\frac{2}{\pi(1-t)t^3}\right)^{\frac{1}{2}} \exp\left(-\frac{y^2}{2}\right) \int_0^{\infty} (x+y)\exp\left(-\frac{(x+(1-t)y)^2}{2t(1-t)}\right) dx \\
&= \sqrt{\frac{2}{\pi}\frac{1}{t}} \exp\left(-\frac{y^2}{2}\right) \int_{\sqrt{\frac{1-t}{t}}y}^{\infty} \left(\sqrt{t(1-t)}z + ty\right)\exp\left(-\frac{z^2}{2}\right) dz \\
&= \sqrt{\frac{2}{\pi}}\sqrt{\frac{1-t}{t}} \exp\left(-\frac{y^2}{2t}\right) + 2y\exp\left(-\frac{y^2}{2}\right)\Phi\left(-\sqrt{\frac{1-t}{t}}y\right),
\end{aligned}
$$

where, in the last equality, Φ is the distribution function of the centered normal distribution with variance 1. Finally, from (6.15.2), the density of $\int_0^t \frac{dL_v}{1+v}$ is given by:

$$h_{\frac{t}{1+t}}(y) = \sqrt{\frac{2}{\pi t}} \exp\left(-\frac{(1+t)y^2}{t}\right) + 2y\exp\left(-\frac{y^2}{2}\right)\Phi\left(-\frac{y}{\sqrt{t}}\right).$$

Solution to Exercise 6.16

1. Let (t_n) be any sequence of positive real numbers which tends to 0. From the scaling property and the hypothesis, for any n, $P(X_{t_n}/t_n^\alpha \geq a) = P(X_1 \geq a) > 0$. Therefore, $P(\limsup_n\{X_{t_n}/t_n^\alpha \geq a\}) \geq P(X_1 \geq a) > 0$. But since the event $\limsup_n\{X_{t_n}/t_n^\alpha \geq a\}$ belongs to the σ-field $\cap_{t>0}\sigma\{X_s, s \leq t\}$ which is trivial, its probability is 1. We deduce that for any $a \geq 0$, $\limsup_n X_{t_n}/t_n^\alpha \geq a$, almost surely.

2. Since B satisfies the hypothesis of question 1, then $\limsup_{t\downarrow 0} B_t/\sqrt{t} = +\infty$, almost surely, hence the random time $g_a = \sup\{t \leq 1 : B_t = a\sqrt{t}\}$ is almost surely positive, $P(0 < g_a < 1) = 1$. Now, set $\tilde{B} = (tB_{1/t}, t > 0)$, $\tilde{B}_0 = 0$, then from the time inversion property of Brownian motion, $\tilde{B}$ has the same law as B and $g_a \stackrel{(\text{law})}{=} \tau_a^{-1}$, where $\tau_a \stackrel{(\text{def})}{=} \inf\{t \geq 1 : \tilde{B}_t = a\sqrt{t}\}$. We also have:

$$(g_a^{-1/2}B_{g_a t}, 0 < t \leq 1) \stackrel{(\text{law})}{=} (t\tau_a^{-1/2}\tilde{B}_{\tau_a/t}, 0 < t \leq 1).$$

Note that τ_a is a stopping time in the filtration generated by $\tilde{B}$, so if we set $\overline{B} \stackrel{\text{(def)}}{=}$ $(\tilde{B}_{\tau_a+t} - a\sqrt{\tau_a}, t \geq 0)$, then $\overline{B}$ is a standard Brownian motion and we have, from above,

$$(g_a^{-1/2} B_{g_a t}, 0 < t \leq 1) \stackrel{\text{(law)}}{=} (t\tau_a^{-1/2}[\overline{B}_{\frac{1-t}{t}\tau_a} + a\sqrt{\tau_a}], 0 < t \leq 1).$$

Note also that $\overline{B}$ is independent of τ_a, so that from the scaling property,

$$(g_a^{-1/2} B_{g_a t}, 0 < t \leq 1) \stackrel{\text{(law)}}{=} (t\overline{B}_{\frac{1-t}{t}} + at, 0 < t \leq 1).$$

From Exercise **6.13**, the process on the right hand side of the last equality is a standard Brownian bridge from 0 to a.

Solution to Exercise 6.17

1. From the time inversion invariance property of Brownian motion, the process $\tilde{B}_t \stackrel{\text{def}}{=} tB_{\frac{1}{t}}, t \geq 0$ is a Brownian motion and we have

$$\frac{1}{\Lambda} = \inf\{u : \tilde{B}_u = 1\} \stackrel{\text{def}}{=} \tilde{T}_1.$$

As is well known, and follows from the scaling property and the reflection principle (see question 1 of Exercise **6.12**), one has: $\tilde{T}_1 \stackrel{\text{(law)}}{=} \tilde{B}_1^{-2}$, which gives (6.17.1).

It also appears in Exercise **5.12** that $\tilde{T}_1$ has a unilateral stable $(1/2)$ distribution (see the comments at the end). So, the result follows from question 1 of Exercise **4.17**.

2. Reversing the time in (6.17.2), we see that this equation is equivalent to

$$\left(\tilde{B}_t - 1, t \geq \tilde{T}_1\right) \stackrel{\text{(law)}}{=} \left(t\sqrt{\Lambda}b\left(\frac{1}{t\Lambda}\right), t \geq \frac{1}{\Lambda}\right).$$

Now, the proof of the above identity may be further reduced by making the changes of variables $t = \tilde{T}_1 + u$ and $t = \frac{1}{\Lambda} + u$, to showing:

$$\left[\tilde{T}_1, \left(\tilde{B}_{u+\tilde{T}_1} - 1, u \geq 0\right)\right] \stackrel{\text{(law)}}{=} \left[\frac{1}{\Lambda}, \left(\frac{1+\Lambda u}{\sqrt{\Lambda}}b\left(\frac{1}{1+\Lambda u}\right), u \geq 0\right)\right].$$

Applying the strong Markov property of Brownian motion at time $\tilde{T}_1$ and using the fact that it has time homogeneous and stationary increments, this also may be written as

$$\left[\tilde{T}_1, (B'_u, u \geq 0)\right] \stackrel{\text{(law)}}{=} \left[\frac{1}{\Lambda}, \left(\frac{1+\Lambda u}{\sqrt{\Lambda}}b\left(\frac{1}{1+\Lambda u}\right), u \geq 0\right)\right],$$

where B' is a standard Brownian motion which is independent of $\tilde{T}_1$. Finally it follows from the scaling property of Brownian motion, and question 5 of Exercise **6.15** that the process

$$\left(\frac{1+\Lambda u}{\sqrt{\Lambda}} b\left(\frac{1}{1+\Lambda u}\right), u \geq 0\right)$$

is a standard Brownian motion which is independent of Λ. This proves the result.

3. (i) and (ii). The proofs of (6.17.3) and (6.17.4) are exactly the same as the proofs of questions 1 and 2, once we notice that

$$\frac{1}{\Lambda_\mu} = \inf\{u : \tilde{B}_u = \mu\}.$$

From the scaling property of Brownian motion, for any $\lambda > 0$, the re-scaled process $\left(\sqrt{\lambda} b\left(\frac{t}{\lambda}\right), 0 \leq t \leq \lambda\right)$ is distributed as a Brownian bridge starting from 0, ending at 0 and with length λ. Hence, the second part of the question follows directly from the first.

Solution to Exercise 6.18

1. First recall that the process $B' = (tB_{\frac{1}{t}}, t > 0)$ has the same law as $(B_t, t > 0)$. Moreover, the process $b = \left(B'_s - \frac{s}{1+m} B'_{1+m}, s \in [0, 1+m]\right)$ has the law of a standard Brownian bridge $P_{0\to0}^{(1+m)}$. From the Markov property of Brownian bridge applied at time 1, conditionally on $b_1 = a$, the process

$$(b_{1+t}, t \in [0,m]) = \left((1+t)\left(B_{\frac{1}{1+t}} - B_{\frac{1}{1+m}}\right), t \in [0,m]\right)$$

has law $P_{a\to0}^{(m)}$. Finally, we deduce the result from the identity in law

$$\left((1+t)\left(B_{\frac{1}{1+t}} - B_{\frac{1}{1+m}}\right), t \in [0,m]\right) \overset{(\text{law})}{=} \left((1+t)\left(B_{\frac{1}{1+t} - \frac{1}{1+m}}\right), t \in [0,m]\right).$$

2. From above, using the fact that if $(X_u, u \leq m)$ is distributed as $P_{a\to0}^{(m)}$, then $(X_{m-u}, u \leq m)$ is distributed as $P_{a\to0}^{(m)}$. This leads to the identity in law:

$$\left(\int_0^m dt\, \varphi(t) B_t^2 \mid B_m = a\right) \overset{(\text{law})}{=} \left(\int_0^m \varphi(m-t)(1+t)^2 B^2_{\frac{1}{1+t} - \frac{1}{1+m}} dt \mid B_{\frac{m}{1+m}} = a\right),$$

and from the change of variable $\frac{s}{1+m} = \frac{1}{1+t} - \frac{1}{1+m}$ in the right hand side, it follows:

$$\left(\int_0^m dt\, \varphi(t) B_t^2 \mid B_m = a\right) \overset{(\text{law})}{=} \left(\int_0^m \varphi\left(\frac{(1+m)s}{1+s}\right) \frac{(m+1)^3}{(s+1)^4} B^2_{\frac{s}{1+m}} ds \mid B_{\frac{m}{1+m}} = a\right).$$

Recall that from the scaling property of Brownian motion,

$$\left(\sqrt{m+1} B_{\frac{s}{m+1}}, 0 \leq s \leq m\right) \overset{(\text{law})}{=} (B_s, 0 \leq s \leq m),$$

which yields, from above:

$$\left(\int_0^m dt\, \varphi(t) B_t^2 \mid B_m = a \right)$$

$$\stackrel{\text{(law)}}{=} \left((1+m)^2 \int_0^m \frac{ds}{(s+1)^4} B_s^2 \varphi\left(\frac{(1+m)s}{1+s} \right) \mid B_m = a\sqrt{1+m} \right).$$

One means to compute the expression $E\left[\exp\left(-\lambda \int_0^m dt\, B_t^2 \right) \mid B_m = a \right]$, is to determine the Laplace transform

$$I_{\alpha,b} \stackrel{\text{def}}{=} E\left[\exp\left(-\alpha B_m^2 - \frac{b^2}{2} \int_0^m dt\, B_t^2 \right) \right],$$

as a consequence of Girsanov's transformation. One will find a detailed proof in the book referred to in the statement of the exercise.

Solution to Exercise 6.19

1. First observe that from the condition (6.19.1), the semimartingale X converges almost surely, so that X_T and $\langle X \rangle_T$ are well defined for any stopping time T (taking possibly the value ∞). We also point out that from Itô's formula, we have

$$X_t^2 - \langle X \rangle_t = 2 \int_0^t X_s\, dX_s$$

$$= 2 \int_0^t X_s\, dM_s + 2 \int_0^t X_s\, dV_s. \qquad (6.19.\text{a})$$

Assume that (6.19.2) holds for every stopping time T. Then it is well known that the process $(X_t^2 - \langle X \rangle_t, t \geq 0)$ is a martingale. From (6.19.a) we deduce that $(\int_0^t X_s\, dV_s, t \geq 0)$ is also a martingale. Since this process has finite variation, it necessarily vanishes. This is equivalent to (6.19.3).

Assume now that (6.19.3) holds. Then the process $(\int_0^t X_s\, dV_s, t \geq 0)$ vanishes and from (6.19.a), $(X_t^2 - \langle X \rangle_t, t \geq 0)$ is a martingale. Therefore, (6.19.2) holds for every bounded stopping time. If T is any stopping time, then we may find a sequence of (bounded) stopping times T_n, $n \geq 1$ which converges almost surely towards T and such that for every $n \geq 1$, $X_{t \wedge T_n}^2 - \langle X \rangle_{t \wedge T_n}$, $t \geq 0$, is a bounded martingale. Hence,

$$\lim_{n \to \infty} \left(X_{T_n}^2 - \langle X \rangle_{T_n} \right) = X_T^2 - \langle X \rangle_T, \quad \text{a.s., and} \qquad (6.19.\text{b})$$

$$E[X_{T_n}^2] = E[\langle X \rangle_{T_n}]. \qquad (6.19.\text{c})$$

From the Burkholder–Davis–Gundy inequalities, there exists a constant C such that:

$$E\left[\left(\sup_{s \geq 0} |X_s| \right)^2 \right] \leq CE\left[\langle M \rangle_\infty + \left(\int_0^\infty |dV_s| \right)^2 \right], \qquad (6.19.\text{d})$$

hence we obtain (6.19.2) for T from (6.19.b), (6.19.c), (6.19.1) and Lebesgue's Theorem of dominated convergence.

2. X^+ and X^- are semimartingales such that for any time $t \geq 0$, $\langle X^+ \rangle_t + \langle X^- \rangle_t = \langle X \rangle_t$ and $(X_t^+)^2 + (X_t^-)^2 = X_t^2$, almost surely. Hence, if X^+ and X^- satisfy (6.19.2), then so does X.

Suppose that X satisfies (6.19.2). From question 1, X also satisfies (6.19.3). We will check that X^+ and X^- satisfy (6.19.1) and (6.19.3) (so that they satisfy (6.19.2)), that is

$$E\left[\langle M^{(+)} \rangle_\infty + \left(\int_0^\infty |dV_s^{(+)}|\right)^2\right] < \infty, \; E\left[\langle M^{(-)} \rangle_\infty + \left(\int_0^\infty |dV_s^{(-)}|\right)^2\right] < \infty,$$

$$\tag{6.19.e}$$

$$\int_0^\infty \mathbb{1}_{\{X_s^+ \neq 0\}} |dV_s^{(+)}| = 0, \quad \text{and} \quad \int_0^\infty \mathbb{1}_{\{X_s^- \neq 0\}} |dV_s^{(-)}| = 0, \tag{6.19.f}$$

where $M^{(+)}$ and $M^{(-)}$ are the martingale parts of X^+ and X^-, respectively, and $V^{(+)}$ and $V^{(-)}$ are their finite variation parts.

From Tanaka's formula and (6.19.3), $V^{(+)}$ and $V^{(-)}$ are given by

$$V_t^{(+)} = \int_0^t \mathbb{1}_{\{X_s > 0\}} dV_s + \frac{1}{2} \int_0^t dL_s^X = \frac{1}{2} L_t^X$$

$$V_t^{(-)} = -\int_0^t \mathbb{1}_{\{X_s \leq 0\}} dV_s + \frac{1}{2} \int_0^t dL_s^X = -\int_0^t \mathbb{1}_{\{X_s = 0\}} dV_s + \frac{1}{2} L_t^X,$$

where L^X is the local time at 0 of X. Since V and L^X have zero variation on the set $\{t : X_t \neq 0\}$ which contains both $\{t : X_t^+ \neq 0\}$ and $\{t : X_t^- \neq 0\}$, then $V^{(+)}$ (resp. $V^{(-)}$) has zero variation on $\{t : X_t^+ \neq 0\}$ (resp. $\{t : X_t^- \neq 0\}$). This proves (6.19.f).

Now we check (6.19.e). Since $L_t^X = 2\left(X_t^+ - \int_0^t \mathbb{1}_{\{X_s > 0\}} dV_s - \int_0^t \mathbb{1}_{\{X_s > 0\}} dM_s\right)$, there exists a constant D such that

$$E[(L_\infty^X)^2] \leq DE\left[\left(\sup_{s \geq 0} |X_s|\right)^2 + \left(\int_0^\infty |dV_s|\right)^2 + \langle M \rangle_\infty\right].$$

From (6.19.d), the right hand side of the above inequality is finite. Finally, we have $\langle M^{(+)} \rangle_\infty = \int_0^\infty \mathbb{1}_{\{X_s > 0\}} d\langle M \rangle_s \leq \langle M \rangle_\infty$ and $\langle M^{(-)} \rangle_\infty = \int_0^\infty \mathbb{1}_{\{X_s \leq 0\}} d\langle M \rangle_s \leq \langle M \rangle_\infty$, almost surely. This ends the proof of (6.19.e).

3. (i) We easily check that the semimartingale $X = -M + S^M$ satisfies (6.19.3), hence it satisfies (6.19.2).

(ii) Developing the square of the semimartingale $X = M + L^M$, we obtain:

$$E[X_T^2] = E[M_T^2 + 2M_T L_T^M + (L_T^M)^2]$$
$$= E[M_T^2] + E[(L_T^M)^2]$$
$$= E[\langle M \rangle_T] + E[(L_T^M)^2].$$

Since $\langle X \rangle_T = \langle M \rangle_T$, X satisfies (6.19.2) if and only if $L_t^M = 0$, for all t; but since $E[|M_t|] = E[L_t^M]$ for every t, $L_t^M = 0$ is equivalent to $M_t = 0$.

4. Tanaka's formula yields:

$$|X_t| - L_t = \int_0^t \text{sgn}(X_s) \, dX_s \, .$$

If (6.19.4) holds for every stopping time T then $(|X_t| - L_t)$ is a martingale and so is $X_t = \int_0^t \text{sgn}(X_s) \, d(|X_s| - L_s)$.

If (X_t) is a martingale, then (6.19.4) holds for every bounded stopping time. But as in question 1, the condition (6.19.1) allows us to prove that (6.19.4) holds for every stopping time.

Solution to Exercise 6.20

Let $A_t = \int_0^t ds \, \exp(2B_s)$. First we prove the identity

$$\left(\exp(B_t), \int_0^t \exp(B_s) dW_s; \, t \geq 0 \right) = (R(A_t), \beta(A_t); \, t \geq 0) \, , \qquad (6.20.a)$$

where R is a two-dimensional Bessel process started at 1 and β is a Brownian motion, independent of R.

Since $\left(\int_0^t \exp(B_s) dW_s; \, t \geq 0 \right)$ is a martingale, whose quadratic variation is A_t, the process β is nothing but its Dambis–Dubins–Schwarz (DDS) Brownian motion, and the identity

$$\left(\int_0^t \exp(B_s) dW_s; \, t \geq 0 \right) = (\beta(A_t); \, t \geq 0)$$

follows. To prove the other identity, write from Itô's formula:

$$\exp(2B_t) = 1 + 2 \int_0^t \exp(2B_s) \, dB_s + 2 \int_0^t \exp(2B_s) \, ds \, . \qquad (6.20.b)$$

Changing the time on both sides of this equation with the inverse A_u^{-1} of A_t we get

$$\exp(2B_{A_u^{-1}}) = 1 + 2 \int_0^u \exp(B_{A_s^{-1}}) \, d\gamma_s + 2u \, ,$$

where γ is the (DDS) Brownian motion defined by $\int_0^t \exp(B_s) \, dB_s = \gamma(A_t)$. This shows that $\exp(2B_{A_u^{-1}})$ is the square of a two-dimensional Bessel process started at 1 and the first identity of (6.20.a) is proved. The independence between R and β follows from that of B and W: after time changing, γ, the driving Brownian motion of R is orthogonal to β, hence these Brownian motions are independent.

Finally, since $\sqrt{R^2 + \beta^2}$ is transient (it is a three-dimensional Bessel process) and $\lim_{t \to +\infty} A_t = +\infty$, it is clear that the norm of the process $(R(A_t), \beta(A_t); t \geq 0)$ converges almost surely towards ∞ as $t \to +\infty$.

Note that instead of using equation (6.20.b), we might also have simply developed $\exp(B_t)$ with Itô's formula, and found R instead of R^2.

Solution to Exercise 6.21

To present the solution, we find it easier to work with the process $I_t \overset{\text{def}}{=} \int_0^t ds\, X_s$, rather than with $\overline{X}_t = \frac{1}{t} I_t$. Let f be a real valued function defined on $\mathbb{R}$ which is C^1, with compact support. Differentiating $f(I_t)$ with respect to t gives $\frac{d}{dt}\left(f(I_t)\right) = X_t f'(I_t)$, so that $f(I_t) = f(0) + \int_0^t ds\, X_s f'(I_s)$, and

$$E[f(I_t)] = f(0) + E\left[\int_0^t ds\, X_s f'(I_s)\right].$$

On the other hand, using the scaling property of the process X, we obtain that for fixed t, $f(I_t) \overset{(\text{law})}{=} f(t I_1) = f(0) + \int_0^t ds\, f'(s I_1) I_1$, hence:

$$E[f(I_t)] = f(0) + \int_0^t ds\, E[f'(s I_1) I_1].$$

Comparing the two expressions obtained for $E[f(I_t)]$ and again using the scaling property of X, we get, for every s:

$$E[X_s f'(I_s)] = E[I_1 f'(s I_1)] = \frac{1}{s} E[I_s f'(I_s)],$$

which yields the result.

Solution to Exercise 6.22

1. Suppose that there exists another process $(Y_t,\, t \geq 0)$ (rather than $(X'_t,\, t \geq 0)$) such that $X_t - \int_0^t ds\, Y_s$ is an $\mathcal{F}_t$-martingale. Subtracting $X_t - \int_0^t ds\, X'_s$, we see that $\int_0^t ds\,(X'_s - Y_s)$ is an $\mathcal{F}_t$-martingale. Since $\int_0^t ds\,(X'_s - Y_s)$ has finite variations, it necessarily vanishes. This proves that X'_t and Y_t are indistinguishable.

2. It is clear that formula (6.22.1) holds for $n = 1$. Suppose that formula (6.22.1) holds up to order $n - 1$. An integration by parts gives:

$$\frac{t^n}{n!} X_t^{(n)} = \int_0^t \frac{s^{n-1}}{(n-1)!} X_s^{(n)}\, ds + \int_0^t \frac{s^n}{n!}\, dX_s^{(n)}.$$

Since $dX_s^{(n)} = dM_s^{X^{(n)}} + X_s^{(n+1)} \, ds$, we have:

$$(-1)^{n-1} \int_0^t \frac{s^{n-1}}{(n-1)!} X_s^{(n)} \, ds = (-1)^{n-1} \left[\frac{t^n}{n!} X_t^{(n)} - \int_0^t \frac{s^n}{n!} \, dM_s^{X^{(n)}} - \int_0^t \frac{s^n}{n!} X_s^{(n+1)} \, ds \right].$$

$$(6.22.\text{a})$$

Plugging (6.22.a) into formula (6.22.1) taken at order $n-1$, we obtain (6.22.1) at order n.

3. We will prove a more general result. If (X_t) is a Markov process, taking values in $(E, \mathcal{E})$, its extended infinitesimal generator L is an operator acting on functions $f : E \longrightarrow \mathbb{R}$ such that $f(X_t)$ is $(\mathcal{F}_t)$ differentiable, where $\mathcal{F}_t = \sigma\{X_s, s \leq t\}$. Then one can show that $(f(X_t))' = g(X_t)$, for some function g. One denotes $g = Lf$ and $f \in D(L)$. Now it follows directly from the previous question that if $f \in D(L^{n+1})$, then:

$$f(X_t) - tLf(X_t) + \frac{t^2}{2} L^2 f(X_t) + \cdots + (-1)^n \frac{t^n}{n!} L^n f(X_t) - \int_0^t \frac{s^n}{n!} L^{n+1} f(X_s) \, ds$$

is a martingale.

The result of this question simply follows from the above property applied to Brownian motion whose infinitesimal generator L satisfies $Lf = \frac{1}{2} f''$, ($f \in C_c^2$) and to $f(x) = x^k$.

4. A formal proof goes as follows. Write, for $s < t$, the martingale property,

$$E\left[\exp\left(aB_t - \frac{a^2 t}{2} \right) \Big| \mathcal{F}_s \right] = \exp\left(aB_s - \frac{a^2 s}{2} \right),$$

and develop both sides as a series in a; the martingale property for each process $(H_k(B_t, t), \, t \geq 0)$ follows. Now, justify fully those arguments!

Comment. The notion of extended infinitesimal generator for a Markov process (X_t) which, to our knowledge, is due to

H. KUNITA: Absolute continuity of Markov processes and generators. *Nagoya Math. J.*, **36**, 1–26 (1969).

H. KUNITA: Absolute continuity of Markov processes. *Séminaire de Probabilités X*, 44–77, *Lecture Notes in Mathematics*, **511**, Springer, Berlin, 1976

is very convenient to compute martingales associated with (X_t), especially when the laws of X are characterized via a martingale problem, à la Stroock and Varadhan, for which the reader may consult:

D.W. STROOCK AND S.R.S. VARADHAN: *Multidimensional Diffusion Processes.*
Grundlehren der Mathematischen Wissenschaften, 233. Springer-Verlag, Berlin–
New York, (Second Edition), 1997.

Solution to Exercise 6.23

For every $n \in \mathbb{N}$, there exists a function φ_n of $(n+1)$ arguments, such that $M_n = \varphi_n(W_0, \ldots, W_n)$. From (ii), we have:

$$E[M_\gamma \,|\, m = j] = E\left[\varphi_{\sigma(j)}(W_0, W_1, \ldots, W_{\sigma(j)})\right]$$
$$= E[M_{\sigma(j)}] = E[M_0],$$

where the last equality follows from the optional stopping theorem. Consequently, we obtain:

$$E\left[M_\gamma \mathbb{1}_{\{m=j\}}\right] = E\left[E[M_\gamma \,|\, m = j]\mathbb{1}_{\{m=j\}}\right] = E[M_0]P(m = j).$$

Thus, a simple explanation of this result is that, once we condition with respect to $\{m = j\}$, γ becomes a stopping time.

Solution to Exercise 6.24

1. The equivalence between (6.24.1) and (6.24.2) is straightforward.

2. Since Brownian motion (which we shall denote here by X) is at the same time a Lévy process, and a Gaussian process, we know that there exists, for fixed s, t, T, two reals u and v such that

$$E[X_t \,|\, \mathcal{F}_{s,T}] = uX_s + vX_T,$$

which we write equivalently as

$$E[X_t - X_s \,|\, \mathcal{F}_{s,T}] = ((u - 1) + v)X_s + v(X_T - X_s). \tag{6.24.a}$$

From (6.24.a), we can deduce easily that

$$E[X_s(X_t - X_s)] = 0 = (u - 1) + v$$
$$E[(X_t - X_s)(X_T - X_s)] = t - s = v(T - s),$$

which yields the right values of u and v.

3. Any centered Markovian Gaussian process may be represented as $X_t = u(t)\beta_{v(t)}$, with β a Brownian motion and u and v two deterministic functions, v being a

monotone. Call $(\mathcal{B}_t)$ the natural filtration of β and assume that v is increasing (for simplicity). Then from question 1, we have

$$
\begin{aligned}
E[X_t \mid \mathcal{F}_{s,T}] &= u(t)E[\beta_{v(t)} \mid \mathcal{B}_{v(s),v(t)}]\\
&= u(t)\left(\frac{v(T)-v(t)}{v(T)-v(s)}\beta_{v(s)} + \frac{v(t)-v(s)}{v(T)-v(s)}\beta'_{v(T)}\right)\\
&= u(t)\left(\frac{v(T)-v(t)}{v(T)-v(s)}\frac{X_s}{u(s)} + \frac{v(t)-v(s)}{v(T)-v(s)}\frac{X(T)}{u(T)}\right),
\end{aligned}
$$

which yields the right values of α and β. Now, for X to be an affine process, we should have

$$
\frac{v(T)-v(t)}{v(T)-v(s)}\frac{u(t)}{u(s)} = \frac{T-t}{T-s}
$$

$$
\frac{v(t)-v(s)}{v(T)-v(s)}\frac{u(t)}{u(T)} = \frac{t-s}{T-s}.
$$

4. The identity presented in the hint is easily deduced from the following

$$
E[\exp\left(i(\lambda X_a + \mu X_b)\right)] = \exp(-a\psi(\lambda+\mu) - (b-a)\psi(\mu)),
$$

where ψ denotes the Lévy exponent associated with X, and $a < b$, $\lambda, \mu \in \mathbb{R}$. Then we have

$$
iE[X_a \exp(i\mu X_b)] = \frac{d}{d\lambda}E[\exp i(\lambda X_a + \mu X_b)]_{|\lambda=0} = -a\psi'(\mu)\exp(-b\psi(\mu)),
$$

which implies the desired result. It follows that

$$
E\left[\frac{X_a}{a} \,\middle|\, \mathcal{F}_b^+\right] = \frac{X_b}{b},
$$

and with the help of the homogeneity of the increments of X, we obtain:

$$
E\left[\frac{X_t - X_s}{t-s} \,\middle|\, \mathcal{F}_{s,T}\right] = \frac{X_T - X_s}{T-s},
$$

which shows that X is an affine process.

5. The proof is the same after replacing the characteristic function by the Laplace transform.

6. Assume that X is differentiable. By letting $t' \downarrow t$ in (6.24.2), we obtain $E[Y_t \mid \mathcal{F}_{s,u}]$, which does not depend on $t \in (s, u)$; but we also have $Y_s = \lim_{\varepsilon \to +0}\frac{X_s - X_{s-\varepsilon}}{\varepsilon}$, in L^1 and the same is true for Y_u. Thus Y_s and Y_u are $\mathcal{F}_{s,u}$-measurable, hence they are both equal to $\frac{X_u - X_s}{u-s}$; the proof of this question is now easily ended.

7. It is easily shown that the property (6.24.2) is satisfied for t and t' of the form $s + \frac{k}{n}(u-s)$, $0 \le k \le n$; hence, it also holds for all $t, t' \in [s, u]$, using the L^1 continuity of X.

Solution to Exercise 6.25

1. We first compute $E[M_u^{(t)} \mid \mathcal{F}_{a,b}] \overset{(\text{def})}{=} m(u,t)$, for $a < u < t < b$:

$$
\begin{aligned}
m(u,t) &= E\left[X_u - \int_0^u ds \frac{X_t - X_s}{t-s} \,\Big|\, \mathcal{F}_{a,b}\right] = X_a + E[X_u - X_a \mid \mathcal{F}_{a,b}] \\
&\quad - \int_0^a ds \frac{E[X_t - X_s \mid \mathcal{F}_{a,b}]}{t-s} - \int_a^u ds\, E\left[\frac{X_t - X_s}{t-s} \,\Big|\, \mathcal{F}_{a,b}\right] \\
&= X_a - \frac{X_b - X_a}{b-a}(u-a) - \int_0^a \frac{ds}{t-s}\big((X_a - X_s) \\
&\quad + E[X_t - X_a \mid \mathcal{F}_{a,b}]\big) - \int_a^u ds \frac{X_b - X_a}{b-a} \\
&= X_a - \int_0^a ds \frac{X_a - X_s}{t-s} - \int_0^a \frac{ds}{t-s}\frac{t-a}{b-a}(X_b - X_a). \qquad (6.25.\text{a})
\end{aligned}
$$

When $t = b$, one easily checks that this expression is reduced to $M_a^{(b)}$, so that $(M_t^{(T)})$ is a martingale with respect to the variable t.

2. We want to show that, for $s_0 < s < t < T$,

$$
E\left[\frac{X_t - X_s}{t-s} \,\Big|\, \mathcal{F}_{s_0,T}\right] = \frac{X_t - X_{s_0}}{T - s_0}. \qquad (6.25.\text{b})
$$

But it suffices to obtain:

$$
E\left[\frac{X_t - X_s}{t-s} \,\Big|\, \mathcal{F}_{s,T}\right] = \frac{X_t - X_s}{T-s}. \qquad (6.25.\text{c})
$$

Indeed, if (6.25.c) holds then for any triplet $s < t < T$, one can write

$$
E\left[\frac{X_t - X_s}{t-s} \,\Big|\, \mathcal{F}_{s_0,T}\right] = E\left[\frac{X_T - X_s}{T-s} \,\Big|\, \mathcal{F}_{s_0,T}\right], \qquad (6.25.\text{d})
$$

since $\mathcal{F}_{s_0,T} \subset \mathcal{F}_{s,T}$. Then the right hand side of (6.25.d) is equal to:

$$
\begin{aligned}
\frac{X_T - X_{s_0}}{T-s} + E\left[\frac{X_{s_0} - X_s}{T-s} \,\Big|\, \mathcal{F}_{s_0,T}\right] &= \frac{X_T - X_{s_0}}{T-s} - \frac{1}{T-s}(X_T - X_{s_0})\frac{s - s_0}{T - s_0} \\
&= \frac{X_T - X_{s_0}}{T - s_0},
\end{aligned}
$$

which proves (6.25.b).

Now, we prove that (6.25.c) follows from the hypothesis of question 2. To this aim, write:

$$
X_t - X_s = M_t^{(T)} - M_s^{(T)} + \int_s^t du \frac{X_T - X_u}{T - u}.
$$

We deduce from this identity that:

$$E[X_t - X_s \mid \mathcal{F}_{s,T}] = \int_s^t du\, E\left[\frac{X_T - X_u}{T - u} \mid \mathcal{F}_{s,T}\right]$$

$$= \int_s^t \frac{du}{T - u}(X_T - X_s)$$

$$- \int_s^t \frac{du}{T - u} E[X_u - X_s \mid \mathcal{F}_{s,T}].$$ (6.25.e)

Fix s and T and put $\Phi(t) = E[X_t - X_s \mid \mathcal{F}_{s,T}]$ and $a(t) = (X_T - X_s)\int_s^t \frac{du}{T-u}$. From (6.25.e), we have

$$\Phi(t) = a(t) - \int_s^t \frac{du}{T - u}\Phi(u).$$ (6.25.f)

This is a first-order linear equation. Moreover, we have $\Phi(s) = 0$. Thus, the unique solution is given by

$$\Phi(t) = (t - s)\frac{X_T - X_s}{T - s}.$$ (6.25.g)

Joint solution to Exercises 6.26 and 6.27

We note that it suffices to prove Exercise **6.27**, since then Exercise **6.26** will follow, by considering

$$H_t = \int_0^t du\, \alpha(u).$$

To prove Exercise **6.27**, we consider $s < t < u$, and we write

$$\alpha(s) = E\left[\frac{H(u) - H(s)}{u - s} \mid \mathcal{F}_s\right]$$

$$= E\left[E\left[\frac{H(u) - H(s)}{u - s} \mid \mathcal{F}_t\right] \mid \mathcal{F}_s\right]$$

$$= \frac{u - t}{u - s}E[\alpha(t) \mid \mathcal{F}_s] + \alpha(s)\frac{t - s}{u - s}.$$

Comparing the extreme terms, we obtain

$$\alpha(s) = E[\alpha(t) \mid \mathcal{F}_s].$$

To prove question 2 of Exercise **6.27**, write $\alpha(s) = E\left[\frac{1}{t-s}\int_s^t du\,\alpha_u \mid \mathcal{F}_s\right]$, hence $E[H_t - H_s \mid \mathcal{F}_s] = E\left[\int_s^t du\,\alpha_u \mid \mathcal{F}_s\right]$, which allows us to deduce that $H_t - \int_0^t du\,\alpha_u$ is a martingale.

Solution to Exercise 6.28

1. Let $c < a < b < d$, then applying the Harness relation (6.25.1) which is recalled at the beginning of Exercise **6.25**, we obtain

$$E(M_{a,b} \mid \mathcal{F}_{c,d}) = \int_0^c \varphi_-(s)\, dB_s + \int_d^\infty \varphi_+(s)\, dB_s + C_{a,b} \frac{b-a}{d-c}(B_d - B_c)$$

$$+ E\left(\int_c^a \varphi_-(s)\, dB_s \mid \mathcal{F}_{c,d}\right) + E\left(\int_b^d \varphi_+(s)\, dB_s \mid \mathcal{F}_{c,d}\right).$$

Using the fact that $\beta_s^{(d)} = B_s - \int_0^s du\, \frac{B_d - B_u}{d-u}$, $s \le d$ is a Brownian motion in the filtration $(\mathcal{F}_s \vee \sigma(B_d))$, we may write

$$E\left(\int_c^a \varphi_-(s)\, dB_s \mid \mathcal{F}_{c,d}\right) = E\left(\int_c^a \varphi_-(s)\, ds\, \frac{B_d - B_s}{d-s} \mid \mathcal{F}_c \vee \sigma(B_d)\right)$$

$$= \left(\int_c^a \varphi_-(s)\, ds\right) \frac{B_d - B_c}{d-c}.$$

Doing the same for $E\left(\int_b^d \varphi_+(s)\, dB_s \mid \mathcal{F}_{c,d}\right)$, we have

$$E(M_{a,b} \mid \mathcal{F}_{c,d}) = \int_0^c \varphi_-(s)\, dB_s + \int_d^\infty \varphi_+(s)\, dB_s + C_{a,b} \frac{b-a}{d-c}(B_d - B_c)$$

$$+ \left(\int_c^a \varphi_-(s)\, ds\right) \frac{B_d - B_c}{d-c} + \left(\int_b^d \varphi_-(s)\, ds\right) \frac{B_d - B_c}{d-c}.$$

So, for $M_{a,b}$ to be a $\mathcal{F}_{a,b}$-martingale we must have

$$(b-a)C_{a,b} + \int_c^a \varphi_-(s)\, ds + \int_b^d \varphi_+(s)\, ds = (d-c)C_{c,d}.$$

This identity may also be written as follows

$$(b-a)C_{a,b} + \int_0^a \varphi_-(s)\, ds + \int_b^\infty \varphi_+(s)\, ds = (d-c)C_{c,d} + \int_0^c \varphi_-(s)\, ds + \int_d^\infty \varphi_+(s)\, ds.$$

Hence the left hand side of this equality does not depend on (a, b). Let us denote this value by C. It follows from the above expression that

$$C_{a,b} = \frac{C}{b-a} - \frac{1}{b-a}\left(\int_0^a \varphi_-(s)\, ds + \int_b^\infty \varphi_+(s)\, ds\right),$$

so that $M_{a,b}$ can be written as follows

$$M_{a,b} = \int_0^a \varphi_-(s)\, dB_s + \int_b^\infty \varphi_+(s)\, dB_s + \frac{C}{b-a}(B_b - B_a)$$

$$- \frac{B_b - B_a}{b-a}\left(\int_0^a \varphi_-(s)\, ds + \int_b^\infty \varphi_+(s)\, ds\right).$$

2. First write

$$E\left(\exp\left(\int_0^\infty \varphi(s)\,dB_s\right) \mid \mathcal{F}_{a,b}\right) = \qquad\qquad (6.28.a)$$

$$\exp\left(\int_0^a \varphi(s)\,dB_s + \int_b^\infty \varphi(s)\,dB_s\right) E\left(\exp\left(\int_a^b \varphi(s)\,dB_s\right) \mid \mathcal{F}_{a,b}\right),$$

and define Φ by

$$\int_a^b \varphi(s)\,dB_s = \frac{B_b - B_a}{b - a}\int_a^b \varphi(s)\,ds + \Phi.$$

Then since Φ is a Gaussian variable that is independent from $\mathcal{F}_{a,b}$, we obtain

$$E\left(\exp\left(\int_a^b \varphi(s)\,dB_s\right) \mid \mathcal{F}_{a,b}\right) = \exp\left(\frac{B_b - B_a}{b - a}\int_a^b \varphi(u)\,du\right)\exp\left(\frac{1}{2}E\left[\Phi^2\right]\right).$$

But from the definition of Φ, we have

$$E\left[\Phi^2\right] = \int_a^b \varphi^2(s)\,ds - \frac{1}{b - a}\left(\int_a^b \varphi(s)\,ds\right)^2,$$

so that with (6.28.a), we obtain

$$E\left(\exp\left(\int_0^\infty \varphi(s)\,dB_s\right) \mid \mathcal{F}_{a,b}\right)$$

$$= \exp\left(\int_0^a \varphi(s)\,dB_s + \int_b^\infty \varphi(s)\,dB_s + C_{a,b}^\varphi(B_b - B_a) + D_{a,b}^\varphi\right),$$

where

$$C_{a,b}^\varphi = \frac{1}{b - a}\int_a^b \varphi(s)\,ds \quad\text{and}$$

$$D_{a,b}^\varphi = \frac{1}{2}\left(\int_a^b \varphi^2(s)\,ds - \frac{1}{b - a}\left(\int_a^b \varphi(s)\,ds\right)^2\right).$$

3. Let $c < a < b < d$ and define Φ_- and Φ_+ by

$$\int_c^a \varphi_-(s)\,dB_s = \frac{B_d - B_c}{d - c}\int_c^a \varphi_-(s)\,ds + \Phi_- \qquad (6.28.b)$$

$$\int_b^d \varphi_-(s)\,dB_s = \frac{B_d - B_c}{d - c}\int_b^d \varphi_+(s)\,ds + \Phi_+. \qquad (6.28.c)$$

Note that, from this definition, Φ_- and Φ_+ are Gaussian variables, independent from $\mathcal{F}_{c,d}$ (and so, from $B_d - B_c$). On the other hand, from the Harness property of B,

$$B_b - B_a = \frac{b - a}{d - c}(B_d - B_c) + \Psi, \qquad (6.28.d)$$

where Ψ is independent from $\mathcal{F}_{c,d}$. Then let us write,

$$E\left[\mathcal{E}_{a,b} \mid \mathcal{F}_{c,d}\right] = E\left[\exp\left(\int_0^a \varphi_-(s)\,dB_s + \int_b^\infty \varphi_+(s)\,dB_s + C_{a,b}^{\hat\varphi}(B_b - B_a) + D_{a,b}^{\hat\varphi}\right) \mid \mathcal{F}_{c,d}\right]$$
$$= I \times II, \qquad\qquad (6.28.\text{e})$$

where

$$I = \exp\left(D_{a,b}^{\hat\varphi} + \int_0^c \varphi_-(s)\,dB_s + \int_d^\infty \varphi_+(s)\,dB_s\right)$$

and

$$II = E\left[\exp\left(\int_c^a \varphi_-(s)\,dB_s + \int_b^d \varphi_+(s)\,dB_s + C_{a,b}^{\hat\varphi}(B_b - B_a)\right) \mid \mathcal{F}_{c,d}\right].$$

Now using (6.28.b), (6.28.c) and (6.28.d), we obtain:

$$II = III \times IV,$$

where

$$III = \exp\left[\frac{B_d - B_c}{d - c}\left(\int_c^a \varphi_-(s)\,ds + \int_b^d \varphi_+(s)\,ds + C_{a,b}^{\hat\varphi}(b - a)\right)\right],$$

and

$$IV = \exp\left(\frac{1}{2}E\left[(\Phi_- + \Phi_+ + \Psi)^2\right]\right).$$

It now remains to compute $E\left[(\Phi_- + \Phi_+ + \Psi)^2\right]$. To this aim, let us set:

$$X = \int_c^a \varphi_-(s)\,dB_s, \quad Y = \int_b^d \varphi_-(s)\,dB_s, \quad Z = B_b - B_a, \quad T = \frac{B_d - B_c}{d - c}$$
$$x = \int_c^a \varphi_-(s)\,ds, \quad y = \int_b^d \varphi_+(s)\,ds, \quad z = b - a.$$

Then from (6.28.b), (6.28.c) and (6.28.d), we may state:

$$\begin{cases} X = xT + \Phi_- \\ Y = yT + \Phi_+ \\ Z = zT + \Psi \end{cases},$$

where X, Y, and Z are independent. These equalities correspond to the orthogonal decomposition of X, Y and Z with respect to T, so that we may write:

$$E\left[(\Phi_- + \Phi_+ + \Psi)^2\right] = E[X^2 + Y^2 + Z^2] - (x + y + z)^2 E[T^2].$$

Thus, we have obtained:

$$E\left[(\Phi_- + \Phi_+ + \Psi)^2\right] = \int_c^a \varphi_-(s)^2\,ds + \int_b^d \varphi_+(s)^2\,ds + (b - a)^2 -$$
$$\frac{1}{d - c}\left(\int_c^a \varphi_-(s)\,ds + \int_b^d \varphi_+(s)\,ds + b - a\right)^2.$$

Plugging these expressions in (6.28.e), yields:

$$E\left[\mathcal{E}_{a,b} \,|\, \mathcal{F}_{c,d}\right] = I \times III \times IV$$
$$= \exp\left[D_{a,b}^{\hat{\varphi}} + \frac{1}{2}\int_c^a \varphi_-(s)^2\,ds + \frac{1}{2}\int_b^d \varphi_+(s)^2\,ds + \frac{1}{2}(b-a)^2\right.$$
$$-\frac{1}{2(d-c)}\left(\int_c^a \varphi_-(s)\,ds + \int_b^d \varphi_+(s)\,ds + b - a\right)^2$$
$$+\int_0^c \varphi_-(s)\,dB_s + \int_d^\infty \varphi_+(s)\,dB_s$$
$$\left.+\frac{B_d - B_c}{d-c}\left(\int_c^a \varphi_-(s)\,ds + \int_b^d \varphi_+(s)\,ds + C_{a,b}^{\hat{\varphi}}(b-a)\right)\right],$$

so that

$$C_{c,d}^{\hat{\varphi}} = \frac{1}{d-c}\left(\int_c^a \varphi_-(s)\,ds + \int_b^d \varphi_+(s)\,ds + C_{a,b}^{\hat{\varphi}}(b-a)\right)$$
$$D_{c,d}^{\hat{\varphi}} = D_{a,b}^{\hat{\varphi}} + \frac{1}{2}\int_c^a \varphi_-(s)^2\,ds + \frac{1}{2}\int_b^d \varphi_+(s)^2\,ds + \frac{1}{2}(b-a)^2$$
$$-\frac{1}{2(d-c)}\left(\int_c^a \varphi_-(s)\,ds + \int_b^d \varphi_+(s)\,ds + b - a\right)^2.$$

In conclusion, by taking $a = b$, we find:

$$C_{c,d}^{\hat{\varphi}} = \frac{1}{d-c}\left(\int_c^d (\varphi_-(s) + \varphi_+(s))\,ds\right)$$
$$D_{c,d}^{\hat{\varphi}} = \frac{1}{2}\left[\int_c^d (\varphi_-(s)^2 + \varphi_+(s)^2)\,ds - \frac{1}{d-c}\left(\int_c^d (\varphi_-(s) + \varphi_+(s))\,ds\right)^2\right],$$

and this choice of $C_{a,b}^{\hat{\varphi}}$ and $D_{a,b}^{\hat{\varphi}}$ ensures that $(\mathcal{E}_{a,b})$ is a $\mathcal{F}_{a,b}$-martingale.

Solution to Exercise 6.29

1. We deduce from (6.29.1) that, for any simple function $\varphi : \mathbb{R}_+ \to \mathbb{R}$, one has:

$$\int \varphi(u)\,dB_u \overset{\text{(law)}}{=} \int \varphi(u)\,dM_u^{(1)} + \cdots + \int \varphi(u)\,dM_u^{(k)}.$$

From Cramer's theorem about additive decompositions of a Gaussian variable,

$$\int \varphi(u)\,dM_u^{(i)}, \quad i = 1, 2, \ldots, k$$

are independent Gaussian variables. Hence, for $i = 1, 2, \ldots, k$ $(M_t^{(i)}, t \geq 0)$ is a Gaussian process. Furthermore, there is the inequality: $E\left[(M_t^{(i)} - M_s^{(i)})^2\right] \leq t - s$. From the Gaussian character of $M^{(i)}$, this reinforces as:

$$E\left[(M_t^{(i)} - M_s^{(i)})^{2m}\right] \leq g_m(t-s)^m,$$

with $g_m = E[G^{2m}]$. Hence, $M^{(i)}$ admits a continuous version (from Kolmogorov's theorem); in fact $M^{(i)}$ is "as continuous as Brownian motion".

Considering now the increasing processes of the $M^{(i)}$'s, we obtain, again as a consequence of (6.29.1):

$$t = \sum_{i=1}^{k} \langle M^{(i)} \rangle_t. \tag{6.29.a}$$

Now, since $M^{(i)}$ is a Gaussian process, $\langle M^{(i)} \rangle_t$ is a deterministic function, in fact, from (6.29.a), there exists a deterministic function $(m_s^{(i)}, s \geq 0)$ such that:

$$\langle M^{(i)} \rangle_t = \int_0^t ds \, (m_s^{(i)})^2,$$

and from (6.29.a), we deduce: $1 = \sum_{i=1}^{k} (m_s^{(i)})^2$, ds-a.e.

2. (a) It is well known that if $W^{(1)}$ and $W^{(2)}$ are two independent Brownian motions, then so are $(W^{(1)} + W^{(2)})/\sqrt{2}$ and $(W^{(1)} - W^{(2)})/\sqrt{2}$. Hence, (6.29.1) is satisfied with $M^{(1)} = (W^{(1)} + W^{(2)})/2$, $M^{(2)} = (W^{(1)} - W^{(2)})/2$ $B^{(1)} = (W^{(1)} + W^{(2)})/\sqrt{2}$, $B^{(2)} = (W^{(1)} - W^{(2)})/\sqrt{2}$ and $m^{(1)} = m^{(2)} \equiv 1/\sqrt{2}$. In particular, (6.29.2) is not satisfied.

2. (b) For $i \neq j$, since $\langle M^{(i)}, M^{(j)} \rangle = \langle N^{(i)}, B \rangle \equiv 0$, a.s., we obtain $\langle M^{(i)}, M^{(j)} \rangle_t = \int_0^t \mu_s^{(i)} \mu_s^{(j)} ds + \langle N^{(i)}, N^{(j)} \rangle_t = 0$, so that

$$\langle N^{(i)}, N^{(j)} \rangle_t = - \int_0^t \mu_s^{(i)} \mu_s^{(j)} ds.$$

3. From an application of Itô formula to $\log(L_t)$, we derive that each strictly positive martingale $L^{(i)}$ may be written as $L^{(i)} = \exp\left(M^{(i)} - \frac{1}{2}\langle M^{(i)} \rangle\right)$, where $M_t^{(i)} = \int_0^t \frac{dL_s^{(i)}}{L_s^{(i)}}$. Moreover from (i), (ii) and (iii), the martingales $M^{(1)}, M^{(2)}, \ldots, M^{(k)}$ are independent and satisfy $M_0^{(1)} = \cdots = M_0^{(k)} = 0$ and $M_t^{(1)} + M_t^{(2)} + \ldots + M_t^{(k)} = B_t$, so the martingales $M^{(i)}$ (and hence $L^{(i)}$) may be characterized through question 1. (a).

4. We shall solve this question under the following restrictive assumptions: assume that (M_t) and (N_t) are independent martingales with respect to the natural filtration of B and that $M_t N_t = B_t$. Then there exist a real c (possibly, a priori, equal to 0) and two predictable processes (m_s) and (n_s) such that:

$$M_t = c + \int_0^t m_s \, dB_s, \quad N_t = \int_0^t n_s \, dB_s. \tag{6.29.b}$$

Furthermore, we deduce from our assumptions that:

$$E[M_t^2] < \infty \quad \text{and} \quad E[N_t^2] < \infty, \text{ for all } t \geq 0.$$

Consequently, (m_s) and (n_s) satisfy:

$$E\left[\int_0^t ds\, m_s^2\right] < \infty \quad \text{and} \quad E\left[\int_0^t ds\, n_s^2\right] < \infty, \tag{6.29.c}$$

for all $t \geq 0$.

Since (M_t) and (N_t) are independent, they are, a fortiori, orthogonal, hence:

$$m_s n_s = 0, \quad ds\, dP \text{ a.e.} \tag{6.29.d}$$

Furthermore, the increasing processes $(\langle M \rangle_t)$ and $(\langle N \rangle_t)$ are also independent and so are (m_s^2) and (n_s^2). Consequently, we deduce from (6.29.d) that:

$$E[m_s^2]E[n_s^2] = 0, \quad ds \text{ a.e.} \tag{6.29.e}$$

Now, introduce the deterministic sets:

$$\mathcal{P}_m \overset{\text{def}}{=} \{s : P(m_s \neq 0) > 0\} = \{s : E(m_s^2) > 0\} \text{ and } \mathcal{O}_m \overset{\text{def}}{=} \{s : P(m_s = 0) = 1\},$$

as well as $\mathcal{P}_n$ and $\mathcal{O}_n$. It follows from (6.29.e) that if $s \in \mathcal{P}_m$, then $s \in \mathcal{O}_n$, i.e. $\mathcal{P}_m \subset \mathcal{O}_n$, and by symmetry, $\mathcal{P}_n \subset \mathcal{O}_m$.

Developing $M_t N_t$ with Itô's formula and using (6.29.b), we obtain:

$$M_s n_s + N_s m_s = 1, \quad ds\, dP \text{ a.e.} \tag{6.29.f}$$

We now take $s \in \mathcal{P}_m$, and later $s \in \mathcal{P}_n$. If $s \in \mathcal{P}_m$, then $s \in \mathcal{O}_n$ and (6.29.f) becomes:

$$N_s m_s = 1, \quad \text{hence} \quad N_s^2 m_s^2 = 1, \quad dP \text{ a.s.} \tag{6.29.g}$$

But since N_s^2 and m_s^2 are independent, (6.29.g) can only hold if N_s^2 and m_s^2 are constant, i.e.: $N_s^2 = C_s$ and $m_s^2 = 1/C_s$. Likewise, if $s \in \mathcal{P}_n$, then $M_s^2 = D_s$ and $n_s^2 = 1/D_s$.

Consequently, the integral representations (6.29.b) for (M_t) and (N_t) take the form:

$$M_t = c + \int_0^t \frac{\mathbb{1}_{\{s \in \mathcal{P}_m\}}}{\sqrt{C_s}} \mu_s\, dB_s, \quad N_t = \int_0^t \frac{\mathbb{1}_{\{s \in \mathcal{P}_n\}}}{\sqrt{D_s}} \nu_s\, dB_s,$$

with (μ_s) and (ν_s) two predictable processes valued in $\{-1, +1\}$. In particular, (M_t) and (N_t) are two (independent) Gaussian martingales.

It now follows from the equality: $B_t = M_t N_t$, by a simple Fourier-Laplace computation that: $M_t^2 E(N_t^2) = t$, hence M_t must be a constant, this constant is equal to c (as in (6.29.b)) which must be different from 0. Finally it turns out that almost surely, $\mathcal{P}_m$ is empty; $\mathcal{P}_n = \mathbb{R}_+$; $\nu_s \equiv 1$; $D_s = c^2$. The proof is ended.

Nonetheless, we do not see how to study the more general equation with: $B_t \overset{(\text{law})}{=} M_t N_t$, and M and N two independent martingales.

Where is the notion N discussed ?

In this book	In the literature
Monotone Class Theorem Chapter 1	Meyer ([60], T19, T20; pp. 27, 28); Dacunha-Castelle and Duflo ([21], chapter 3)
Uniform integrability Ex. 1.3, Ex. 1.4	Meyer ([60], pp. 35–40); Durrett ([26], section 4.5); Grimmett and Stirzaker ([38], pp. 353)
Convergence of r.v.s Chapter 5	Fristedt and Gray ([33], chapter 12); Billingsley [6]; Chung [17]
Independence Chapter 2	Dacunha-Castelle and Duflo ([21], chapter 5); Kac [45]
Conditioning Chapter 2	Meyer ([60], chapter 2, section 4); Williams [98]; Fristedt and Gray ([33], chapter 21, chapter 23)
Gaussian space Chapter 3	Neveu [64]; Janson [41]; Lifschits [54]
Laws of large numbers Chapter 5	Williams ([97], p. 103); Durrett ([26], chapter 1); Feller ([28], chapter VII)
Central Limit Theorems Ex. 5.9	Williams ([97], p. 156); Durrett ([26], chapter 2); Feller ([28], chapter VIII)
Large deviations Ex. 5.6	Toulouse ([93], chapter 3); Azencott (see ex. 3.11); Durrett ([26], 1.9)
Characteristic functions Chapter 4, Chapter 5	Lukacs [55]; Williams ([97], p. 166); Fristedt and Gray ([33], chapter 13)
Laplace transform Chapter 4	Feller ([28], chapter VII, 6) Chung ([17], 66); Meyer [60]
Mellin transform Ex. 1.14, Ex. 4.23	Zolotarev [103]; Widder [96]; Patterson [66]
Infinitely divisible laws Ex. 1.13, Ex. 5.12	Fristedt and Gray ([33], chapter 16); Feller ([28], chapter 6); Durrett ([26], 2.8)
Stable laws Ex. 4.19 to 4.21, Ex. 5.12	Zolotarev [103]; Feller [28] Fristedt and Gray ([33], chapter 17);
Domains of attraction Ex. 5.17	Feller ([28], chapter IX); Petrov [68]
Ergodic Theorems Ex. 1.9	Durrett ([26], chapter 6); Billingsley (see ex. 1.9)
Martingales Ex. 1.6, Chapter 6	Baldi, Mazliak and Priouret [2]; Neveu ([62], section B); Williams [98]; Grimmett and Stirzaker ([38], section 7.7, 7.8, chapter 12)

Final suggestions: how to go further ?

The reader who has had the patience and/or interest to remain with us until now may want to (and will certainly) draw some conclusions from this enterprise.... If you are really "hooked" (in French slang, "accro"!) on exercises, counterexamples, etc., you may get into [2], [19], [22], [25], [30], [37], [52], [78],[86], [90].

We would like to suggest two complementary directions.

(i) When, in the middle of a somewhat complex research program, involving a highly sophisticated probability model, it is often comforting to check one's assertions by "coming down" to some consequences involving only one – (or finite) – dimensional random variables. As explained in the foreword of this book, most of our exercises have been constructed in this way, mainly by "stripping" the Brownian set-up.

(ii) The converse attitude may also be quite fruitful; namely to view a one-dimensional model as embedded in an infinitely-dimensional one. An excellent example of the gains one might draw in "looking at the big picture" is Itô's theory of (Brownian) excursions: jointly (as a process of excursions), they constitute a big Poisson process; this theory allowed us to recover most of Lévy's results for the individual excursion, and indeed many more....

To illustrate, here is a discussion relative to the arc-sine law of P. Lévy: in his famous 1939 paper, P. Lévy noticed that the law of $\frac{1}{t} \int_0^t ds\, \mathbb{I}_{\{B_s > 0\}}$, for fixed t, is also the same as that of $\frac{1}{\tau_h} \int_0^{\tau_h} ds\, \mathbb{I}_{\{B_s > 0\}}$, where, for fixed h, $\tau_h = \inf\{t : l_t > h\}$, $h \geq 0$, is the inverse of the local time process l. This striking remark motivated Pitman and Yor to establish the following infinite-dimensional reinforcement of Lévy's remark: for fixed t and h, both sequences: $\frac{1}{t}(M_1(t), \ldots, M_n(t), \ldots)$ and $\frac{1}{\tau_h}(M_1(\tau_h), \ldots, M_n(\tau_h), \ldots)$ have the same distribution, where $M_1(t) > M_2(t) > \ldots$ is the decreasing sequence of lengths of Brownian excursions over the interval $(0, t)$.

We could not find any better way to conclude on this topic, and with this book, than by simply reproducing one sentence in Professor Itô's Foreword to his *Selected Papers* (Springer, 1987):

"After several years, it became my habit to observe even finite-dimensional facts from the infinite-dimensional viewpoint."

So, let us try to imitate Professor Itô !!

References

[1] G.E. ANDREWS, R. ASKEY AND R. ROY: *Special Functions*. Encyclopedia of Mathematics and its Applications, 71. Cambridge University Press, Cambridge, 1999.

[2] P. BALDI, L. MAZLIAK AND P. PRIOURET: *Martingales et Chaînes de Markov*. Collection Méthodes, Hermann, 1998. English version: Chapman & Hall/CRC, Oxford, 2002.

[3] PH. BARBE AND M. LEDOUX: *Probabilité*. De la Licence à l'Agrégation. Éditions Espaces 34, Belin, 1998.

[4] O.E. BARNDORFF-NIELSEN AND A. SHIRYAEV: *Change of Time and Change of Measure*. Advanced Series on Statistical Science & Applied Probability, 13. World Scientific Publishing Co. Pte. Ltd., Hackensack, NJ, 2010.

[5] J. BERTOIN: *Lévy Processes*. Cambridge University Press, Cambridge, 1996.

[6] P. BILLINGSLEY: *Convergence of Probability Measures*. Second edition. John Wiley & Sons, Inc., New York, 1999.

[7] P. BILLINGSLEY: *Probability and Measure*. Third edition. John Wiley & Sons, Inc., New York, 1995.

[8] P. BILLINGSLEY: *Ergodic Theory and Information*. Reprint of the 1965 original. Robert E. Krieger Publishing Co., Huntington, NY, 1978.

[9] N.H. BINGHAM, C.M. GOLDIE AND J.L. TEUGELS: *Regular Variation*. Cambridge University Press, Cambridge, 1989.

[10] A.N. BORODIN AND P. SALMINEN: *Handbook of Brownian Motion – Facts and Formulae*. Second edition. Probability and its Applications. Birkhäuser Verlag, Basel, 2002. Third edition in preparation (September 2011).

[11] M. BRANCOVAN AND TH. JEULIN: *Probabilités – Niveau M1*, Vol. 2, Ellipses, in preparation (September 2011).

[12] M. BRANCOVAN AND TH. JEULIN: *Probabilités – Niveau M1,* Ellipses, 2006.

[13] L. BREIMAN: *Probability.* Addison-Wesley Publishing Company, Reading, Mass.-London-Don Mills, Ont., 1968.

[14] P. BRÉMAUD: *An Introduction to Probabilistic Modeling.* Corrected reprint of the 1988 original. Springer-Verlag, New York, 1994.

[15] L. CHAUMONT: *An Introduction to Self-similar Processes.* In preparation (March 2012).

[16] Y.S. CHOW AND H. TEICHER: *Probability Theory. Independence, Interchangeability, Martingales.* Third edition. Springer-Verlag, New York, 1997.

[17] K.L. CHUNG: *A Course in Probability Theory.* Third edition. Academic Press, Inc., San Diego, CA, 2001.

[18] E. ÇINLAR: *Probability and Stochastics.* Graduate Texts in Mathematics, 261. Springer, New York, 2011.

[19] D. DACUNHA-CASTELLE, M. DUFLO AND V. GENON-CATALOT: *Exercices de Probabilités et Statistiques. Tome 2. Problèmes à temps mobile.* Masson, 1984.

[20] D. DACUNHA-CASTELLE AND M. DUFLO: *Probabilités et Statistiques. Tome 2. Problèmes à temps mobile.* Masson, Paris, 1983.

[21] D. DACUNHA-CASTELLE AND M. DUFLO: *Probabilités et Statistiques. Tome 1. Problèmes à temps fixe.* Masson, Paris, 1982.

[22] D. DACUNHA-CASTELLE, D. REVUZ AND M. SCHREIBER: *Recueil de Problèmes de Calcul des Probabilités.* Deuxième édition. Masson, Paris, 1970.

[23] C. DELLACHERIE AND P.A. MEYER: *Probabilités et Potentiel.* Chapitres I à IV. Hermann, Paris, 1975.

[24] J.L. DOOB: *Stochastic Processes.* A Wiley-Interscience Publication. John Wiley & Sons, Inc., New York, 1990.

[25] A.Y. DOROGOVTSEV, D.S. SILVESTROV, A.V. SKOROKHOD AND M.I. YADRENKO: *Probability Theory: Collection of Problems.* Translations of Mathematical Monographs, 163. American Mathematical Society, Providence, RI, 1997.

[26] R. DURRETT: *Probability: Theory and Examples.* Fourth edition. Cambridge University Press, Cambridge, 2010.

[27] P. EMBRECHTS AND M. MAEJIMA: *Selfsimilar Processes*. Princeton University Press, Princeton, NJ, 2002.

[28] W. FELLER: *An Introduction to Probability Theory and its Applications*. Vol. II. Second edition. John Wiley & Sons, Inc., New York–London–Sydney, 1971.

[29] X. FERNIQUE: *Fonctions Aléatoires Gaussiennes, Vecteurs Aléatoires Gaussiens*. Université de Montréal, Centre de Recherches Mathématiques, Montréal, QC, 1997.

[30] D. FOATA AND A. FUCHS: *Processus Stochastiques, Processus de Poisson, Chaînes de Markov et Martingales. Cours et exercices corrigés*. Dunod, Paris, 2002.

[31] D. FOATA AND A. FUCHS: *Calcul des Probabilités*. Second edition. Dunod, 1998.

[32] D. FREEDMAN: *Brownian Motion and Diffusion*. Second edition. Springer-Verlag, New York–Berlin, 1983.

[33] B. FRISTEDT AND L. GRAY: *A Modern Approach to Probability Theory*. Probability and its applications. Birkhäuser Boston, Inc., Boston, MA, 1997.

[34] L. GALLARDO: *Mouvement Brownien et Calcul d'Itô – Cours et Exercices Corrigés*. Hermann, 2008.

[35] H.-O. GEORGII: *Gibbs Measures and Phase Transitions*. Second edition. de Gruyter Studies in Mathematics, 9. Walter de Gruyter & Co., Berlin, 2011.

[36] G.R. GRIMMETT: *Probability on Graphs. Random Processes on Graphs and Lattices*. Cambridge University Press, Cambridge, 2010.

[37] G.R. GRIMMETT AND D.R. STIRZAKER: *One Thousand Exercises in Probability*. The Clarendon Press, Oxford University Press, New York, 2001.

[38] G.R. GRIMMETT AND D.R. STIRZAKER: *Probability and Random Processes*. Second edition. The Clarendon Press, Oxford University Press, New York, 1992.

[39] T. HIDA AND M. HITSUDA: *Gaussian processes*. Translations of Mathematical Monographs, 120. American Mathematical Society, Providence, RI, 1993.

[40] J. JACOD AND PH. PROTTER: *Probability Essentials*. Second edition. Universitext. Springer-Verlag, Berlin, 2002.

[41] S. JANSON: *Gaussian Hilbert Spaces*. Cambridge Tracts in Mathematics, 129. Cambridge University Press, Cambridge, 1997.

[42] N.L. JOHNSON, S. KOTZ AND N. BALAKRISHNAN: *Continuous Univariate Distributions.* Vol. 2. Second edition. John Wiley & Sons, Inc., New York, 1995.

[43] M. KAC: *Statistical Independence in Probability, Analysis and Number Theory.* The Carus Mathematical Monographs, No. 12. Distributed by John Wiley and Sons, Inc., New York 1959.

[44] O. KALLENBERG: *Foundations of Modern Probability.* Second edition. Probability and its Applications. Springer-Verlag, New York, 2002.

[45] G. KALLIANPUR: *Stochastic Filtering Theory.* Applications of Mathematics, 13. Springer-Verlag, New York-Berlin, 1980.

[46] I. KARATZAS AND S.N. SHREVE: *Brownian Motion and Stochastic Calculus.* Second edition. Springer-Verlag, New York, 1991.

[47] D. KHOSHNEVISAN: *Probability.* Graduate Studies in Mathematics, 80. American Mathematical Society, Providence, RI, 2007.

[48] A. KLENKE: *Probability Theory. A Comprehensive Course.* Springer-Verlag, London, Ltd., London, 2008.

[49] A.N. KOLMOGOROV: *Foundations of the Theory of Probability.* Chelsea Publishing Company, New York, 1950.

[50] M. LEDOUX AND M. TALAGRAND: *Probability in Banach Spaces. Isoperimetry and Processes.* Results in Mathematics and Related Areas (3), 23. Springer-Verlag, Berlin, 1991.

[51] N.N. LEBEDEV: *Special Functions and their Applications.* Unabridged and corrected republication. Dover Publications, Inc., New York, 1972.

[52] G. LETAC: *Exercises and Solutions Manual for Integration and Probability.* Springer-Verlag, New York, 1995. French edition: Masson, Second edition, 1997.

[53] M.A. LIFSCHITS: *Gaussian Random Functions.* Kluwer Academic Publishers, 1995.

[54] T.M. LIGGETT: *Continuous Time Markov Processes. An Introduction.* Graduate Studies in Mathematics, 113. American Mathematical Society, Providence, RI, 2010.

[55] E. LUKACS: *Developments in Characteristic Function Theory.* Macmillan Co., New York, 1983.

[56] H.P. MC KEAN: *Stochastic Integrals*. Reprint of the 1969 edition, with errata. AMS Chelsea Publishing, Providence, RI, 2005.

[57] P. MALLIAVIN AND H. AIRAULT: *Integration and Probability*. Graduate Texts in Mathematics, 157. Springer-Verlag, New York, 1995. French edition: Masson, 1994.

[58] R. MANSUY AND M. YOR: *Aspects of Brownian Motion*. Universitext. Springer-Verlag, Berlin, 2008.

[59] S. MÉLÉARD: *Aléatoire. Introduction à la Théorie et au Calcul des Probabilités*. Éditions de l'École Polytechnique, 2010.

[60] P.A. MEYER: *Probability and Potentials*. Blaisdell Publishing Co. Ginn and Co., Waltham, Mass.–Toronto, Ont.–London, 1966. French edition: Herman (1966).

[61] P. MÖRTERS AND Y. PERES: *Brownian Motion*. With an Appendix by Oded Schramm and Wendelin Werner. Cambridge University Press, Cambridge, 2010.

[62] J. NEVEU: *Discrete-parameter Martingales*. Revised edition. North-Holland Mathematical Library, Vol. 10., Amsterdam–Oxford–New York, 1975.

[63] J. NEVEU: *Mathematical Foundations of the Calculus of Probability*. Holden-Day, Inc., 1965. French edition: Masson, Second edition, 1970.

[64] J. NEVEU: *Processus Aléatoires Gaussiens*. Séminaire de Mathématiques Supérieures, No. 34. Les Presses de l'Université de Montréal, 1968.

[65] B. ØKSENDAL: *Stochastic Differential Equations. An Introduction with Applications*. Sixth edition. Universitext. Springer-Verlag, Berlin, 2003.

[66] S.J. PATTERSON: *An Introduction to the Theory of the Riemann Zeta-function*. Cambridge Studies in Advanced Mathematics, 14. Cambridge University Press, Cambridge, 1988.

[67] K. PETERSEN: *Ergodic Theory*. Cambridge Studies in Advanced Mathematics 2. Cambridge University Press, Cambridge, 1983.

[68] V.V. PETROV: *Limit Theorems of Probability Theory. Sequences of Independent Random Variables*. Oxford University Press, New York, 1995.

[69] J. PITMAN: *Probability*. Springer-Verlag, 1993.

[70] CH. PROFETA, B. ROYNETTE AND M. YOR: *Option Prices as Probabilities. A New Look at Generalized Black–Scholes Formulae*. Springer Finance. Springer-Verlag, Berlin, 2010.

[71] P.E. PROTTER: *Stochastic Integration and Differential Equations.* Second edition. Applications of Mathematics, 21. Springer-Verlag, Berlin, 2004.

[72] A. RÉNYI: *Calcul des Probabilités.* Collection Universitaire de Mathématiques, No. 21, Dunod, Paris, 1966.

[73] D. REVUZ: *Probabilités.* Hermann, Paris, 1998.

[74] D. REVUZ: *Intégration.* Hermann, Paris, 1997.

[75] D. REVUZ AND M. YOR: *Continuous Martingales and Brownian Motion.* Third edition. Springer-Verlag, Berlin, 1999.

[76] L.C.G. ROGERS AND D. WILLIAMS: *Diffusions, Markov Processes, and Martingales. Vol. 1. Foundations.* Reprint of the second (1994) edition. Cambridge Mathematical Library. Cambridge University Press, Cambridge, 2000.

[77] L.C.G. ROGERS AND D. WILLIAMS: *Diffusions, Markov Processes, and Martingales. Vol. 2. Itô Calculus.* Second edition. Cambridge University Press, Cambridge, 2000.

[78] J.P. ROMANO AND A.F. SIEGEL: *Counterexamples in Probability and Statistics.* Wadsworth & Brooks/Cole Advanced Books & Software, Monterey, CA, 1986.

[79] S.M. ROSS: *Stochastic Processes.* Second edition. John Wiley & Sons, Inc., New York, 1996.

[80] G. SAMORODNITSKY AND M.S. TAQQU: *Stable Non-Gaussian Random Processes.* Chapman & Hall, New York, 1994.

[81] K. SATO: *Lévy Processes and Infinitely Divisible Distributions.* Cambridge University Press, Cambridge, 1999.

[82] A.N. SHIRYAEV: *Probability.* Second edition. Graduate Texts in Mathematics, 95. Springer-Verlag, New York, 1996.

[83] Y.G. SINAÏ: *Probability Theory. An Introductory Course.* Springer-Verlag, Berlin, 1992.

[84] A.V. SKOROKHOD: *Basic Principles and Applications of Probability Theory.* Springer-Verlag, Berlin, 2005.

[85] F.W. STEUTEL AND K. VAN HARN: *Infinite Divisibility of Probability Distributions on the Real Line.* Monographs and Textbooks in Pure and Applied Mathematics, 259. Marcel Dekker, Inc., New York, 2004.

[86] J.M. STOYANOV: *Counterexamples in Probability.* Second edition. Wiley Series in Probability and Mathematical Statistics. John Wiley & Sons, Ltd., Chichester, 1997.

[87] D.W. STROOCK: *Probability Theory, an Analytic View.* Second edition. Cambridge University Press, Cambridge, 2011.

[88] D.W. STROOCK: *Partial Differential Equations for Probabilists.* Cambridge Studies in Advanced Mathematics, 112. Cambridge University Press, Cambridge, 2008.

[89] D.W. STROOCK: *Markov Processes from K. Itô's Perspective.* Annals of Mathematics Studies, 155, Princeton University Press, Princeton, NJ, 2003.

[90] G.J SZÉKELY: *Paradoxes in Probability Theory and Mathematical Statistics.* Mathematics and its Applications (East European Series), 15. D. Reidel Publishing Co., Dordrecht, 1986.

[91] M. TALAGRAND: *Mean Field Models For Spin Glasses.* Volume I. Basic examples. 54. Springer-Verlag, Berlin, 2011.

[92] M. TALAGRAND: *Spin Glasses: a Challenge for Mathematicians. Cavity and Mean Field Models.* Results in Mathematics and Related Areas. 3rd Series, 46. Springer-Verlag, Berlin, 2003.

[93] P.S. TOULOUSE: *Thèmes de Probabilités et Statistiques. Agrégation de Mathématiques.* Dunod, 1999.

[94] V.V. UCHAIKIN AND V.M. ZOLOTAREV: *Chance and Stability. Stable Distributions and their Applications.* Modern Probability and Statistics. VSP, Utrecht, 1999.

[95] P. WHITTLE: *Probability via Expectation.* Fourth edition. Springer Texts in Statistics. Springer-Verlag, New York, 2000.

[96] D.V. WIDDER: *The Laplace Transform.* Princeton Mathematical Series, v. 6. Princeton University Press, Princeton, NJ, 1941.

[97] D. WILLIAMS: *Weighing the Odds. A Course in Probability and Statistics.* Cambridge University Press, Cambridge, 2001.

[98] D. WILLIAMS: *Probability with Martingales.* Cambridge Mathematical Textbooks. Cambridge University Press, Cambridge, 1991.

[99] D. WILLIAMS: *Diffusions, Markov Processes, and Martingales.* Vol. 1. Foundations. Probability and Mathematical Statistics. John Wiley & Sons, Ltd., Chichester, 1979.

[100] M. YOR: *Small and Big Probability Worlds.* Banach Center Publications, Vol. 95, Institute of Mathematics Polish Academy of Sciences Warszawa, 2011, pp. 261–271.

[101] M. YOR: *Some Aspects of Brownian Motion,* Part I. Some special functionals. Lectures in Mathematics ETH Zürich. Birkhäuser Verlag, Basel, 1992.

[102] M. YOR, M.E. VARES AND T. MIKOSCH: A tribute to Professor Kiyosi Itô. *Stochastic Process. Appl.,* **120** (2010), no. 1, 1.

[103] V.M. ZOLOTAREV: *One-dimensional Stable Distributions.* Translations of Mathematical Monographs, 65. American Mathematical Society, Providence, RI, 1986.

Index

Printed in the United States
By Bookmasters